# Preface

Identification of unknown organic compounds and mixtures forms an important part of the curriculum in Chemistry at graduate and postgraduate levels. Though time saving spectroscopic techniques can be used for structure elucidation, the traditional (chemical) methods of analysis are widely used to train the students of organic chemistry to approach a given problem in a systematic and logical way.

This book has been written for the students of B.Sc. (General), B.Sc. (Honours) and M.Sc. courses in Chemistry at different universities in India. Both chemical and spectral methods used for identification of compounds and mixtures have been discussed. A systematic scheme used for analysis has been given in detail. It includes preliminary examination, determination of melting and boiling points, detection of extra elements, unsaturation, functional groups and preparation of derivatives. Details of chemistry, important precautions to be observed and some relevant notes have been given in addition to the procedure used for each experiment. A section on recrystallisation has been included, keeping in mind the difficulties faced by the students in the crystallisation of crude derivatives.

Spectral methods used for analysis have not been ignored. Detection of functional groups using IR and NMR are given in detail along with the chemical tests. A separate section on IR, NMR and UV includes the principle, the method used for recording the spectrum and the important correlation data.

A detailed scheme for the separation of the components of the mixtures is given. A comprehensive list of various types of organic compounds, their derivatives and specific tests has been provided in the tables.

For the convenience of the readers, a summary of the scheme used for analysis, along with page numbers, has also been provided. A comprehensive index appended at the end will facilitate easy location and reference.

The authors express their thanks to Prof. G.B.V. Subramanian, Prof. M.R. Parthasarthy, Dr. (Ms.) Renu Agarwal, Ms. Pooja Sharma, Suman Dudeja, Bindu Goel, Madhu Bala and Aruna Dahiya for the help rendered during the preparation of the manuscript.

**V K Ahluwalia**
**Sunita Dhingra**

# Contents

# Safety in the Laboratory

Your own safety and that of the other students in the laboratory is largely determined by your work practices. Accidents in the laboratory can be prevented to a large extent by working carefully, with full concentration, neatly and unhurriedly. Accidents in the laboratory can have serious consequences and one must believe in the fact that *prevention is better than cure*. Here are some precautions that you must follow.

1. Never work alone in the laboratory.
2. Always wear a cotton laboratory coat.
3. Use safety glasses for the protection of your eyes and wear gloves for the safety of your hands.
4. All chemicals in the laboratory are hazardous in some way or the other. Handle them carefully.
5. All organic solvents are inflammable. Never heat them directly on a naked flame. While heating them, avoid working near the flame. Solvent ether is highly inflammable. Do not even open a bottle of ether in a laboratory where other students are using a flame.
6. Check your burner, the rubber tube and leakage of gas before lighting the burner.
7. Check all glass apparatus before use. Never use ordinary glass apparatus or cracked apparatus for any reaction.
8. In case of an accident inform your supervisor immediately.
9. If you are not sure about handling or disposal of certain reagents, consult your supervisor before performing the experiment.
10. Use the fume chamber whenever necessary.
11. Follow all the instructions given by the supervisor.

Refer to the following for safety practices.


1. The BDH (Merck) hazard data sheet. Merck Ltd. 1990
2. Guide to safe practices in the chemical laboratory. Royal Society of Chemistry. 1986

# Introduction

Qualitative organic analysis of unknown organic compounds is interesting as well as challenging for a student of organic chemistry as a very large number of natural and synthetic compounds are reported in the literature and a large number are added to the list every year. These compounds contain a variety of functional groups. Many of these compounds are being used as chemotherapeutic agents, cleansing agents, dyes, cosmetics, for making polymers, plastics, fibres and elastomers etc. Proteins, carbohydrates lipids, nucleic acids, enzymes, vitamins etc. are some of very vital natural organic compounds. It is therefore important to train a student of organic chemistry in the methods used for the identification of organic compounds.

The identity of a given unknown compound can be established by carrying out a systematic study of its physical, chemical and spectral properties.

The compound used for analysis must be pure as an impure compound may respond to a number of chemical tests which may not be dependable. An impure solid may be purified by crystallisation and an impure liquid by distillation. The pure compound is then subjected to the following examination.

1. Preliminary examination includes

(i) Physical state
(ii) Colour
(iii) Odour
(iv) Flame test
(v) Action of alkali
(vi) Action of concentrated sulphuric acid
(vii) Ferric chloride test
(viii) Potassium permaganate test
(ix) Bromine water test
(x) Solubility test
(xi) Detection of extra elements
(xii) Determination of melting point/boiling point
(xiii) Detection of unsaturation

2. Detection of functional group which may be done by chemical or spectroscopic methods.

On the basis of above steps one knows the type of compound i.e. whether it is an alcohol, an aldehyde, a ketone or a carbohydrate etc. From the Melting Point/Boiling Point the choice of compound is limited to two or three. The final confirmation of the structure is done by preparation of the derivatives.

3. Preparation of the derivative and comparison of the melting points with those given in the literature enable the identification of the given compound.

A student is advised to carry out all the tests systematically and avoid short cuts to save time. The observations must be recorded immediately after the test is performed. The inferences drawn from the test should be analysed before reporting the result.

# 1

# Preliminary Examination

## 1.1 Physical State

Note whether the compound is a solid or a liquid. If solid, whether it is crystalline or amorphous.

## 1.2 Colour

Note the colour of the compound. The colour of the compound gives information which may be helpful for its identification.

| *Observation* | *Inference* |
|---|---|
| Colourless | Carbohdyrates, aldehydes, ketones, carboxylic acids, esters, alcohols, ethers and some simple hydrocarbons. |
| Colourless turning reddish or pinkish (due to oxidation by air) | Phenols, amines, aminophenols etc. |
| Reddish orange | Nitroamines, azo compounds, naphthaquinones etc. |
| Yellow | Nitro compounds, quinones, iodoform etc. |
| Greenish yellow | *p*-Nitroso compounds. |

**Note** The above information will be useful only if the compound is pure. Many compounds show colour due to impurities.

## 1.3 Odour

Some organic compounds have a very characteristic odour.

| *Observation* | *Inference* |
|---|---|
| Odourless | Carbohydrates, aromatic acids etc. |
| Bitter almond odour | Nitrobenzene, benzaldehyde, benzonitrile etc. |
| Cinnamom-like | Cinnamaldehyde |
| Pleasant-fruity | Esters |
| Sweet odour | Chloroform, alcohols |
| Garlic-like | Thiophenols, thioalcohols |
| Phenolic odour | Phenols |
| Irritating pungent odour | Lower acids, aldehydes and acid halides etc. |
| Vinegar-like | Acetic acid |

| | |
|---|---|
| Pyridine-like | Heterocyclic bases |
| Fish-like | Aliphatic amines, side chain aromatic amines |

**Note** Do not inhale the vapours of the compound directly.

## 1.4 Flame Test

Heat the organic compound (50 mg) on a nickel spatula over a low flame, raising the temperature gradually.

| *Observation* | *Inference* |
|---|---|
| Non-sooty flame | Aliphatic compounds |
| Sooty flame | Aromatic compounds |
| Ammonical odour | Urea, thiourea etc. |
| Charring with smell of burnt sugar | Carbohydrates, tartrates, citrates etc. |

**Note** Some aliphatic compounds like chloroform, carbon tetrachloride etc. also produce a sooty flame.

## 1.5 Action of Alkali

Treat the organic compound (100 mg) with aqueous sodium hydroxide (2–3 ml, 20%) first in the cold and then heat for 1–2 minutes.

| *Observation* | *Inference* |
|---|---|
| Smell of ammonia | Primary amides |
| Smell of chloroform | Chloral |
| Smell of amine on prolonged boiling | Anilides |

## 1.6 Action of Concentrated Sulphuric Acid

Treat the organic compound (100 mg) in a dry test tube with concentrated sulphuric acid (1 ml) first at room temperature and then heat.

| *Observation* | *Inference* |
|---|---|
| Evolution of carbon monoxide | Foramtes, citrates |
| Evolution of carbon monoxide and carbon dioxide | Oxalates |
| Rapid charring with effervescence | Carbohydrates, tartrates |
| Evolution of halogens | N-halogeno compounds |

**Note** $CO_2$ turns lime water milky, CO burns with a blue flame and halogens ($I_2$, $Br_2$, $Cl_2$) are detected by the colour of the vapours.

## 1.7 Ferric Chloride Test

To a solution of the organic compound (50 mg) in alcohol or water (5 ml) add one drop of neutral ferric chloride solution through the sides of the test tube. Note the colour produced carefully. Shake the test tube and add one more drop of neutral ferric chloride solution (note any change).

| *Observation* | *Inference* |
|---|---|
| Intense purple green, blue or red colour | Phenols and enols |
| Buff coloured precipitate | Some aromatic acids |
| Green precipitate | Quinol |
| Red colour (decolourised on addition of dilute hydrochloric acid) | Amino acids, lower fatty acids |
| Intense reddish brown colour or precipitate (no change on addition of hydrochloric acid) | o- and m-phenylenediamines |

**Note** The neutral ferric chloride solution is prepared by adding a dilute ammonium hydroxide solution dropwise to the ferric chloride solution until a faint precipitate is formed, which does not dissolve on shaking. The solution is filtered and the filtrate is used for the test.

## 1.8 Potassium Permanganate Test

To a solution of the organic compound (100 mg) in water or acetone add dilute sodium carbonate solution (0.5 ml, 5%) and then treat dropwise with dilute potassium permanganate solution (2–3 drops, 1–2%). Disappearance of the purple colour of potassium permanganate indicates the presence of unsaturation or the presence of an easily oxidisable group like – CHO in the organic compound (for details, see page 14).

## 1.9 Bromine Test

To a suspension or a solution of the organic compound (100 mg) in water or carbon tetrachloride (1–2 ml), dropwise add a solution of bromine in the same solvent (2%). Rapid disappearance of colour without evolution of hydrogen bromide indicates unsaturation. If on the other hand, the colour disappears with the evolution of hydrogen bromide it indicates the substitution of bromine, e.g., as in phenol or aniline. (for details, see page 14).

## 1.10 Solubility Test

This test gives useful information about the nature of the organic compound. The solubility of the compound should be checked in water, dilute sodium bicarbonate solution, dilute sodium hydroxide solution, dilute hydrochloric acid, concentrated sulphuric acid and ether. On the basis of their solubility, the compounds are divided into three categories.

The compound may be soluble in water as well as in ether (category-1). The compound may be soluble in water but insoluble in ether (category-2). Lastly, the compound may be insoluble in water (category-3). In categories 1 and 2 note whether the aqueous solution is neutral, acidic or alkaline to litmus.

### *(a) Category-1*

Compounds soluble in water and ether.

| *Observation* | *Inference* |
| --- | --- |
| The aqueous solution is acidic to litmus | Phenols, aliphatic acids, armoatic acid (more soluble in hot water). |
| The aqueous solution is alkaline to litmus | Amines. |
| The aqueous solution is neutral to litmus | Aliphatic alcohols, aldehydes, ketones, nitriles, esters. |

### *(b) Category-2*

Compounds soluble in water but insoluble in ether.

| *Observation* | *Inference* |
| --- | --- |
| The aqueous solution is acidic to litmus | Hydroxy acids, sulphonic acids, salts of amines, amino acids. |
| The aqueous solution is alkaline to litmus | Amino alcohols. |
| The aqueous solution is neutral to litmus | Carbohydrates, polyhydric alcohols. |

**Note** The ether used for checking the solubility must be dry, otherwise in presence of moisture the carbohydrates may also dissolve.

### *(c) Category-3*

If the compound is insoluble in water, it may be an aromatic acid, phenol, ester, aldehyde, ketone, ether, amine, nitro compound or a hydrocarbon.

Check the solubility in dilute sodium bicarbonate, dilute sodium hydroxide and dilute hydrochloric acid.

| *Observation* | *Inference* |
| --- | --- |
| Soluble in sodium bicarbonate solution | Carboxylic acids, polynitro phenols. |
| Soluble in dilute sodium hydroxide (5%) but insoluble in sodium bicarbonate | Phenols. |
| Soluble in dilute hydrochloric acid | Amines. |

If the compound does not respond to any of the above solubility tests, check its solubility in concentrated sulphuric acid.

| *Observation* | *Inference* |
| --- | --- |
| Soluble | Neutral compounds like ethers, alcohols, esters, aldehydes and ketones etc. |
| Insoluble | Inert compounds like benzene, chlorobenzene, toluene, xylenes etc. |

The solubility tests are summarised in Chart 1.1.

## 1.11 Detection of Extra Elements

The elements carbon, hydrogen and oxygen are assumed to be present in an organic compound. Nitrogen, sulphur and the halogens (chlorine, bromine and iodine) are elements which may also be present in an organic compound. The presence of these extra elements may be confirmed by the following tests.

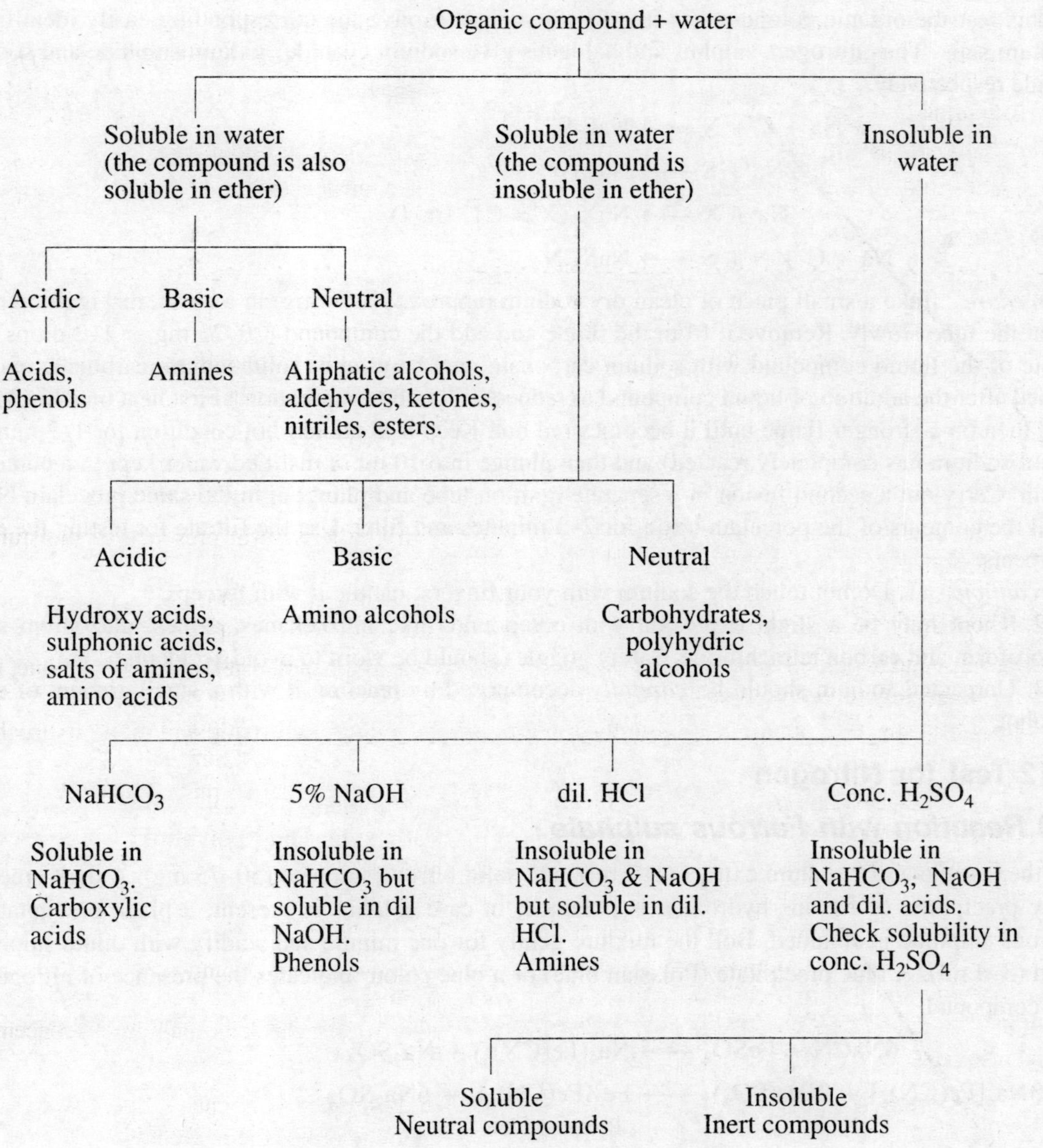


**Chart 1.1 Solubility tests**

***Lassaigne's test*** This is the most widely used test for the detection of nitrogen, sulphur and halogens. In this test the organic compound is fused with sodium to give the corresponding easily identifiable sodium salts. Thus nitrogen, sulphur and halogens give sodium cyanide, sodium sulphide and sodium halide respectively.

$$Na + C + N \longrightarrow NaCN$$

$$2Na + S \longrightarrow Na_2S$$

$$Na + X \longrightarrow NaX\ (X = Cl, Br, I)$$

$$Na + C + N + S \longrightarrow NaSCN$$

*Procedure* Take a small piece of clean dry sodium (approx. 50–75 mg) in a clean, dry ignition tube. Heat the tube slowly. Remove it from the flame and add the compound (20–25 mg or 1–2 drops). (A paste of the liquid compound with sodium carbonate may be used or solid sodium carbonate may be added after the addition of liquid compound to reduce its loss by evaporation.) First heat on a low flame and then on a stronger flame until it becomes red hot. Keep it in the red-hot condition for 1–2 minutes (until sodium has completely reacted) and then plunge into 10 ml of distilled water kept in a porcelain basin. Carry out a second fusion in a separate ignition tube and plunge it in the same procelain basin. Boil the contents of the porcelain basin for 2–3 minutes and filter. Use the filtrate for testing the extra elements.

*Precautions* 1. Do not touch the sodium with your fingers; handle it with forceps.

2. There may be a slight explosion with compounds like nitroalkanes, azides, diazonium salts, chloroform and carbon tetrachloride. Safety goggles should be worn to avoid accidents.

3. Unreacted sodium should be *carefully* decomposed by reacting it with a small amount of ethyl alcohol.

## 1.12 Test for Nitrogen

### *(a) Reaction with Ferrous sulphate*

To the lassaigne's or sodium extract (1–2 ml), add solid ferrous sulphate (50–75 mg). A dark greenish grey precipitate of ferrous hydroxide is obtained; in case sulphur is present, a black precipitate of ferrous sulphide is obtained. Boil the mixture gently for one minute and acidify with dilute sulphuric acid (3–4 ml). A blue precipitate (Prussian blue) or a blue colour indicates the presence of nitrogen in the compound.

$$6NaCN + FeSO_4 \longrightarrow Na_4[Fe(CN)_6] + Na_2SO_4$$

$$3Na_4[Fe(CN)_6] + 2Fe_2(SO_4)_3 \longrightarrow \underset{\text{Prussian blue}}{Fe_4[Fe(CN)_6]_3} + 6Na_2SO_4$$

$Fe_2(SO_4)_3$ is obtained in situ by oxidation of alkaline ferrous salt solution in presence of atmospheric oxygen.

### *(b) Benzidine-Copper Sulphate Test*

Acidify the sodium extract (1–2 ml) with acetic acid (3–4 drops). Add freshly prepared benzidine solution (1% in 50% acetic acid, 2–3 drops) to the above solution and shake the mixture. Treat with copper sulphate solution (1%, 1–2 drops). Appearance of a blue colour or a precipitate indicates the presence of nitrogen. If iodine is also present along with nitrogen, a greenish precipitate is obtained in this test.

## *(c) Reaction with PNB and ODNB*

Nitrogen can also be detected in the Lassaigne's extract by treating it with p-nitrobenzaldehyde (PNB) and o-dinitrobenzene (ODNB). Development of a purple colour indicates a positive test.

Purple colour is obtained due to the formation of o-nitrophenyl hydroxylamine from ODNB and its conversion to the o-quinonoid form (A) in alkaline medium.

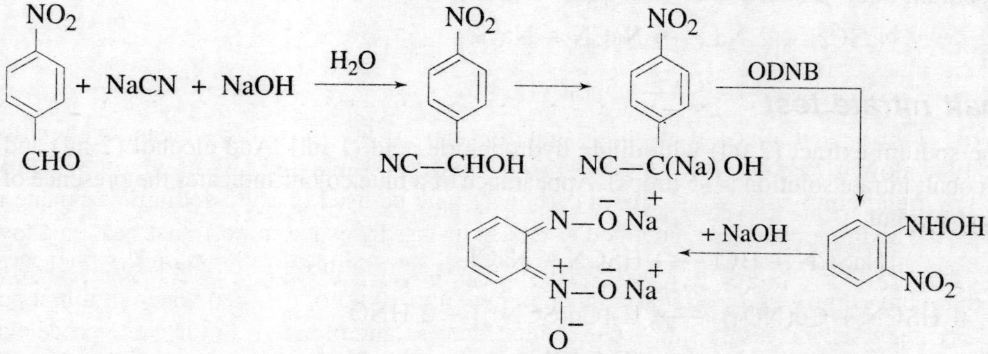


[A] Purple

*Procedure* To 1 ml of the alkaline Lassaigne's extract (add a few drops of NaOH solution, if the extract is not alkaline) add p-nitrobenzaldehyde solution (4–5 drops) and o-dinitrobenzene solution (4–5 drops). Development of a purple colour indicates the presence of nitrogen.

**Note** Use 0.1 M solution of PNB and ODNB in methyl cellosol (2-methoxyethanol).

## 1.13 Test for Sulphur

### (a) Lead acetate test

Acidify the sodium extract (2 ml) with acetic acid (2 ml) and add a few drops (4-5 drops) of lead acetate solution. Appearance of a black precipitate of lead sulphide indicates the presence of sulphur.

$$Na_2S + (CH_3COO)_2\,Pb \longrightarrow \underset{\text{black}}{PbS} \downarrow + 2\,CH_3COONa$$

### (b) Sodium Nitroprusside Test

To the sodium extract (1–2 ml) add a freshly prepared aqueous solution of sodium nitroprusside (1–2 drops). A deep reddish violet coloration indicates the presence of sulphur.

$$Na_2S + \underset{\text{Sodium nitroprusside}}{Na_2\,[Fe(CN)_5NO]} \longrightarrow \underset{\text{Sodium sulphonitroprusside Reddish violet}}{Na_4\,[Fe(CN)_5NOS]}$$

## 1.14 Test for Nitrogen and Sulphur Present Together

### *(a) Test with $FeCl_3$*

Acidify the sodium extract (2 ml) with dilute hydrochloric acid (5–6 drops) and treat with ferric chloride solution (2–3 drops). The appearance of a blood red colour indicates the presence of nitrogen and sulphur.

$$3\ NaSCN + FeCl_3 \longrightarrow \underset{\text{Blood red}}{\underset{\text{Ferric thiocyanate}}{Fe(SCN)_3}} + 3\ NaCl$$

**Note** If the organic compound containing both nitrogen and sulphur is fused with an excess of sodium for the preparation of Lassaigne's extract, sodium cyanide and sodium sulphide are formed instead of sodium thiocyanate, and the test for NaCN and $Na_2S$ can be performed.

$$NaSCN + 2\ Na \longrightarrow NaCN + Na_2S$$

### (b) Cobalt nitrate test

Acidify the sodium extract (2 ml) with dilute hydrochloric acid (1 ml). Add alcohol (2 ml) and then treat with cobalt nitrate solution (4–5 drops). Appearance of a blue colour indicates the presence of both nitrogen and sulphur.

$$NaSCN + HCl \longrightarrow HSCN + NaCl$$

$$4\ HSCN + Co(NO_3)_2 \longrightarrow \underset{\text{Blue}}{\underset{\text{Cobaltithiocyanic acid}}{H_2[Co(SCN)_4]}} + 2\ HNO_3$$

## 1.15 Tests for Halogens

### (a) Beilstein Test

Heat a copper wire having a small loop at its end in a flame until the flame is no longer coloured. Cool, dip the loop in the organic compound and heat on the edge of a clear flame. A green coloured flame indicates the presence of a halogen (fluorine does not give this test). This test is extremely sensitive since minute traces of halogen containing impurities give this test. This test may also fail for very volatile liquids, since these evaporate before the copper wire is heated sufficiently to cause decomposition of the given compound. It is to be noted that certain compounds not containing halogen (e.g., derivatives of quinoline, pyridine, organic acids and urea) impart a green colour to the flame.

### (b) Silver nitrate test

*Procedure,* Acidify the sodium extract (2 ml) with dilute nitric acid (2-3 ml). Boil the solution gently for 1-2 minutes to expel any hydrogen cyanide or hydrogen sulphide that may be present due to the presence of nitrogen or sulphur in the organic compound. Add silver nitrate solution (0.5 ml) to the above solution. A white precipitate readily soluble in ammonium hydroxide indicates the presence of chlorine. A pale yellow precipitate soluble in excess of ammonium hydroxide solution indicates the presence of bromine. In case a yellow precipitate insoluble in ammonium hydroxide is formed, the presence of iodine is indicated.

$$NaX + AgNO_3 \longrightarrow AgX \downarrow + Na\ NO_3$$

**Note** 1. If the given compound does not contain N or S as extra elements, the extract need not be boiled with nitric acid.

2. The colour of the silver halides and their solubility in ammonium hydroxide are *not dependable* observations for inferring the presence of a specific halogen (Cl, Br, I) in an organic compound because the amount of sodium halide formed by fusion of the organic compound with sodium is

very small. Even silver iodide which is present in a very small amount will be soluble in excess of ammonium hydroxide solution. Thus a positive silver nitrate test will only indicate the presence of a halogen. To find out which halogen is present, the layer test should be performed.

### (c) Layer test

In this test, the sodium halide is oxidised to the corresponding halogen which dissolves in the organic solvent. For eg., NaBr on oxidation gives $Br_2$ which dissolves in the organic solvent to impart an orange brown colour to the organic layer.

*Procedure* To the sodium extract (2 ml) add concentrated nitric acid (1 ml). Heat the mixture for a few seconds. Add carbon tetrachloride or chloroform (1 ml) and shake well. If the organic layer attains a violet colour, the presence of iodine is indicated; On the other hand if, the organic layer attains an orange brown colour the presence of bromine is indicated; in case both iodine and bromine are absent and in the silver nitrate test a white precipitate is obtained, the presence of chlorine is indicated.

**Note** 1. Do not heat the extract with nitric acid for a long time, as the halogen formed due to oxidation of sodium halide may escape.

2. This test is helpful in case only one halogen (chlorine, bromine or iodine) is present in the organic compound.
3. Chlorine water or acidified potassium permanganate may be used as an oxidising agent instead of nitric acid in the above test.

### (d) Schiff's reagent test

*Procedure* Acidify the sodium extract (2 ml) with glacial acetic acid (1 ml) and heat to boiling with lead dioxide (0.1 – 0.2 g). Place a piece of filter paper moistened with Schiff's reagent on the mouth of the test tube. If the paper turns violet—the presence of bromine is indicated, a blue colour indicates the presence of iodine.

Alternatively, a piece of filter paper dipped in a 1% alcoholic solution of fluorescein is kept over the mouth of the test tube in place of Schiff's reagent. Appearance of a rose-pink colour indicates the presence of bromine and the appearance of brown colour indicates the presence of iodine.

### (e) Test for halogens (when present together)

Acidify the sodium extract (2 ml) with dilute sulphuric acid. Boil for a minute, cool and add carbon tetrachloride (2 ml). Now add a solution of sodium nitrite (20%) while shaking the test tube cautiously. A purple colour in the organic layer indicates the presence of *iodine*. If iodine is indicated, add more of sodium nitrite and boil until all the iodine (violet vapours) is removed. Boil for some more time to expel the nitrous fumes. Cool the remaining solution, add an oxidising agent ($KMnO_4$) and carbon tetrachloride (1 ml) and shake well. An organe-brown colour in the organic layer indicates *bromine*. Boil this solution with glacial acetic acid (1 ml) and lead dioxide (0.5 gm) to remove all the bromine. After the removal of bromine, acidify the remaining solution with dilute nitric acid and then add silver nitrate solution (1 ml). Appearance of a white precipitate indicates the presence of *chlorine*.

### (f) Test for fluorine

Acidify the sodium extract (2 ml) with acetic acid, boil and cool. Place One drop of the above solution on a piece of zirconium-alizarin test paper. A yellow colour on the red paper indicates the presence of fluorine. The test paper is prepared by dipping a piece of filter paper into a solution obtained by mixing ethanolic alizarin solution (3 ml, 1%) and zirconium chloride (or nitrate) solution (2 ml, 0.4%). The red filter paper is dried before use and moistened with a drop of acetic acid (50%).

## DETERMINATION OF MELTING POINT AND BOILING POINT

### 1.16 Melting Point

Determination of the melting point is extremely valuable for identification of an organic compound. Melting point of a solid is the temperature at which it begins to change into a liquid; the change from solid to the liquid state being quite sharp (within 0.5°). The melting point is considerably influenced by the presence of other substances (impurities) and is therefore an important criteria of purity.

The correct melting point can be determined by using a calibrated thermometer. The calibration is done by determining the melting or boiling points of a few selected organic compounds such as p-dichlorobenzene (53°), benzoic acid (122°), salicylic acid (159°) and anisic acid (184°) as reference compounds. Acetone (B.P 56°) and water (B.P. 100°) are also useful reference compounds.

*Procedure* Powder the organic compound (100 mg) with the help of a spatula on a clean, dry watch glass. Seal one end of a capillary tube (~ 3" long) and fill it (3–5 mm) with the finely powdered sample to form a compact column. Attach the capillary to the thermometer and place it in the Kjeldahl flask (100 ml capacity) containing the bath liquid (~ 50 ml) (concentrated sulphuric acid or paraffin wax) as shown in the diagram (Fig. 1.1). Heat the Kjeldahl flask gently at a uniform rate (1-2° rise in temperature per minute). Note the temperature at which the solid melts.

**Note** 1. It is advisable to determine the approximate melting point first. Repeat to determine the correct melting point. In the second determination, the rate of heating should be carefully controlled (specially near the melting point of the compound.)

2. The organic compound should be finely powdered.
3. The compound should form a compact column at the sealed end of the capillary tube. This can be done either by dropping the capillary tube through a glass tube (18–20" long) or tapping the lower end of capillary.
4. A Thiele's tube may be used in place of a Kjeldhal flask (see Fig. 1.1).
5. Many organic compounds undergo a change in crystalline structure just before melting, usually because of release of solvent of crystalisation. This may lead to shrinkage of sample in the capillary tube. This should not be taken as the beginning of the melting process.

### 1.17 Boiling Point

Like the melting point of a solid, boiling point is useful for identification of a liquid organic compound. The boiling point of a liquid (at a given pressure) is defined as the temperature at which its vapour pressure is equal to the atmospheric (surrounding) pressure. Though a pure liquid has a sharp boiling point—the converse is not always true. A sharp boiling point may sometimes be caused by a constant boiling mixture (azeotrope) of two or more liquids.

*Procedure* Take the liquid (20–25 ml) in a round bottomed flask (100 ml). Add a few pieces of pumice stones and set up the distillation apparatus as shown is Fig. 1.2 a.

Heat the liquid gently at a uniform rate. Discard first fraction of distillate (2-3 ml). Note the temperature at which the liquid boils (constant temperature) and collect the distillate in a clean dry flask.

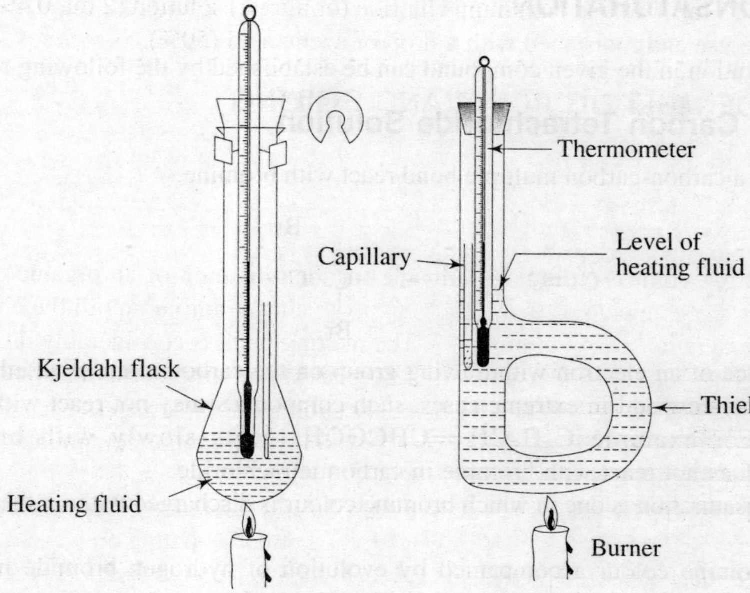


**Fig. 1.1** Thiele's tube

In case only a small amount of the liquid is available, determine the boiling point by *Siwoloboff's method*. Take the organic liquid (0.5-1 ml) in an ignition tube (4-5 mm in diameter and 80–100 mm long). Introduce a capillary tube sealed at the upper end into the liquid (as shown Fig. 1.2b) Attach the ignition tube with the help of a rubber band to the thermometer and fit this into a melting point bath. Heat the bath gently. The temperature at which a rapid and continuous stream of bubbles first emerges from the capillary tube is the boiling point of the liquid. At this temperature the bubbling ceases and the liquid begins to rise in the capillary.

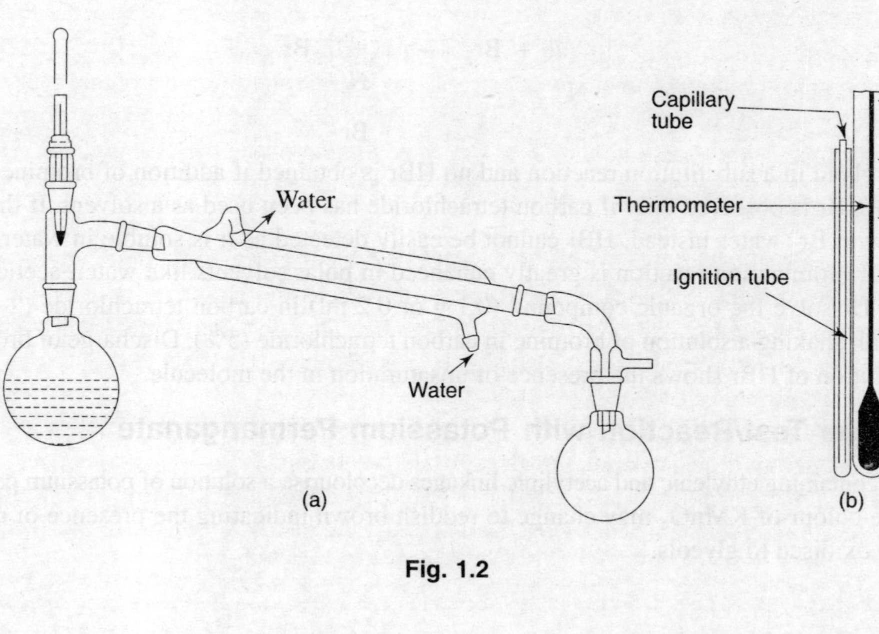


**Fig. 1.2**

## DETECTION OF UNSATURATION

The presence of unsaturation in the given compound can be established by the following reactions.

### 1.18 Bromine in Carbon Tetrachloride Solution

Compounds containing a carbon-carbon multiple bond react with bromine.

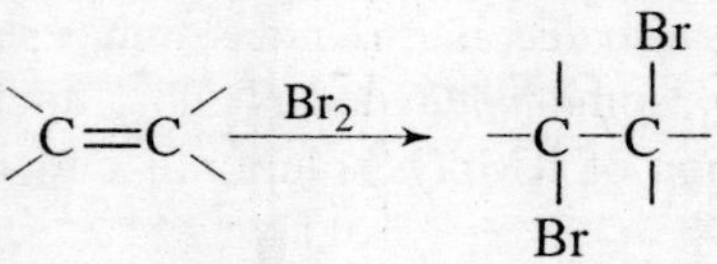

However, the presence of an electron withdrawing group on the carbon atom of an ethylenic bond causes the addition to be slow and in extreme cases, such compounds may not react with bromine at room temperature. For example $C_6H_5CH{=}CHCOOH$ reacts slowly with bromine and $(C_6H_5)_2C = C(C_6H_5)_2$ does not react with bromine in carbon tetrachloride.

A positive test for unsaturation is one in which bromine colour is discharged without the evolution of hydrogen bromide.

Discharge of the bromine colour accompanied by evolution of hydrogen bromide indicates that substitution has occurred. Many aromatic compounds which have electron donating groups present on the ring undergo this type of reaction. Some typical examples are phenols and amines.

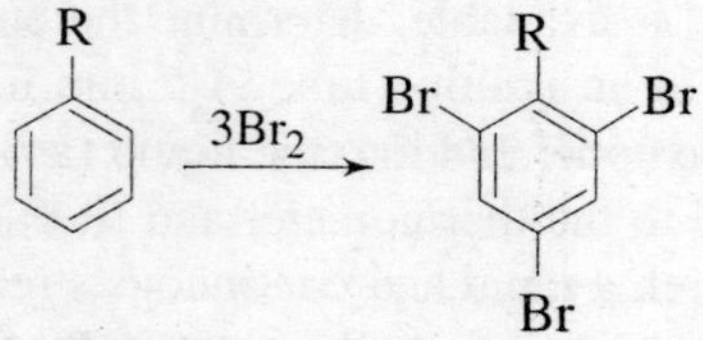

R=OH, $NH_2$

Certain heterocyclic compounds containing tertiary nitrogen and aliphatic amines add bromine. For example

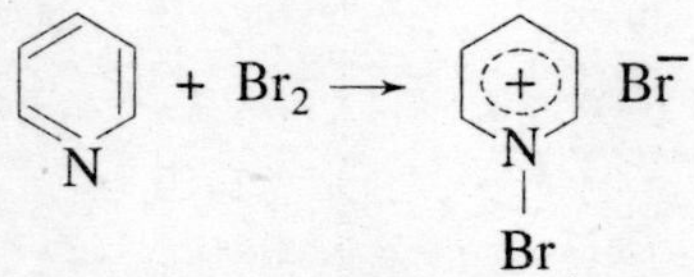

HBr is evolved in a substitution reaction and no HBr is obtained if addition of bromine takes place. Evolution of HBr is observed only if carbon tetrachloride has been used as a solvent. If the reaction is carried out with $Br_2$ water instead, HBr cannot be easily detected as it is soluble in water.

The rate of bromination reaction is greatly enhanced in polar solvents like water, acetic acid etc.

*Procedure* Dissolve the organic compound (0.1 g or 0.2 ml) in carbon tetrachloride (2 ml) and add drop wise with shaking a solution of bromine in carbon tetrachloride (5%). Discharge of bromine colour without evolution of HBr shows the presence of unsaturation in the molecule.

### 1.19 Baeyer Test/Reaction with Potassium Permanganate

Compounds containing ethylenic and acetylinic linkages decolourise a solution of potassium permanganate, or the purple colour of $KMnO_4$ may change to reddish brown indicating the presence of unsaturation. Olefines are oxidised to glycols.

$$3\left(\rangle C{=}C\langle\right) + 2KMnO_4 + 4H_2O \longrightarrow 3\left(-\underset{OH}{\overset{|}{C}}-\underset{OH}{\overset{|}{C}}-\right) + 2MnO_2 + 2KOH$$

Acetylinic compounds are generally oxidised to a mixture of carboxylic acids.

$$RC{\equiv}CR' \xrightarrow{[O]} RCOOH + R'COOH$$

All easily oxidisable compounds also decolourise potassium permanganate solution even if no olefinic or acetylinic bonds are present, e.g., some aldehydes, phenols, amines, formic acid and its esters, thiols, thioesters etc. So the decolourisation of $KMnO_4$ solution is either due to unsaturation or oxidation of easily oxidisable groups.

*Procedure* Dissolve the organic compound (0.1 g or 0.2 ml) in water (2 ml) or acetone if the compound is insoluble in water. Add potassium permanganate solution (2%) dropwise with shaking. Decolourisation of a few drops of potassium permanganate solution is a positive test.

# 2

# Detection of Functional Groups

Identification of the functional groups is a very important part of qualitative organic analysis. Chemical and spectral methods can be used to identify functional groups. The following important points should be kept in mind while performing these tests.

(i) A given unknown compound may contain more than one functional group and hence tests for all the functional groups should be systematically performed (e.g. if nitrogen is not present in the compound, functional groups containing nitrogen are obviously absent). Note down the observations immediately e.g. the colour of the precipitate or solution etc.

(ii) Some of the colour tests may be positive for several different functional groups, e.g. $FeCl_3$ test, a test recommended for phenols is also positive for some amines. In such cases results of preliminary tests should be kept in mind before making derivatives.

(iii) Short cuts should be avoided and a systematic and a logical approach should be adopted to save time and labour.

## 2.1 Compounds Containing C, H and O

### *(a) Carbohydrates*

Polyhydroxy aldehydes or ketones or compounds which yield these on hydrolysis are called carbohydrates, sugars or saccharides. The simplest type of carbohydrates are called monosaccharides and complex polymers are called polysaccharides. Carbohydrates having an aldehydic group are called aldoses and those having a keto group are ketoses. These may be either reducing (which reduce Fehling, Tollen's solution etc.) or non-reducing in nature.

Simple carbohydrates are water soluble, give a characteristic smell of burnt sugar on dry heating, give positive ceric ammonium nitrate test (page 26) but do not give 2, 4-dinitrophenylhydrazine test. The melting points of carbohydrates and glycosides are of little value in the identification as these compounds generally melt over a range of temperature and often decompose on heating. They respond to the following tests.

(i) **Molisch's test.** An aqueous solution of a carbohydrate on treatment with $\alpha$-naphthol followed by addition of conc. $H_2SO_4$ gives a violet coloured ring. This is a general test for identification of carbohydrates.

The reaction is based on the formation of furfural or substituted furfural from the carbohydrate in presence of conc. $H_2SO_4$, which condenses with $\alpha$-naphthol to produce a violet ring, e.g. glucose undergoes cyclisation and dehydration to produce 5-hydroxymethylfurfural, which gives a violet colour with $\alpha$-naphthul.

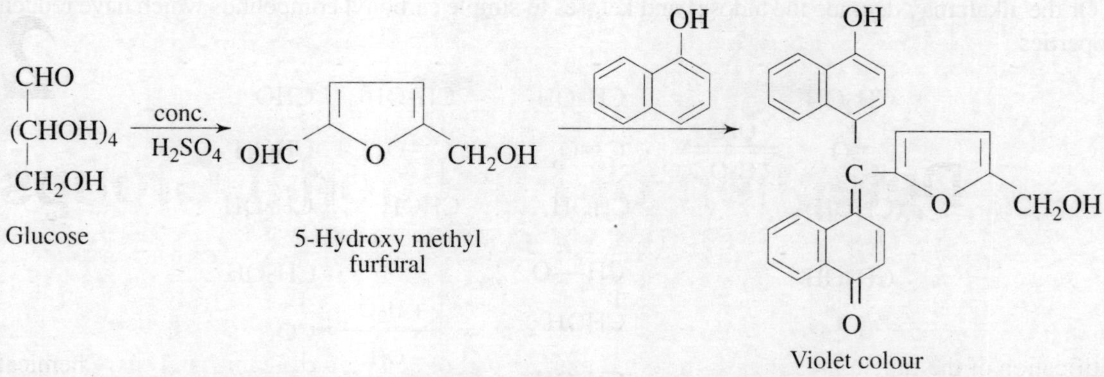


*Procedure* Add an alcoholic solution of $\alpha$-naphthol (10%, 1 ml) to an aqueous solution of carbohydrate (1-2 ml). Then pour conc. $H_2SO_4$ (1 ml) carefully along the sides in the test tube. A violet ring at the junction of two layers indicates the presence of a carbohydrate.

**Note** Though this is a general test for carbohydrates, compounds like furfural and substituted furfural also give this test.

**(ii) Barfoed test.** This test helps to distinguish a monosaccharide from a disaccharide. Monosaccharides react at a faster rate with the Barfoed reagent as compared to disaccharides. A red precipitate of $Cu_2O$ is formed within two minutes with a monosaccharide.

$$\underset{\text{}}{CHO{-}(CHOH)_4{-}CH_2OH} + Cu^{+2} + 2\,H_2O \longrightarrow COOH{-}(CHOH)_4{-}CH_2OH + H^+ + \underset{\text{Red ppt.}}{Cu_2O}$$

*Procedure* Heat an aqueous solution of the compound (1 ml) with the Barfoed reagent (1 ml, freshly prepared) in a boiling water bath for 1-2 minutes. Formation of a red precipitate of $Cu_2O$ indicates the presence of a monosaccharide.

**(iii) Test with Fehling or Benedict solution**. It is a general method employed for distinction between reducing and non-reducing saccharides. Copper hydroxide can oxidise some carbohydrates, itself getting reduced to $Cu_2O$ (red precipitate) in the process. However the rate of reaction is slow because of the low solubility of $Cu(OH)_2$ in alkaline solution. Its reactivity is enhanced by the formation of soluble complexes of $Cu^{+2}$ with either tartarate ions (Fehling solution) or citrate ions (Benedict solution).

Aldoses reduce Fehling solution or Benedict solution forming a red precipitate of $Cu_2O$. Ketoses also reduce these solutions though simple ketones do not. Ketoses give these tests because of the following base catalysed reaction.

$$\underset{\text{Ketose}}{CH_2OH{-}C{=}O{-}CHOH{-}} \xrightarrow{\overset{\ominus}{O}H} CHOH{=}C{-}OH{-}CHOH{-} \xrightarrow{\overset{\ominus}{O}H} \underset{\text{Aldose}}{CHO{-}CHOH{-}CHOH{-}}$$

Or the alkali may degrade the aldoses and ketoses to simple carbonyl compounds which have reducing properties.

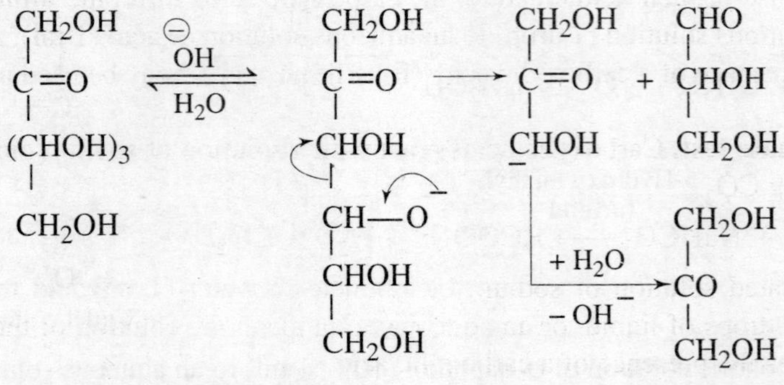

*Procedure* Heat an aqueous solution of the compound (1-2 ml) with an equal amount of Benedict or Fehling solution (mix equal amounts of Fehling A and B). Formation of a red precipitate indicates the presence of a reducing sugar.

**(iv) Seliwanoff test.** This is a test for ketoses and helps to distinguish them from aldoses and is based on the fact that ketoses are more readily dehydrated to furfural and its derivatives as compared to aldoses. Ketoses on condensation with resorcinol in presence of concentrated hydrochloric acid give a red-blue colouration within two minutes.

The structure of the product obtained is similar to that obtained in the Molisch's test.

*Procedure* Add Seliwanoff's reagent (2 ml) to an aqueous solution of the compound (0.1 g in 2 ml water) in a test tube. Development of cherry red or blue colour indicates the presence of a ketose.

**(v) Bial test.** Pentoses can be distinguished from hexoses. Pentoses give blue green colour on reaction with Bial's reagent and hexoses give a muddy brown to grey solution. Bial's solution contains orcinol which condenses with furfural or substituted furfural to produce a characteristic colour.

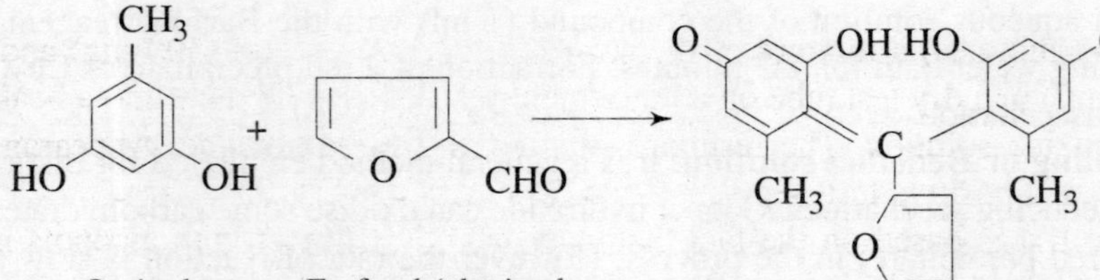

Orcinol　　Furfural (obtained from a pentose)　　Blue green

*Procedure* Heat an aqueous dilute solution of the compound (4-5 drops) with Bial's reagent (2 ml) for 2 minutes in a boiling water bath. Development of a green to blue colour indicates a pentose and a brown or a grey colour a hexose.

**Note** Heptoses and trioses give blue to green colour in the above test.

## (b) Carboxylic acids

The acids do not show the typical reaction of carboynl groups because of the following resonance.

However, they have a tendency to donate protons and act as acids. It is this property which is helpful in the identification of a —COOH group.

**(i) Litmus test.** Because of their acidic nature, the carboxylic acids turn blue litmus red.

*Procedure* Add blue litmus solution (1 drop) to an aqueous solution of acid (1 ml). Appearance of a red colour indicates the presence of a carboxylic acid. (Blue litmus paper may be used in place of a solution of blue litmus).

**(ii) Sodium bicarbonate test.** Carboxylic acids react with a solution of sodium bicarbonate with brisk effervescence evolving $CO_2$.

$$RCOOH + NaHCO_3 \longrightarrow RCOONa + H_2O + CO_2 \uparrow$$

*Procedure* To a saturated solution of sodium bicarbonate in water (1 ml) add the given compound (pinch of solid or 2-3 drops of liquid, or an aqueous or an alcoholic solution of the compound). Brisk effervescence indicates the presence of a carboxylic acid.

**Note** 1. Use acid free alcohol for the test.
2. If alcohol is used as a solvent, perform a blank test also.

**(iii) Fluorescein test.** This test should be performed only if the above tests for—COOH group (i and ii) are positive.

It is a test given by dicarboxylic acids which can form an anhydride on heating (e.g. phthalic, succinic acid etc.). When a dicarboxylic acid or its anhydride is heated with resorcinol in presence of concentrated sulphuric acid, a fluorescent dye called fluorescein is obtained.

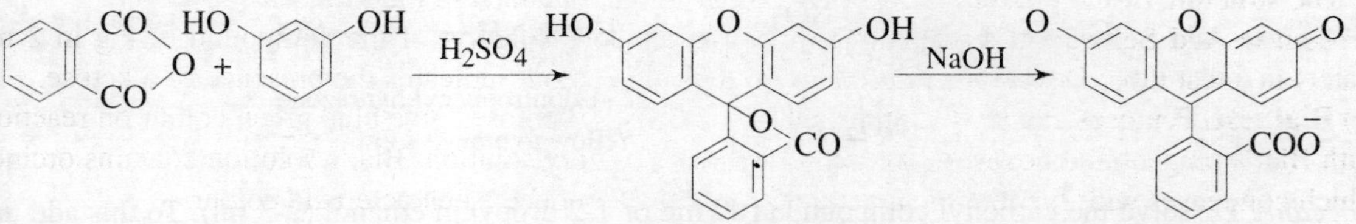


Fluorescein

*Procedure* Heat (1-2 minutes) a mixture of the acid (100 mg), resorcinol (100 mg) and concentrated sulphuric acid (0.5 ml) in a dry test tube on a bunsen burner. Pour the mixture into a beaker containing dilute sodium hydroxide solution. The resultant solution should be alkaline. Appearance of a green fluorescent solution is a positive test.

**(iv) Ester test.** The test is based on the fact that carboxylic acids react with alcohols in presence of concentrated sulphuric acid producing sweet smelling esters.

$$RCOOH + R'OH \xrightarrow[\Delta]{\text{Conc. } H_2SO_4} RCOOR' + H_2O$$

*Procedure* Heat (2-3 minutes) a mixture of the acid (100 mg), ethanol (1 ml) and concentrated sulphuric acid (2-3 drops) in a dry test tube in a water-bath. Pour the reaction mixture carefully into a beaker containing water (~ 25 ml). Neutralise the excess sulphuric acid. Formation of a sweet smelling substance indicates the presence of an acid.

**(v) Spectroscopy.**

IR—Carboxylic acids show two broad absorption bands at 1725–1690 $cm^{-1}$ and 3300–2500 $cm^{-1}$ due to $>C{=}O$ and O—H streching respectively.

NMR—The hydrogen of — COOH appears between 10–13 $\delta$ and can be easily recognised by deuterium exchange. The hydrogens $\alpha$-to a-COOH group absorb between 2–3 $\delta$.

## (c) Aldehydes and Ketones (Carbonyl compounds)

Identification of aldehydes and ketones is based on two types of reactions of the $>C=O$ group, i.e. (a) addition to $>C=O$ and (b) oxidation.

Of the addition reactions, the addition of $NH_3$ and substituted ammonias are more important. The addition is generally followed by elimination resulting into an unsaturated compound.

$$>C=O + ZNH_2 \longrightarrow >C{<}^{OH}_{NHZ} \xrightarrow{-H_2O} >C=NZ$$

These reactions are acid or base catalysed and do not occur in strongly acidic or basic solutions. For each reaction there is an optimum pH, and hence maintenance of pH while carrying out the reaction is very important.

Aldehydes can be easily oxidised to carboxylic acids. Ketones, on the other hand are difficult to oxidise. This difference in their reactivity is the basis for distinction between aldehydes and ketones.

The following qualitative tests may be performed for identification of aldehydes and ketones.

**(i) Reaction with 2, 4-dinitrophenylhydrazine**. Aldehydes and ketones give yellow to orange precipitate when treated with 2, 4-dinitrophenylhydrazine.

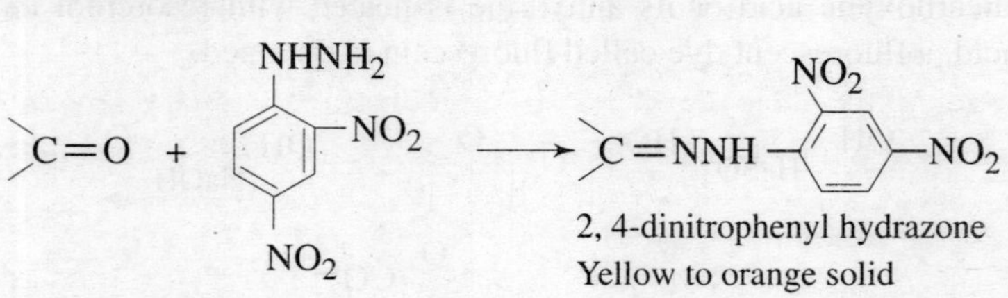

2, 4-dinitrophenyl hydrazone
Yellow to orange solid

*Procedure* Dissolve the carbonyl compound (100 mg or 1-2 drops) in ethanol (2-3 ml). To this add an alcoholic solution of 2, 4-dinitrophenylhydrazine (2 ml) and shake the mixture well. A yellow to orange precipitate indicates the compound to be an aldehyde or a ketone. An orange precipitate is obtained from carbonyl compound in which the C=O groups is conjugated with C=C.

**Note** 1. The test can also be carried out by dissolving the compound in water and adding an aqueous solution of 2, 4-DNP.

2. Brady's reagent can be used for both water soluble and water insoluble carbonyl compounds. It is prepared as follows: Dissolve 2, 4-DNP (1 gm) in concentrated sulphuric acid (5 ml). Add to this, slowly with shaking, a mixture of water (7 ml) and ethanol (25 ml). Filter to remove any suspended impurities.
3. Some aliphatic ketones do not give a precipitate of 2, 4-DNP by addition of 2, 4-DNP reagent to the solution of the compound. In such cases reverse addition, i.e. dropwise addition of the solution of the carbonyl compound to the 2, 4-DNP reagent gives better results.
4. It is advisable to carry out a blank test. A blank test is performed by adding all reagents except the compound.
5. Some amines like aniline etc. may give a precipitate of the salt of the amine with acid used in the preparation of the reagent. However, this precipitated salt is soluble in water.
6. At times, the 2, 4-dinitrophenylhydrazine may precipitate out as an organe solid, which is soluble in dilute sulphuric acid whereas the 2, 4-dinitrophenylhydrazones are insoluble.

Aldehydes can be distinguished from ketones on the basis of the following tests.

**(ii) Tollen's test.** Tollen's reagent consists of silver ammonia complex in ammonia solution. Aldehydes give a grey black precipitate or a silver mirror when treated with a freshly prepared Tollen's reagent. Aldehydes are oxidised to the corresponding acid and silver in tollen's reagent is reduced from + 1 oxidation state to elemental form. Ketones do not reduce Tollen's reagent.

$$RCHO + 2\,[Ag(NH_3)_2]OH \longrightarrow RCOONH_4 + 3\,NH_3 + H_2O + 2Ag$$

*Procedure* To the Tollen's reagent (1 ml), add a solution of the aldehyde (2 drops or 50 mg) dissolved in aldehyde free alcohol (2-3 ml) and warm the solution in a hot water bath. Formation of a grey black precipitate or silver mirror indicates the presence of an aldehyde.

**Note** 1. Tollen's reagent should be freshly prepared for use.

2. Some formates, tartrates, lactates, quinones, $\alpha$-diketones and phenols also give a positive test. Reducing carbohydrates also give this reaction.

3. After performing the test, pour the contents of the test tube into the sink and wash the test tube well with dilute nitric acid, (Silver fluminate, an explosive substance, that might be produced is thus decomposed.)

**(iii) Schiff's test.** Aldehydes restore the colour of the Schiff's reagent. Schiff's reagent is prepared by passing sulphur dioxide into a solution of the dye fuchsin magenta. The solution becomes colourless due to the formation of an addition product. An aldehyde abstracts sulphurous acid from the Schiff's reagent and restores the pink colour.

*Procedure* To a solution of the aldehyde (1-2 drops or 50 mg in 2-3 ml of aldehyde free alcohol) add the Schiff's reagent (2-3 drops). Instant appearance of a violet or a purple colour indicates the presence of an aldehyde.

**Note** 1. Do not heat the solution.

2. Some ketones regenerate a light pink colour of the reagent. Generation of light pink colour is, therefore, not a positive test.

**(iv) Test with Chromic acid.** Aldehydes give a green or a blue precipitate when treated with chromic acid. Ketones do not react.

*Procedure* Add the carbonyl compound (1 drop or 20 mg of solid) to the chromic acid reagent (see reagents page 267) (1 ml). Appearance of a green-blue precipitate indicates the presence of an aldehyde.

**Note** Primary and secondary alcohols also give a positive test with chromic acid but do not give the 2, 4-DNP test. Enols may give a positive test. Phenols give a dark coloured solution entirely different from green-blue colour of a positive test.

**(v) Fehling test.** Fehling solution contains complex of $Cu^{+2}$. When treated with an aldehyde, $Cu^{+2}$ is reduced to $Cu^{+1}$ (as a red precipitate of $Cu_2O$) and the aldehyde is oxidised to the acid. Aromatic aldehydes do not reduce the Fehling solution.

*Procedure* Heat the mixture of the organic compound (50 mg dissolved in 2-3 ml alcohol) and Fehiling solution (2-3 ml, obtained by mixing equal amounts of Fehling solution A + Fehling solution B). Formation of a brick red precipitate indicates the presence of an aldehyde.

**(vi) Iodoform test.** It is based on the fact that when a carbonyl compound containing $\alpha$-hydrogen is treated with a halogen in basic solution, substitution of hydrogen occurs at the $\alpha$-position.

When a compound containing $CH_3CO$—group is treated with a solution of iodine in basic medium, iodoform, a yellow solid separates out.

$$CH_3COR \xrightarrow{I_2 + NaOH} CI_3COR \xrightarrow[\Delta]{NaOH} \underset{\substack{\text{Iodoform} \\ \text{Yellow solid}}}{CHI_3} + RCOONa$$

Positive test is given by $CH_3CHO$, methyl ketones and some alcohols (e.g. $C_2H_5OH$, $CH_3CHOH$ $CH_3$ etc). It is also known as the hypoiodide test.

*Procedure.* Dissolve the compound (2-3 drops or 100 mg) in water (2-3 ml) in a test tube and add sodium hydroxide (1 ml, 10%). To this add a saturated solution of iodine-potassium iodide in water with stirring until the dark colour of iodine persists. Heat the solution in a boiling water bath for 1-2 minutes. Remove the colour of excess iodine by adding a few drops of sodium hydroxide solution. A yellow precipitate of iodoform indicates the presence of the $CH_3CO$— group in the compound.

**Note** 1. Use dioxane to dissolve the compound if it is water insoluble.

2. If the iodoform is dark reddish yellow in colour, dissolve it in 3-4 ml of dioxane and treat it with 1 ml of 10% sodium hydroxide solution, shake until a light yellow solid remains. Dilute with water and filter the solid.

**(v) Spectroscopy.** IR-Carbonyl group is one of the strongest absorbing group in the infrared spectrum, appearing between 1800-1650 $cm^{-1}$.

*Aldehydes*-A strong absorption band is generally observed between 1725-1680 $cm^{-1}$. Two weak bands at 2750 and 2850 $cm^{-1}$ respectively are due to C-H strech of-CHO group. These bands are absent in the spectrum of a ketone

*Ketones*-A strong absorption band due to $>C=O$ strech is generally observed between 1780-1665 $cm^{-1}$

*NMR*-Hydrogens which are alpha to a carbonyl group show absorption between 2-3$\delta$. The H of—CHO absorbs between 9-10$\delta$. Coupling of the absorption signal at 9-10$\delta$ is observed if the aldehyde contains $\alpha$-hydrogen atoms. For example, a singlet is observed for benzaldehyde and a quarlet for acetaldehyde between 9-10$\delta$.

## (d) Phenols

Hydroxy derivatives of aromatic compounds are known as phenols. Phenols are weaker acids than carboxylic acids, and hence dissolve in dilute sodium hydroxide and are generally insoluble in dilute sodium bicarbonate solution. Some phenols (like 2, 4, 6-trinitrophenol, 2, 4-dinitrophenol etc.) which contain strong electron withdrawing groups are stronger acids and dissolve even in sodium bicarbonate. Phenols undergo easy electrophilic substitution reactions, e.g. they decolourise bromine solution at room temperature forming a polybromo substituted phenol. Some phenols like catechol and hydroquinone are powerful reducing agents, because they are easily oxidised to the corresponding quinones.

Following tests may be performed for showing the presence of a phenolic group.

**(i) Solubility in sodium hydroxide.** Most phenols are soluble in dilute sodium hydroxide but insoluble in sodium bicarbonate solution.

**(ii) Ferric chloride test.** Most phenols and enols give a dark coloured solution (blue, green, purple etc.) on reaction with a neutral solution of ferric chloride, e.g. simple phenol reacts with $FeCl_3$ as follows:

$$6\ C_6H_5OH + FeCl_3 \longrightarrow \underset{\text{Purple}}{H_3\,[Fe(OC_6H_5)_6]} + 3\ HCl$$

This test should be performed using a dilute solution of the phenol, observation should be made as the added $FeCl_3$ mixes with the phenol solution. Some phenols do not give this test and so a negative test must not be taken as an indication of the absence of phenol without supporting information (e.g. solubility in dilute sodium hydroxide, $Br_2$ test, Liebermann nitroso reaction and phthalein test).

Some phenols react in a different way with ferric chloride, e.g. Hydroquinone is oxidised to quinone which forms a dark green solid of quinhydrone with unreacted hydroquinone.

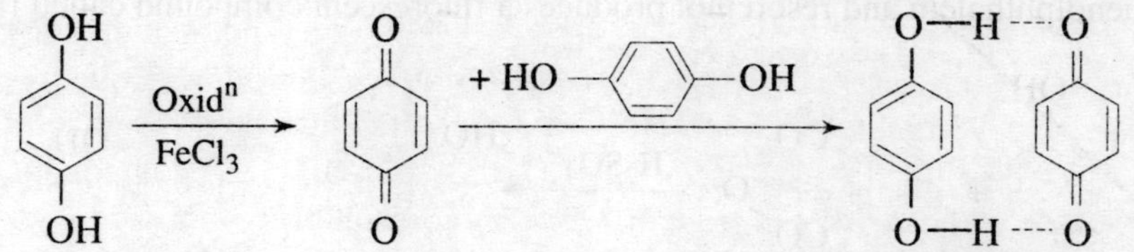


Quinhydron
Dark green solid

$\alpha$-and $\beta$-Naphthol undergo oxidative coupling in the presence of $FeCl_3$.

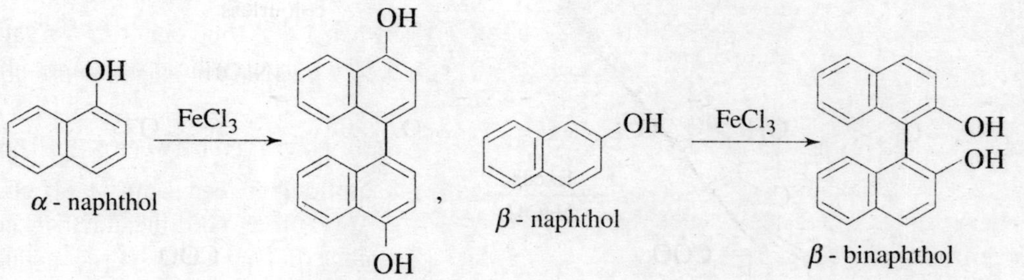


$\alpha$ - naphthol, $\beta$ - naphthol, $\beta$ - binaphthol

$\alpha$ - binaphthol

*Procedure* Dissolve the given compound (50 mg) in water (4-5 ml) and add a neutral solution of $FeCl_3$ slowly (several drops, dropwise) and observe the change in colour. A red, blue, green or purple colouration indicates the presence of a phenol. Enols produce a red, violet or tan colour.

**Note 1.** Use alcohol or chloroform if the compound is insoluble in water.

**2.** Phenols can be distinguished from enols by the following test:

Take the given compound (50 mg) and treat it with a solution of mercurous nitrate in nitric acid. Formation of an immediate grey precipitate shows the presence of an enol. Phenols do not give this test.

**(iii) Liebermann Nitroso Reaction.** Phenols having a free para-position undergo this reaction.

Phenol on heating with concentrated Sulphuric acid and sodium nitrite undergoes the following reaction.

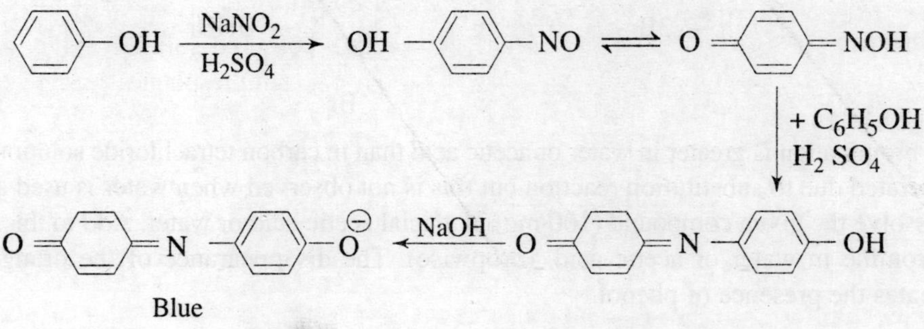


Blue

*Procedure* Take the compound (100 mg) in a dry test tube, add sodium nitrite (few crystals) and concentrated sulphuric acid (1 ml), mix well and heat gently. A blue colour is obtained. The solution turns red on dilution with water and blue on basification with a dilute solution of sodium hydroxide.

**(iv) Phthalein test.** A number of phenols having a free p-position respond to this test. On heating with phthalic anhydride in presence of concentrated sulphuric acid, phenols produce condensation compounds which give a characteristic colour in alkaline solution.

Phenol gives phenolphthalein and resorcinol produces a fluorescent compound called fluorescein.

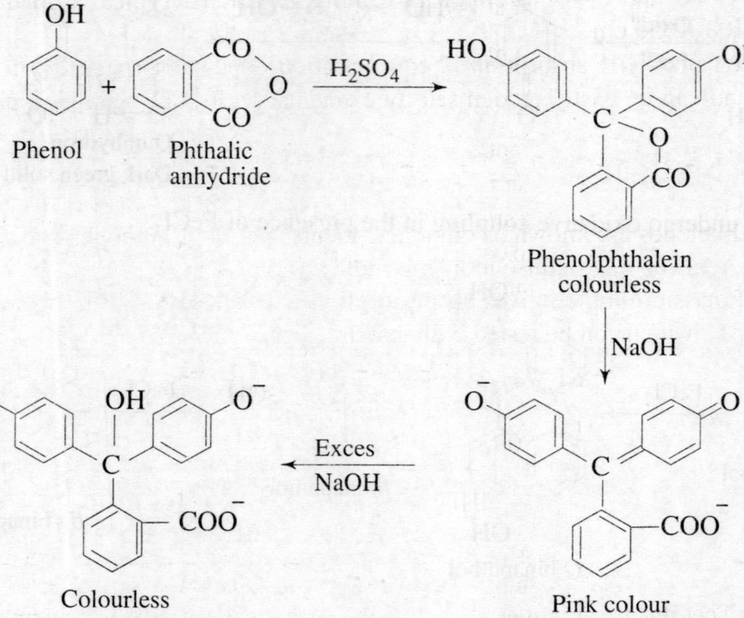


*Procedure* In a dry test tube, heat the given compound (100 mg) with an equal amount of phthalic anhydride and concentrated sulphuric acid (2-3 drops), for 1-2 minutes. Cool and pour into a beaker containing dilute sodium hydroxide solution. Appearance of a pink, blue, green, red etc. colour indicates the presence of a phenol. However, the colour disappears on addition of large excess of concentrated sodium hydroxide solution.

**(v) Bromine test.** Phenols are readily brominated by bromine forming polybromo-substituted phenols. For examples.

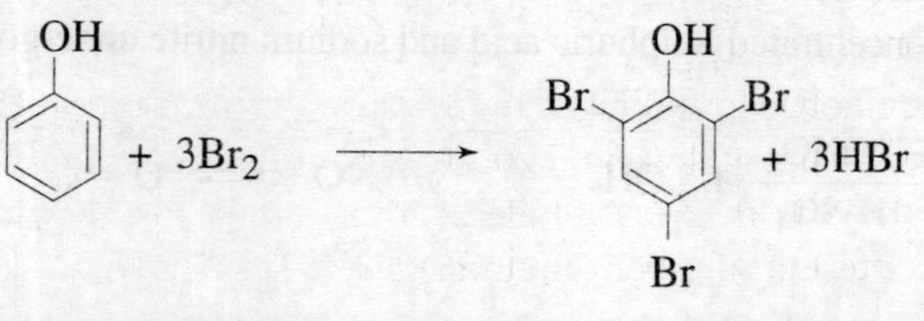


The rate of bromination is greater in water or acetic acid than in carbon tetrachloride solution. Hydrogen bromide is liberated due to substitution reaction but this is not observed when water is used as a solvent.

*Procedure* Dissolve the given compound (100 mg) in glacial acetic acid or water. Add to this, a saturated solution of bromine in water or acetic acid. (dropwise). The disappearance of the orange colour of bromine indicates the presence of phenol.

**Note** 1. Compound may be dissolved in acetic acid or carbon tetrachloride and treated with the solution of bromine dissolved in the same solvent.

2. Amines and other activated aromatic compounds also decolourise bromine water. Solubility and other tests given in the text may be used to distinguish phenols from amines.

3. Compounds containing an olefinic or acetylinic linkage decolourise bromine water without liberating hydrogen bromide.

**(vi) Spectroscopy.** IR—Absorptions at 3600, 1200 and 3100 $cm^{-1}$ in the infra-red spectrum of phenols are due to O—H, C—O and C—H (aromatic) streching vibrations. Typical aromatic ring adsorptions appear between 1400–1650 $cm^{-1}$.

*NMR*—The position of —OH absorption is concentration dependent (page 27) it may be observed between 4—12$\delta$ and can be easily recognised by exchange with $D_2O$. Aromatic protons absorb near 7$\delta$.

## (e) Quinones

$\alpha$, $\beta$-Unsaturated diketones are known as quinones. Identification of quinones is based on the fact that these can be easily reduced and oxidise iodide to iodine.

**(i) Test with KI**. Potassium iodide gives HI on reaction with dil. $H_2SO_4$, the liberated HI reacts with quinone producing $I_2$, which can be tested with starch paper.

$$KI + H_2SO_4 \longrightarrow K_2SO_4 + HI$$

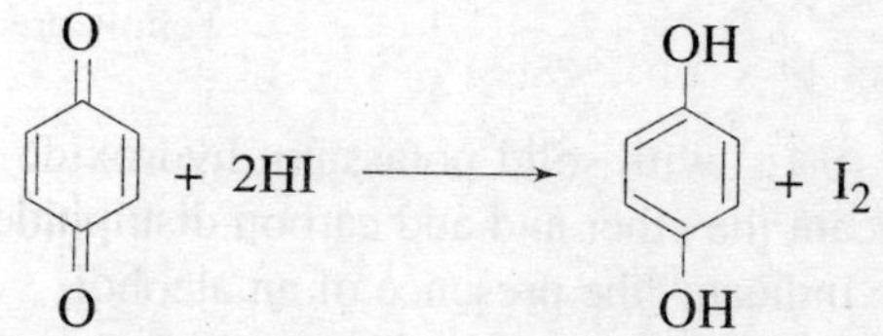

This test is not specific to quinones. Spectral data should be used to confirm the presence of a quinone.

*Procedure* Add KI containing dilute sulphuric acid (1 ml of KI solution containing a few drops of dil. $H_2SO_4$) to the organic compound (0.1 g) in water (2 ml). A pale yellow colour is developed due to the formation of iodine. Add a drop of this solution to a piece of strach paper. Appearance of a blue colour indicates the presence of a quinone.

**(ii) Spectroscopy.** *IR*—Due to conjugation with the olefinic double bond, the C=0 stretching in quinones is shifted to a lower frequency by ~ 30 $cm^{-1}$ as compared to unconjugated ketones. The absorption occurs between 1660 – 1675 $cm^{-1}$. For the same reason, the C=C stretch is shifted to 1600 $cm^{-1}$ and the intensity of absorption is increased.

*UV*—The UV spectrum of quinones is similar to the spectrum of $\alpha$, $\beta$-unsaturated ketones. A strong K band at 245 nm and a weak R band at 435 nm is observed in the spectrum of quinones..

*NMR*—The protons attached to the $\alpha$, $\beta$-unsaturated carbon atoms absorbs at ~ 7.0 $\delta$.

For example, the protons present in p-benzoquinone absorb at $\delta$ 6.8.

## (f) Alcohols

Alcohols are very weak acids and react only with very reactive metals like sodium to form alkoxides. The relative acidity is teritary < secondary < primary < methanol. The following tests can be performed.

for detection of an -OH group and distinction between a primary, secondary and a teritary alcohol. Alcoholic —OH is also present in carbohydrates and other natural compounds like steroids.

**(i) Ceric ammonium nitrate test.** Alcohols replace the nitrate ion of ceric ammonium nitrate, changing the colour of the solution from yellow to red.

$$\underset{\text{Yellow}}{(NH_4)_2\,[Ce(NO_3)_6]} + ROH \longrightarrow \underset{\text{Red}}{(NH_4)_2\,[Ce\,(OR)\,(NO_3)_5]} + HNO_3$$

*Procedure* Dissolve the given organic compound (50 mg) in water (1-2 ml) or dioxane and add ceric ammonium nitrate solution (few drops). Observe the colour change immediately. Appearance of a red colour shows the presence of an alcoholic group.

**Note** 1. A blank test should be performed if dioxane has been used as a solvent.

2. Some alcohols given only a transient red colouration, so look for red colour on mixing.

3. Some amines and phenols also give various colours in this test.

**(ii) Xanthate test.** Alcohols produce a yellow precipitate of xanthate on treatment with $CS_2$ in presence of an alkali.

$$ROH + \underset{\text{(Solid)}}{KOH} \longrightarrow R\overset{-}{O}\overset{+}{K} + H_2O$$

$$\xrightarrow{+\,CS_2} \underset{\text{Yellow precipitate}}{R{-}O{-}\overset{\overset{\displaystyle S}{\|}}{C}{-}\overset{-}{S}\,\overset{+}{K}}$$

*Procedure* Warm the compound (0.1g) with solid potassium hydroxide (one pellet) until it dissolves. Cool, wash with a little ether, decant the ether and add carbon disulphide (2-3 drops). Shake well. The formation of a yellow precipitate indicates the presence of an alcohol.

**(iii) Sodium metal test**. Reactive metals like sodium liberate hydrogen from alcohols.

$$2\,ROH + 2Na \longrightarrow 2RONa + H_2 \uparrow$$

*Procedure* In a dry test tube take the dry compound (1 ml of liquid or a few crystals of solid dissolved in dry benzene). To this add a very small piece of clean, dry sodium. Dissolution of sodium with brisk effervescence indicates the presence of an —OH group.

**Note** 1. A dry sample of the compound should be used.

2. Sodium should be handled carefully.

3. Excess of unreacted sodium should not be discarded in the sink or dust bin. It should be decomposed carefully by adding a small amount of alcohol.

4. A blank test should be performed if the given compound has been dissolved in a solvent.

5. Many other compounds like acetone, acids etc. also give this test.

**(iv) Test with Chromic acid in acetone.** This test is helpful in identifying the presence of the hydroxyl group, provided other easily oxidisable groups like—CHO, carbohydrate etc. are absent.

Both primary and secondary alcohols give a positive test. Tertiary alcohols do not answer the test (for experimental details, see page 21).

**(v) Iodoform test.** Alcohols which contain $CH_3$ CHOHR group are oxidised to $CH_3CO$— group during the reaction, and give a positive iodoform test (For experimental details are page 21)

$$CH_3\ CH_2\ OH \longrightarrow CH_3\ CHO \xrightarrow[NaOH]{I_2} CHI_3$$

$$CH_3\ CHOH\ CH_3 \longrightarrow CH_3COCH_3 \longrightarrow CHI_3$$

**(vi) Lucas test.** This test is used for distinction between primary, secondary and tertiary alcohols. The three types of alcohols can be distinguished by reacting with Lucas reagent (anhydrous zinc chloride in concentrated hydrochloric acid). Zinc chloride, itself being a strong Lewis acid, increases the acidity of the solution. Alcohols react with Lucas reagent at different rates, e.g.

$$\underset{\text{primary}}{RCH_2OH} + HCl \xrightarrow{ZnCl_2} \text{No reaction}$$

$$\underset{\text{secondary}}{R_2CHOH} + HCl \xrightarrow{ZnCl_2} R_2CHCl + H_2O$$

$$\underset{\text{Tertiary}}{R_3COH} + HCl \xrightarrow{ZnCl_2} R_3Cl + H_2O$$

Alcohols are soluble in Lucas reagent but the formed alkyl chlorides are insoluble.

*Procedure* Add Lucas reagent (10 ml) to the given organic compound (1 ml) in a test tube. Shake well and note the time required for the separation of two distinct layers.

| | | |
|---|---|---|
| Tertiary alcohol | – | layers separate immediately |
| Secondary alcohol | – | layer separation takes 5—10 minutes |
| Primary alcohol | – | A clear homogeneous solution is obtained |

Lucas test is applicable to only those alcohols which are soluble in the reagent, as it is based on the appearance of alkyl halides as a second liquid phase.

**(vii) Oxidation.** A primary alcohol gives an aldehyde and a secondary alcohol a ketone on oxidation with chromic acids. Tertiary alcohols do not oxidise under these conditions.

$$RCH_2OH \longrightarrow RCHO,\ R_2CHOH \longrightarrow R_2CO$$

*Procedure* Mix the given compound (1 g), water (5 ml), chromic acid (1 g) and concentrated sulphuric acid (1 ml) in a small distillation flask. Distil the mixture and collect the distillate. Test for the presence of an aldehyde or a ketone (page 20) in the distillate. Positive test with 2.4 DNP indicates the presence of a primary or a secondary alcohol.

**(viii) Spectroscopy.** IR—The characteristic absorption peaks present in alcohols are due to the O—H and C—O stretch. The absorption due to the O—H stretch is broad and highly variable depending mainly on the concentration of the alcohol.

In very dilute solutions, in a non-polar solvent, OH is unassociated (free or non-hydrogen bonded) and shows a sharp absorption band between 3650–3580 $cm^{-1}$.

Intermolecular hydrogen bonding increases with an increase in the concentration resulting in an absorption due to O—H shifting to a lower frequency, i.e. 3550–3200 $cm^{-1}$. The absorption due to the associated O—H is broad. Generally, both the peaks, i.e. at 3650–3580 $cm^{-1}$ and 3550–3200 $cm^{-1}$ are observed in the IR spectrum of an alcohol. If the O—H is intramolecularly associated, then the absorption is broad and independent of concentration. A strong peak at 1300-1000 $cm^{-1}$ is due to the C—O stretch in alcohols.

*NMR*—Oxygen causes a down field shift of $\alpha$-protons as well as the hydrogen attached to oxygen.

The hydrogen attached to oxygen may absorb between 1–5 $\delta$, the exact position depending on the concentration of the solution.

The protons attached to the carbon having the —OH group absorb between 3.3-4.0$\delta$.

| | |
|---|---|
| C – O<u>H</u> | $\delta$ 1.0 – 5.5 |
| <u>H</u>—C—OH | $\delta$ 3.3 – 4.0 |

For a primary alcohol ($RCH_2OH$), the – O<u>H</u> absorption should split into a triplet (n + 1 rule) and for a secondary alcohol ($R_2CHOH$), it should appear as a doublet. This splitting pattern is rarely observed because of the rapid exchange of hydrogen attached to the highly electronegative oxygen. For the same reason, the O—<u>H</u> does not split the neighbouring protons, e.g. in the spectrum of $CH_3CH_2OH$, —O<u>H</u> appears as a singlet and —$CH_3\underline{CH_2}OH$ as a quartet (being split only by the three protons of the $CH_3$ group).

Since the position of —OH absorption is variable, the peak due to —OH is easily recognised by exchange with $D_2O$ (when a small amount of $D_2O$ is added to the solution of alcohol after taking the spectrum, and the spectrum recorded again, the peak due to —O<u>H</u> disappears due to exchange of this proton with D).

## *(g) Esters*

Esters, the alky or aryl derivatives of acids are an important class of organic compounds. Cyclic esters are known as lactones. Esters of glycerol (triglycerides) are important natural compounds. Esters are extensively used in perfumary because of their pleasant smell. The presence of an ester group in a given compound can be detected by the following tests.

**(i) Hydroxamic acid test (Feigl test).** The ester is converted to hydroxamic acid by reaction with hydroxylamine. The formed hydroxamic acid reacts with $FeCl_3$ to produce a coloured complex.

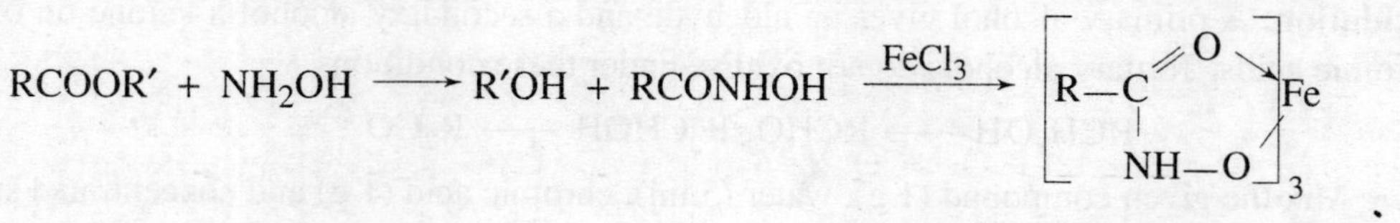


Magenta coloured complex

Carboxylic acids do not give this test. Other derivatives of acids like anhydrides and chlorides give the test. Most imides, some amides and nitriles give a positive test. Primary and secondary nitro compounds also give a positive test because of the presence of the *aci* form i.e.

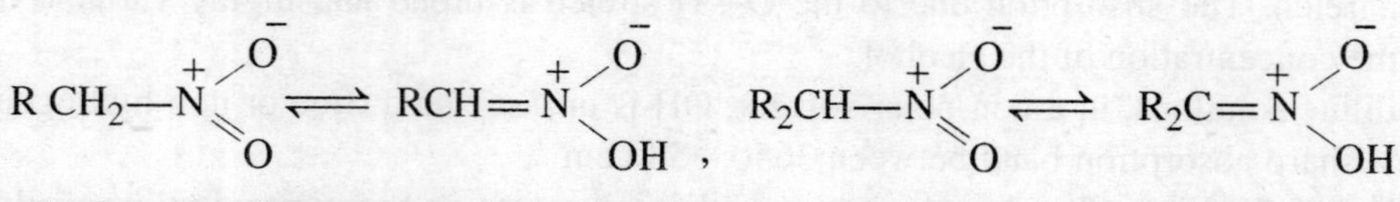


*Procedure* Perform the following preliminary test before carrying out the hydroxamic acid test.

Dissolve the compound (50 mg) in alcohol (1 ml) and add hydrochloric acid (1 M, 1 ml). Add ferric chlorides solution (few drops). If a red, blue, green or a violet colour is produced, the hydroxamic acid test cannot be applied on the given compound. Otherwise proceed as follows.

Mix the compound (40–50 mg) with hydroxylamine hydrochloride (1 ml, 0.5 M solution) and sodium hydroxide (0.2 ml, 6 M) and heat the mixture to boiling. Cool and add hydrochloric acid (2 ml, 1 M). If the solution is cloudy add ethanol (1-2 ml) to make it clear and then add ferric chloride solution (1-2 drops, 0.6 M). Add a little more of ferric chloride if necessary to obtain a permanent colour. Formation of a magneta colour confirms the presence of an ester.

**(ii) Hydrolysis.** An ester on hydrolysis with alkali gives an acid (as its sodium salt).

$$RCOOR' + NaOH \longrightarrow RCOO\,Na + R'OH$$

The formation of acid on hydrolysis can be detected as follows.

*Procedure* Add the given compound (0.1 g) to a dilute solution of sodium hydroxide (1%, 2 ml) containing a drop of phenolphthalein. If the pink colour of the alkaline solution is discharged in cold, add more of dilute sodium hydroxide solution to obtain a pink solution (the colour of the solution is discharged due to the presence of acidic groups). Now boil the solution for a few minutes, discharge of colour indicates the presence of an ester group.

**(iii) Spectroscopy**. IR-Esters show a strong peak at 1735–1750 $cm^{-1}$ due to C═O stretching. In addition, one or two strong bands between 1170–1270 $cm^{-1}$ are also observed in the infrared spectrum of esters due to C—O stretch.

*NMR*–In addition to other signals, the following signals are observed in the NMR spectrum of an ester.

1. The hydrogens attached to an alkyl oxygen absorb between 3.7 – 4.2 $\delta$ (R—C(=O)—O—C$\underline{H}$)
2. The hydrogens $\alpha$ to the carbonyl group absorb between 2 – 2.6 $\delta$ (—C$\underline{H}$—CO—O R)

## *(h) Lactones*

Lactones give positive hydroxamic acid test. On boiling with alkali, the lactone ring is opened up, acidification of the solution regenerates the original compound, whereas an ester is hydrolysed to the sodium salt of an acid, acidification of the solution gives the acid and not the original ester.

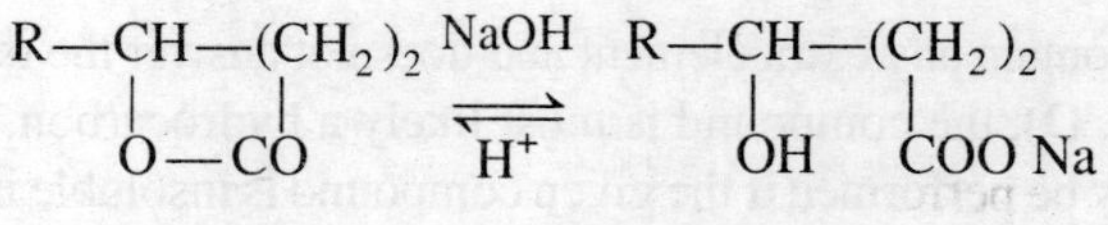

**(i) Spectroscopy.** *IR*—A very strong band in the IR spectrum of lactones at 1735–1750 $cm^{-1}$ is due to the C = O stretch. The stretching frequency depends on the size of the lactone ring, the C═O stretch may be shifted to ~ 1800 $cm^{-1}$ in a highly strained ring structure.

*NMR*—The important absorption positions are:

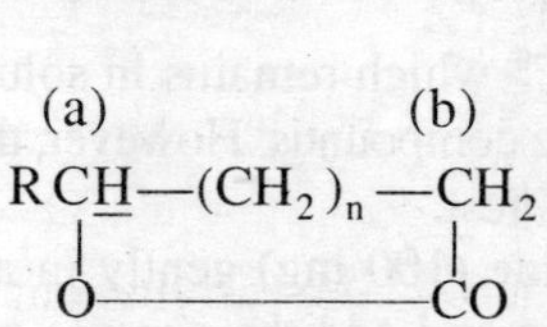

4 – 4.5$\delta$ due to a and 2 – 3.5 $\delta$ due to b protons

### *(i) Acid anhydrides*

Acid anhydrides give positive Feigl test. Anhydrides of some dicarboxylic acids condense with resorcinol in presence of conc. $H_2SO_4$ to give fluorescein (for details see page 19).

**(i) Spectroscopy**. *IR*—Acid anhydrides show two strong absorption peaks due to the C═O stretch at 1800-1870 $cm^{-1}$ and 1740-1790 $cm^{-1}$.

*NMR*—R <u>CH</u>$_2$—CO—O—CO—protons absorb in the range 2-3$\delta$.

### *(j) Ethers*

Most of the ethers are sparingly soluble in water but dissolve in cold concentrated sulphuric acid and are precipitated back on dilution with water (except for some aromatic ethers which may be partially sulphonated and so may be incompletely recovered on dilution).

Ethers, like hydrocarbons, are somewhat unreactive substances and great care is needed when attempting to assign a compound to one of these classes.

**(i)** The following test may be performed to show the presence of an ether group in the compound.

*Procedure* Make a dilute solution of iodine in carbon tetrachloride and to this solution (1 ml), add the given compound (0.5 ml or 0.2 g). A change of colour from purple to tan indicates the presence of an ether.

**Note** 1. A blank test should be performed for comparison.
2. Some aliphatic hydrocarbons also give a positive test.

**(ii) Spectroscopy.** *IR*—Ethers show a strong absorption peak in the region 1050—1300 $cm^{-1}$ due to C–O stretching. Aryl ethers absorb between 1200–1275 $cm^{-1}$ and alkyl ethers between 1050—1150 $cm^{-1}$.

*NMR*—The electronegative oxygen causes the absorption of protons attached to the $\alpha$-carbon atom to shift downfield.

$$R{-}O{-}\overset{|}{\underset{\underline{H}}{C}}{-} \qquad 3.3-4.0\ \delta$$

### *(k) Hydrocarbons*

If a compound does not contain any extra element and does not answer the test for all the above mentioned groups (containing C, H, O), the compound is most likely a hydrocarbon.

The following test may be performed if the given compound is insoluble in concentrated sulphuric acid.

**(i) Test with chloroform.** The test is based on the reaction of aromatic compounds with chloroform in presence of anhydrous aluminium chloride. The compound undergoes a series of Friedal-Crafts reaction to produce a coloured compound.

$$3\ C_6H_6 + CHCl_3 \xrightarrow{AlCl_3} (C_6H_5)_3\ CH + 3HCl$$

The colour is due to the formation of $Ph_3C^{\oplus}$ which remains in solution with $AlCl_4^-$.

The test is positive with many other aromatic compounds. However, the aromatic compounds containing strong deactivating groups do not answer this test.

*Procedure* Heat anhydrous aluminium chloride (100 mg) gently in a pyrex tube by keeping the tube almost horizontal on the flame. Cool the tube and add the organic compound (1 drop or 20 mg) and chloroform (2-3 drops) along the sides of the tube. Appearance of a bright colour indicates the presence of an aromatic compound.

The following colours are obtained with some of the common aromatic hydrocarbons.

| *Compound* | *Colour* |
|---|---|
| Benzene and its homologs | Orange to red |
| Naphthalene | Blue |
| Biphenyl | Purple |
| ArX (X = Cl, Br, I) | Orange to red |

**(ii) Reaction with sulphuric acid.** Shake gently the given organic compound (0.2 g) with fuming sulphuric acid (2 ml, 20%) cautiously. Aromatic hydrocarbons are soluble, while the saturated aliphatic and cyclic compounds are insoluble.

**(iii) Spectroscopy.** *IR*—*(1) Alkanes*—Saturated hydrocarbons show the following main absorptions.

$C_{sp^3}$–H appears below 3000 $cm^{-1}$. $C_{sp^2}$–H and $C_{sp}$–H appears above 3000 $cm^{-1}$. Methyl and methylene also absorb at 1465–1460 $cm^{-1}$. These absorptions are due to methylene scissoring and methyl asymmetric bending. Cycloalkanes show sharp absorption in this region, compared to alkanes. An absorption at 1380 $cm^{-1}$ is due to methyl symmetric bending vibration. It is absent in cycloalkanes lacking a methyl group and is a valuable diagnostic peak used to distinguish an alkane from a cycloalkane. The absorption at 1380 $cm^{-1}$ is split into a doublet in case an isopropyl or a *t*-butyl group is present. For an isopropyl group, the double band is symmetric in intensity and for a *t*-butyl group an asymmetric doublet is produced with the lower frequency peak more intense.

*(2) Alkenes*—Some characteristic peaks which alkenes show in the IR are:

| | |
|---|---|
| =C—H | between 3020 – 3100 $cm^{-1}$ |
| C=C | ~ 1650 $cm^{-1}$ |
| $RCH{=}CH_2$ | ~ 910 – 990 $cm^{-1}$ |
| $R_2C{=}CH_2$ | ~ 890 $cm^{-1}$ |

*(3) Alkynes*—Alkynes show a band between 2100-2260 $cm^{-1}$ due to C≡C stretch. This peak is more intense in the case of terminal alkynes. Terminal alkynes exhibit an intense and sharp peak due to ≡C—H stretch at 3300 $cm^{-1}$. This absorption is absent in disubstituted alkynes

(4) *Aromatic hydrocarbons*—Aromatic hydrocarbons show medium intensity absorption due to C—H stretch at 3010–3095 $cm^{-1}$ and a strong peak at 690–900 $cm^{-1}$ due to out-of-plane bending of C—H bond. The region 690–900 $cm^{-1}$ can also be used to distinguish o-, m- and p-disubstituted benzene derivatives, e.g. for ortho disubstituted, a very strong peak is observed at 750 $cm^{-1}$, for meta disubstituted at 780 and 800 $cm^{-1}$ and for p-disubstituted at 825 $cm^{-1}$.

The C—C stretch causes an absorption between 1500–1600 $cm^{-1}$.

Aromatic peaks are generally sharp and symmetrical whereas the peaks of aliphatic hydrocarbons are often unsymmetrical and broad.

*NMR. (1) Alkanes*—Aliphatic protons absorb at the following values.

| | |
|---|---|
| primary $R{-}C\underline{H}_3$ | 0.9 $\delta$ |
| secondary $R_2C\underline{H}_2$ | 1.3 $\delta$ |
| tertiary $R_3C\underline{H}$ | 1.5 $\delta$ |

The cycloalkanes show the same chemical shift values as alkanes. *(2) Alkenes*.The $\pi$ electron cloud present in alkenes deshields the vinylic protons and the absorption appears downfield.

| | |
|---|---|
| Vinylic C=C—$\underline{H}$ | 4.6 – 5.9 $\delta$ |
| Allylic C=C—C$\underline{H}$ | 1.7 – 2.5 $\delta$ |

*(3) Alkynes.* The chemical shift values for the protons of alkynes are as follows:

| | |
|---|---|
| C≡C—<u>H</u> | 2.0–3.0 $\delta$ |
| — C≡C—C<u>H</u> | 1.8–2.5 $\delta$ |

*(4) Aromatic hydrocarbons.* Due to the ring current produced by the $\pi$ electron cloud of aromatic rings, the absorption is shifted far downfield. Following are the chemical shift values:

| | |
|---|---|
| Aromatic Ar—<u>H</u> | 6–8.5 $\delta$ |
| Benzylic Ar—C—<u>H</u> | 2.2–3.0 $\delta$ |

The 3° benzylic appears most downfield and 1° benzylic the least.

## 2.2 Compounds Containing Halogen

### (a) Alkyl, Vinyl and Aryl halides

If the given compound has been shown to contain a halogen, it may be one of the above mentioned compounds. Their further characterisation is based on the fact that alkyl halides are reactive, the halogen is easily substituted by a nucleophile. The vinyl and aryl halides are unreactive as substitution of halogen does not occur easily.

$$RX + AgNO_3 \xrightarrow[\text{ethanol}]{\Delta} RONO_2 + AgX \downarrow$$

**(i) Alcoholic silver nitrate test**. Tertiary alkyl halides are the most reactive and primary the least. Bromides and iodides react more readily than chlorides. Aryl halides do not react under these conditions. Benzyl halides are the most reactive.

*Procedure* Add the organic compound (1 drop) to an alcoholic solution of silver nitrate (2 ml, 0.1 M). If no reaction is observed within 5 minutes, warm the mixture in a water bath. Formation of a precipitate indicates the presence of RX, a clear solution shows the presence of ArX or vinyl halide.

**(ii) Reaction with Sodium iodide in acetone.** This test depends on the fact that NaCl and NaBr are insoluble in acetone whereas the iodide is soluble. This test can be used for showing the presence of RCl or RBr, when these are treated with a solution of sodium iodide in acetone, a white ppt (if RCl is present) and a pale yellow ppt (if RBr is present) is formed.

$$RCl + NaI \xrightarrow{\text{Acetone}} RI + NaCl \downarrow \text{(White ppt)}$$

$$RBr + NaI \longrightarrow RI + NaBr \downarrow \text{(pale yellow ppt)}$$

*Procedure* Take sodium iodide (1 g) in acetone (1 ml) and add the chloro or bromo compound to it (1-2 drops of liquid or 50 mg of the compound dissolved in acetone), Shake and allow the tube to stand for 3-4 minutes at room temperature. If a precipitate does not separate, warm the mixture for 5-6 minutes. A precipitate indicates RCl or RBr. ArCl and ArBr do not form a precipitate. The rate of the substitution reaction depends on the nature of R, being the fastest with tertiary alkyl chlorides/bromides.

**(iii) Spectroscopy**. IR—In RX and ArX, the C—X absorption position depends on the nature of the halogen present. The absorption for various C—X bonds are:

| | |
|---|---|
| C – I | ~ 500 $cm^{-1}$ |
| C – Br | 500 – 600 $cm^{-1}$ |
| C – Cl | 600 – 800 $cm^{-1}$ |
| C – F | 1000 – 1400 $cm^{-1}$ |

In addition to the above, the absorption peaks due to the hydrocarbon part of the molecule will be observed (page 31).

*NMR*-In RX, the absorption due to the hydrogen attached to the carbon bearing halogen is shifted downfield, the shift depending on the electronegativity of the halogen and the number of halogen atoms present, e.g.

| | |
|---|---|
| $\underline{\underline{H}}$—C—F | 4-4.5 $\delta$ |
| $\underline{\underline{H}}$—C—Cl | 3-4 $\delta$ |
| $\underline{\underline{H}}$—C—Br | 2.5-4 $\delta$ |
| $\underline{\underline{H}}$—C—I | 2-4 $\delta$ |

The number of halogens also affect the position of proton absorption e.g.

| | |
|---|---|
| $C\underline{H}_3Cl$ | 3.0 $\delta$ |
| $C\underline{H}Cl_3$ | 7.24 $\delta$ |

### (b) Acid chlorides

**(i) Test with silver nitrate.** If the compound contains halogen (chlorine) as an extra element and its aqueous solution gives a white precipitate with silver nitrate solution it may be an acid chloride.

*Procedure* Heat the compound (1 g) with dilute sodium hydroxide solution (10 ml, 10%) and acidify the clear solution. Formation of an acid indicates the presence of an acid chloride.

**(ii) Spectroscopy**. *IR*—Acid halides show a very strong absorption band due to the C═O stretch between 1815–1770 $cm^{-1}$.

*NMR*—$R\underline{CH_2}COCl$ absorb between $\delta$ 2.0 – 3.0. For example $C\underline{H}_3COCl$ absorbs at $\delta$ 2.7.

## 2.3 Compounds Containing Nitrogen

### (a) Nitro compounds

Nitro substituted organic compounds are generally yellow in colour. An important reaction of the nitro group is that it can be reduced and the product of reduction depends on the nature of the reducing agent. Some of the products obtained during reduction are:

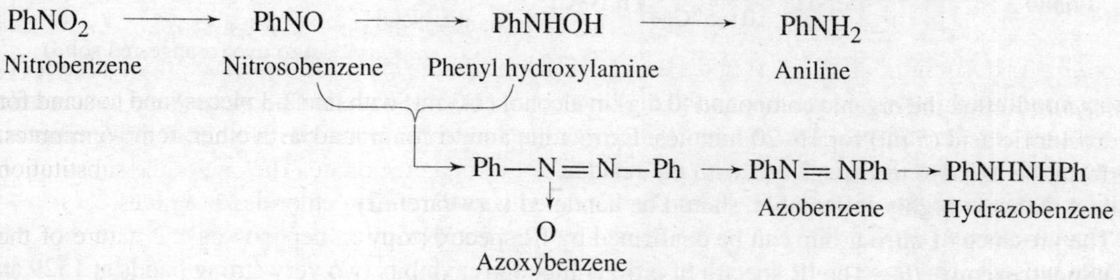

Because of these reactions, the nitro group can act as an oxidising agent.

**(i) Mulliken Barker test**. The test is based on the fact that with a neutral reducing agent, $-NO_2$ is reduced to the —NHOH group. The formed hydroxylamine reduces the Tollen's reagent and itself gets oxidised to a nitroso compound.

$$PhNO_2 + 4H \xrightarrow{Zn + NH_4Cl} PhNHOH + H_2O$$

$$PhNHOH + 2[Ag(NH_3)_2]OH \longrightarrow Ph\,NO + 2H_2O + 4NH_3 + \underset{\text{grey-black}}{2\,Ag\downarrow}$$

*Procedure* Dissolve the given compound (100 mg) in dilute alcohol (1 ml), add a solution of ammonium chloride (0.5 g in 1 ml water) and zinc dust (~ 100 mg). Boil the solution for 2-3 minutes and filter into another tube containing freshly prepared Tollen's reagent. Formation of a grey precipitate turning black indicates the presence of a $-NO_2$ group.

**Note** 1. The test is not applicable if the compounds contains other functional groups like —CHO, carbohydrate etc., which can reduce Tollen's reagent.

2. Freshly prepared Tollen's reagent should be used.

**(ii) Ferrous hydroxide test** Organic compounds that are oxidising agents, oxidise ferrous hydroxide (blue) to ferric hydroxide (brown).

$$PhNO_2 + 6\,\underset{\text{Blue}}{Fe(OH)_2} + 4\,H_2O \longrightarrow Ph\,NH_2 + 6\,\underset{\text{Brown}}{Fe(OH)_3}$$

*Procedure* Take the compound (20 mg or 1 drop) in a test tube, add freshly prepared ferrous ammonium sulphate solution (2 ml, 5%), and a drop of 3 M sulphuric acid. Add potassium hydroxide in methanol (1 ml, 2M). Stopper the tube and shake well. Appearance of a brown precipitate (darkening or greenish colour is a negative test) indicates the presence of a nitro group.

**Note** 1. Perform the reaction in a small tube and fill it with the contents to avoid oxidation of $Fe^{+2}$ to $Fe^{+3}$ by air.

2. Other compounds capable of oxidising $Fe^{+2}$ to $Fe^{+3}$ will also give a positive test.

**(iii) Reduction with tin and hydrochloric acid**. Aromatic nitro compounds may be reduced to amino compounds with tin and HCl. Azo dye test can then be performed on the reduced product. The test is applicable only if the original compound does not undergo the azo dye test.

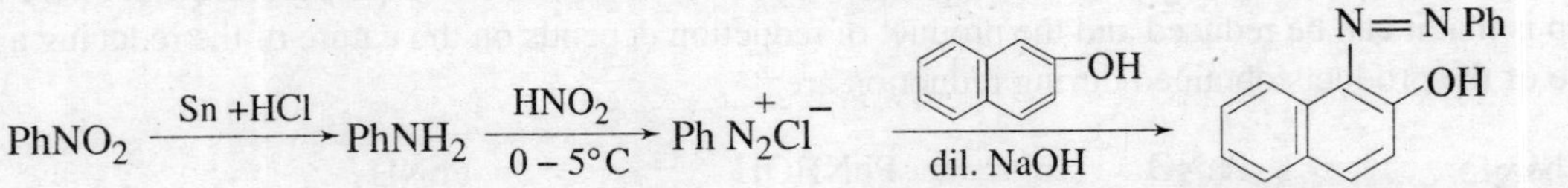


Azo dye (orange-red solid)

*Procedure* Reflux the organic compound (0.5 g) in alcohol (2-3 ml) with tin (2-3 pieces) and concentrated hydrochloric acid (5 ml) for 15–20 minutes. Extract the amino compound with ether, remove ether and perform the azo dye test (page 36) with the residue.

**Note 1.** Ether is highly inflamable, should be handeled very carefully

The presence of nitro group can be confirmed by IR spectroscopy.

**(iv) Spectroscopy**. *IR*—The IR spectra of nitro compounds exhibits two very strong bands at 1529 and 1350 $cm^{-1}$.

*NMR*—$NO_2$ group is a very strong electron withdrawing group and causes the downfield shift of the protons.

In aliphatic nitro compounds, the protons on the $\alpha$ carbon atom absorb between 4.2-4.6 $\delta$. Example, $C\underline{H}_3NO_2$ at 4.3 $\delta$ and $(CH_3)_2\ C\underline{H}NO_2$ at 4.4 $\delta$.

In aromatic nitro compound, the ortho hydrogens to the nitro group are highly deshielded and appear ~ 8.0 $\delta$ as compared to the protons of benzene which appear ~ 7.0 $\delta$.

## (b) Amines (primary, secondary and tertiary)

Alkyl or aryl substituted ammonia derivatives are called amines. These compounds are basic in nature, the strength of the basic character depends on the nature of the alkyl or aryl group. Amines are generally soluble in acids like HCl.

**(i) Action of nitrous acid**. The test can be used for identification of primary, secondary and tertiary amines (the product obtained by treatment with nitrous acid is subjected to further confirmatory tests in all cases).

A primary aromatic amine reacts with nitrous acid to form a diazonium salt. The reaction takes place at about 0–5°C. At a higher temperature the formed diazonium compound undergoes decomposition.

$$C_6H_5NH_2 \xrightarrow[0-5°C]{NaNO_2\ +\ HCl} C_6H_5N{=}\overset{+}{N}\overset{-}{Cl}$$

Benzene diazonium chloride

Since the diazonium compounds are ionic, they remain dissolved in water and hence a clear solution is obtained.

An aliphatic primary amine on reaction with nitrous acid forms an alcohol and gives nitrogen.

$$RNH_2 + HNO_2 \longrightarrow ROH + N_2\uparrow + H_2O$$

Amides also, give nitrogen on reaction with nitrous acid.

$$RCONH_2 + HNO_2 \longrightarrow RCOOH + N_2\uparrow + H_2O$$

Both aliphatic and aromatic secondary amines react with nitrous acid to form N-nitroso derivatives which are liquids, insoluble in the reaction mixture.

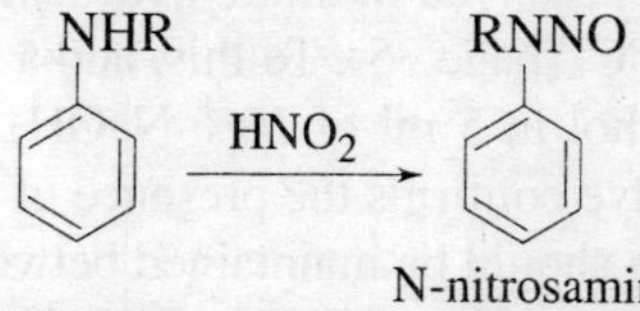


N-nitrosamine

Therefore, when a secondary amine reacts with nitrous acid, an oily N-nitrosamine separates from the reaction mixture.

An aromatic tertiary amine on treatment with nitrous acid forms p-nitroso derivative (if the para position is free), which separates as its hydrochloride from the reaction mixture, as an orange solid.

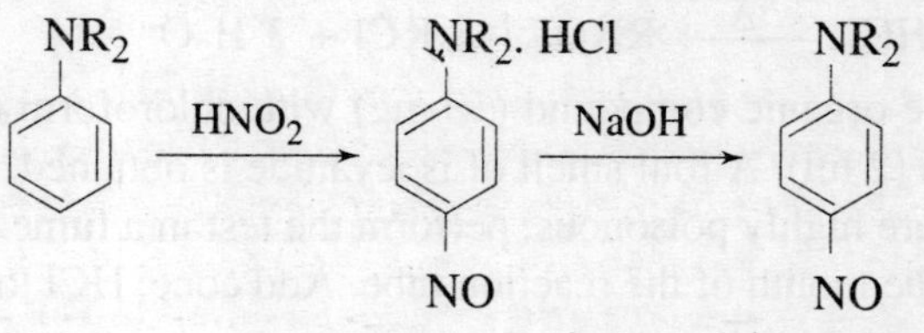


orange solid green solid

An aliphatic tertiary amine forms a salt with $HNO_2$.

$$R_3N + HNO_2 \longrightarrow R_3N.\ HNO_2$$

*Procedure* Dissolve the organic compound (0.2 g) in concentrated hydrochloric acid (2 ml) and water (2 ml) in a boiling tube. Cool the solution in an ice bath to 0–5°C. To this add a cooled (5°C) solution of sodium nitrite (0.2 g in 2–3 ml water) dropwise and mix well. Test for the presence of free nitrous acid (starch iodide paper turns blue when treated with $HNO_2$). If free nitrous acid is not present add some more of the cooled solution of sodium nitrite.

| *Observation* | *Inference* |
|---|---|
| 1. Vigorous evolution of $N_2$ | May be an aliphatic primary amine (confirm by carbylamine test, page 36). |
| 2. Compound dissolves, but reprecipitates on neutralisation of solution | May be a tert. aliphatic or a heterocyclic amine. |
| 3. Clear solution | May be an aromatic primary amine (confirm by azo dye test, page 36). |
| 4. An oily product separates or solution is cloudy | Secondary amine (confirm by Liebermann nitroso test page 37). |
| 5. Deep orange-red solid | Tertiary aromatic amine (confirm by neutralisation). |

**Note** Nitrosamines are highly carcinogenic substances.

**(ii) Azo dye test for confirmation of aromatic primary amines**. Aromatic primary amines form diazonium salts when treated with $HNO_2$. These diazo compounds couple with phenols like $\beta$-naphthol to produce orange-red coloured azo dyes which are water insoluble.

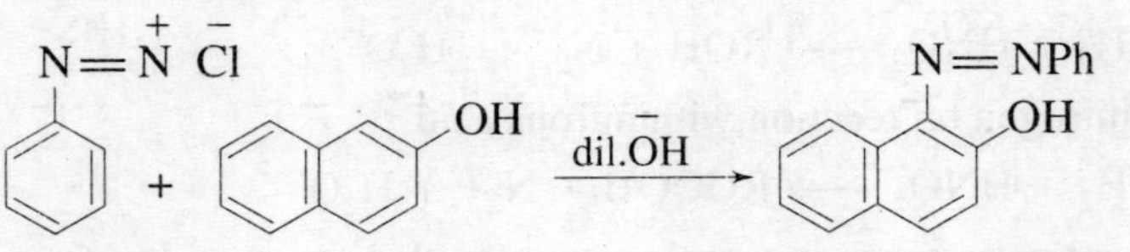


Azo dye

*Procedure* Treat the organic compound dissolved in dilute hydrochloric acid with a cooled solution of $NaNO_2$ in water and cooled to below 5°C, (page 35). To this, add a well cooled solution of $\beta$-naphthol in sodium hydroxide (0.2 g of $\beta$-naphthol in 5 ml of 10% NaOH) dropwise. Stir the mixture with a glass rod. Formation of an orange-red dye confirms the presence of an aromatic primary amine.

**Note** 1. The temperature of the solution should be maintained between 0–5°C throughout the reaction.
2. In certain compounds containing —OH, —COOH, —$SO_3H$ groups etc., the azo dye may remain dissolved.

**(iii) Carbylamine test**. This is also known as isocyanide test, because isocyanides are produced when a primary amine (aliphatic and aromatic) is treated with chloroform in presence of alkali.

$$RNH_2 + 3KOH + CHCl_3 \xrightarrow{\Delta} RNC + 3\ KCl + 3\ H_2O$$

*Procedure* Warm carefully the organic compound (10 mg) with chloroform (1–2 drops) and alcoholic potassium hydroxide solution (2 ml). A foul smell of isocyanide is obtained.

**Note** Since the isocyanides are highly poisonous, perform the test in a fume chamber and do not smell the gas directly from the mouth of the reaction tube. Add conc. HCl to the tube, let it stand for a few minutes in the fume chamber and then discard the contents of the tube into the sink. Hydrochloric acid converts the isocyanide to the amine.

**(iv) Reaction with 5-nitro-salicylaldehyde reagent**. Primary amines react with 5-nitro-salicylaldehyde to produce the Schiff's base or anil, which gives a precipitate of the following complex on reaction with $NiCl_2$.

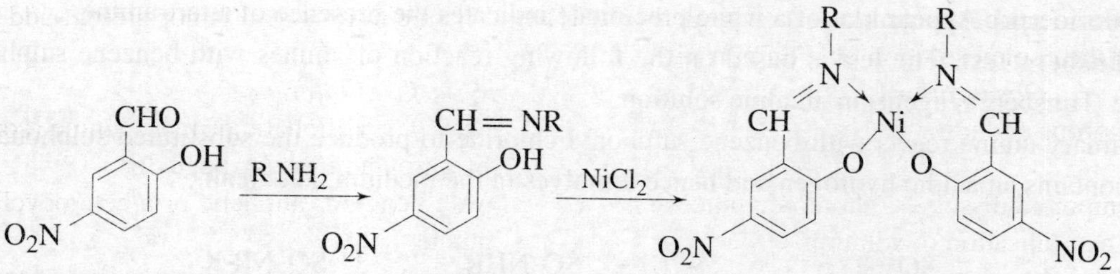


*Procedure* Dissolve the compound (1-2 drops) in water (1 ml) or dil. HCl (4-5 ml). Add 5-nitro-salicylaldehyde reagent (2 ml) (see note below) to the above solution. Aliphatic amines immediately form a copious precipitate. Aromatic amines form a precipitate after 3-4 minutes.

**Note** 5-nitrosalicyldehyde reagent is prepared by dissolving 5-nitrosalicyldehyde (0.48) in a mixture of triethanolamine (12 ml) and water (20 ml) by shaking. To the above solution is added a solution of nickel chloride (0.4g) in water (4 ml). The solution is diluted with water (90 ml.)

**Confirmatory test for secondary amines**

**(v) Liebermann nitroso test.** The N-nitrosamine obtained by reaction of secondary amine with $HNO_2$ is heated with phenol in presence of concentrated sulphuric acid.

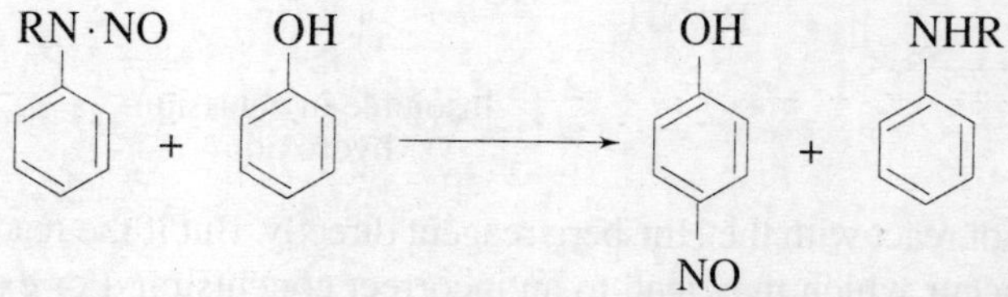


The formed p-nitrosophenol undergoes further reaction (for details of chemistry, see page 23).

*Procedure*-Treat the solution of the secondary amine in HCl with sodium nitrite solution (page 36). Extract the N-nitrosamine with ether. Wash the ether extract with dilute urea solution to remove $HNO_2$. Then dry it over anhydrous $CaCl_2$. Filter and evaporate ether (caution, ether is highly inflammable). Heat the residue with phenol (few crystals), cool, add conc. $H_2SO_4$ (For details of subsequent treatment (see page 23).

**(vi) Confirmatory test for tertiary amine.** On reaction with nitrous acid, tertiary amines form an orange solid. Basification of the solution with dilute sodium hydroxide produces a green solid.

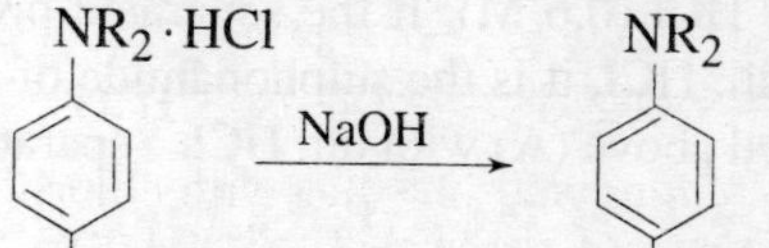


Dark green solid

*Procedure* After the reaction with $HNO_2$, basify the solution with NaOH. A green product is obtained. Extract this with ether, wash and dry the ether extract. Evaporate the ether and recrystallise the formed para-nitroso compound. Determine its melting point.

Chart 2.1 summarises the reaction of amines with nitrous acid

**(vii) Nessler reagent test.** Add Nessler's reagent to the solution of the organic compound in dilute hydrochloric acid. Appearance of a white precipitate indicates the presence of a tert-amine.

**(viii) Hinsberg test.** The test is based on the following reaction of amines with benzene sulphonyl-chloride (Hinsberg reagent) in alkaline solution.

A primary amine reacts with benzene sulphonyl chloride to produce the substituted sulphonamide which contains an acidic hydrogen and hence dissolves in the medium, i.e. alkali.

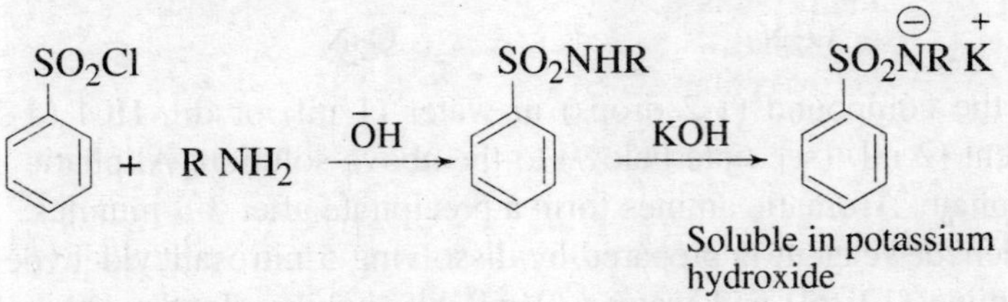


A secondary amine on the other hand forms a substituted sulphonamide which is insoluble in alkali because it does not have an acidic hydrogen.

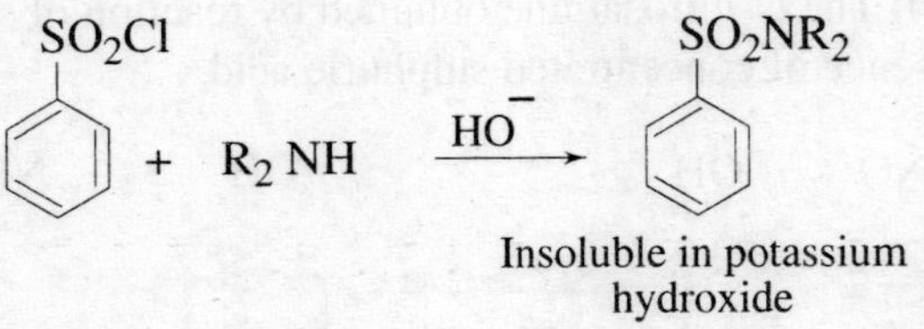


A tertiary amine does not react with the Hinsberg reagent directly. But if the reaction time is prolonged, some side reaction may occur which may lead to an incorrect conclusion. For example:

$$R_3N + Ph\,SO_2Cl \longrightarrow Ph\,SO_2\,\overset{+}{N}R_3\,\overset{-}{Cl} \xrightarrow{R_3N} PhSO_2NR_2 + R_4\overset{+}{N}Cl^-$$

*Procedure* Follow the procedure as closely as possible to avoid any misinterpretation.

Mix aqueous potassium hydroxide (10 ml, 2M), the compound (0.2 ml or 0.2 g) and benzene sulphonyl chloride (0.7 ml) in a test tube. Stopper and shake vigorously until the odour of benzene sulphonyl chloride has gone (cool if necessary). The reaction takes 5-6 minutes. The solution should be alkaline, if necessary add some more potassium hydroxide solution.

If a solid or an oil separates, decant or filter it (keep the filtrate (A) for testing a primary amine). Treat this separated product with dil. HCl (0.6 M). If the separated product is soluble, it is a tertiary amine. If the product is insoluble in dil. HCl, it is the sulphonamide of a secondary amine.

Treat (neutrilise) the filtrate obtained above (A) with dil. HCl. Separation of a solid or an oil indicates the presence of a primary amine.

**(ix) Sodium nitroprusside test.** The Rimini and Simon test involves the reaction of amines with acetone and acetaldehyde respectively. The condensation products obtained give coloured complexes on treatment with sodium nitroprusside solution. These tests have been modified for identification of aromatic amines.

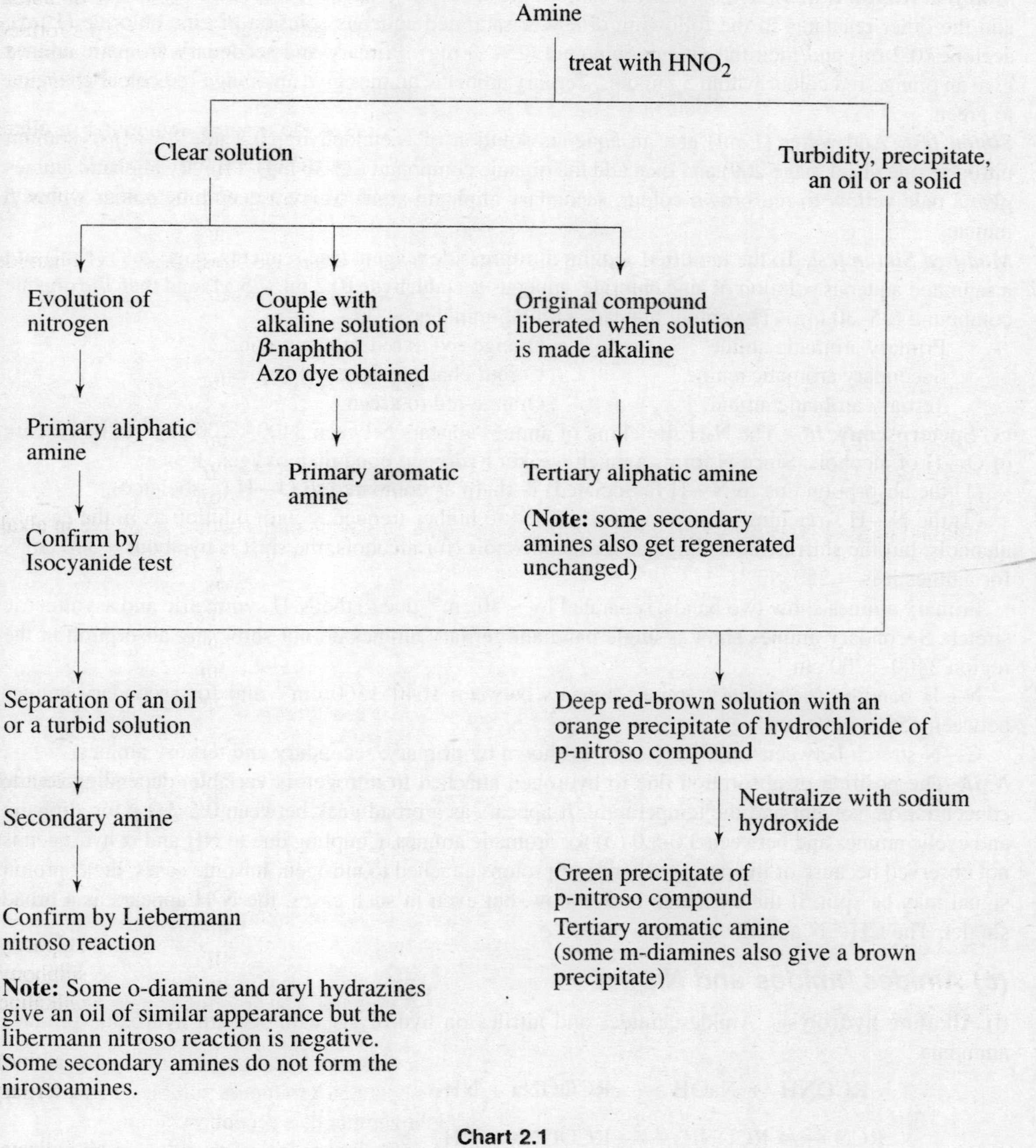


**Chart 2.1**

***Rimini test*.** Add a mixture containing amine (25–30 mg), water (1 ml) and acetone (0.2 ml) to sodium nitroprusside solution (1 ml). Both aliphatic primary and secondary amines give a deep red colour.

***Modified Rimini test.*** Take the modified sodium nitroprusside reagent (1 ml) (see reagents, page 269), add the other reactants in the following order—a saturated aqueous solution of zinc chloride (1 ml), acetone (0.2 ml) and then the given compound (25–30 mg). Primary and secondary aromatic amines give an orange-red colour within 5 minutes. Tertiary aromatic amines give an orange-red colour changing to green.

***Simon test.*** Add water (1 ml) and an aqueous solution of acetaldehyde (0.2 ml, 2.5 M) to sodium nitroprusside (1 ml, page 269) and then add the organic compound (25-30 mg). Primary aliphatic amines give a pale yellow to red-brown colour, secondary aliphatic amines give a deep blue colour within 5 minutes.

***Modified Simon test.*** To the modified sodium nitroprusside reagent (see reagents, page 269) (1 ml) add a saturated aqueous solution of zinc chloride, aqueous acetaldehyde (0.2 ml, 2.5 M) and then the organic compound (25–30 mg). The colour appears within 5 minutes.

| | |
|---|---|
| Primary aromatic amine | Orange-red to red-brown colour |
| Secondary aromatic amine | Colour changes from red to green |
| Tertiary aromatic amine | Orange-red to green |

**(x) Spectroscopy**. *IR*—The N-H stretching of amines appears between 3400–3200 $cm^{-1}$ as in the case of O—H of alcohols. Since N forms a much weaker hydrogen bond than oxygen, hence:

(1) the absorption due to N—H (associated) is sharp as compared to O—H (associated).

(2) the N—H stretching absorptions are shifted to higher frequency with dilution as in the case of alcohols, but the shift is not as large as for the alcohols (for alcohols, the shift is by about ~ 300 $cm^{-1}$, for amines it is ~ 250 $cm^{-1}$).

Primary amines show two bands, separated by ~ 50 $cm^{-1}$ due to the N-H symmetric and asymmetric stretch. Secondary amines show a single band and tertiary amines do not show any absorption in the region 3400–3200 $cm^{-1}$.

N—H bending for primary amines appears between 1640–1560 $cm^{-1}$ and for secondary amines between 1580–1490 $cm^{-1}$.

C—N stretch between 1360-1180 $cm^{-1}$ is shown by primary, secondary and tertiary amines.

*NMR*–The position of absorption due to hydrogen attached to nitrogen is variable, depending on the concentration, solvent and the temperature. It appears as a broad peak between 0.5-3.0 $\delta$ for aliphatic and cyclic amines and between 3.0-5.0 ($\delta$) for aromatic amines. Coupling due to N<u>H</u> and $\alpha$-hydrogen is not observed because of the rapid exchange of protons attached to nitrogen. In some cases, the $\alpha$-proton signal may be split, if the exchange rate is slow, but even in such cases, the N-H appears as a broad singlet. The C<u>H</u>—N absorbs between 2.0-3.0 $\delta$.

## (c) Amides, Imides and Nitriles

**(i) Alkaline hydrolysis.** Amides, imides and nitriles on hydrolysis with sodium hydroxide produce ammonia.

$$RCONH_2 + NaOH \longrightarrow RCOONa + NH_3$$

$$RCN \longrightarrow RCONH_2 \longrightarrow RCOONa + NH_3$$

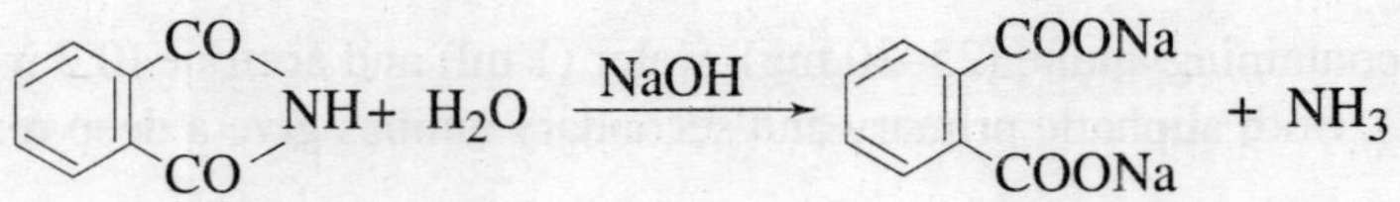

*Procedure* Boil the given organic compound (0.2 g) with NaOH (5 ml, 15%). Evolution of ammonia-which can be tested with a rod dipped in hydrochloric acid or a moist litmus paper or Nessler's reagent, indicates the presence of an amide, nitrile or an imide group.

**(ii) Hydroxamic acid test**. Amides and nitriles on reaction with hydroxyl amine followed by treatment with ferric chloride give a wine-red to violet colour (page 28).

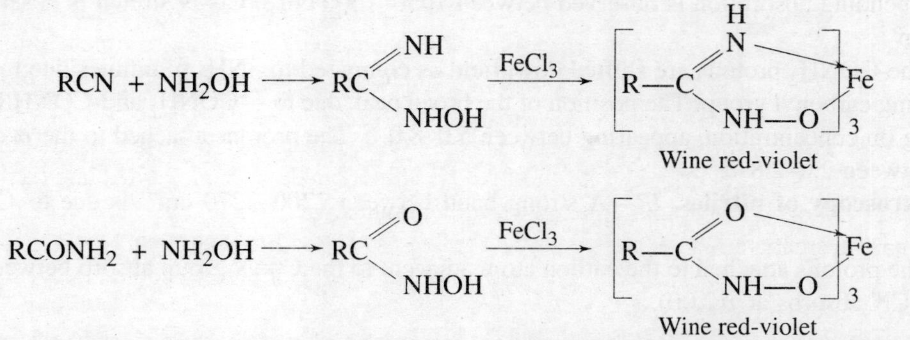


**Note** Esters also give the above test, but the amides, imides and nitriles on heating with alkali give ammonia.

*Procedure* Dissolve hydroxylamine hydrochloride in propylene glycol (1 M solution). Mix this solution (2 ml) with the amide or nitrile dissolved in propylene glycol (20-30 mg in a minimum amount of the solvent). Add a solution of potassium hydroxide (1M, 2 ml). Heat the mixture to boiling for two minutes, cool to room temperature and add alcoholic ferric chloride solution (a few drops, 0.5 M). A red to violet colour is a positive test.

**(iii) Imides.** Add a saturated solution of potassium hydroxide to a saturated solution of the compound. A white precipitate of the potassium salt of the imide is formed.

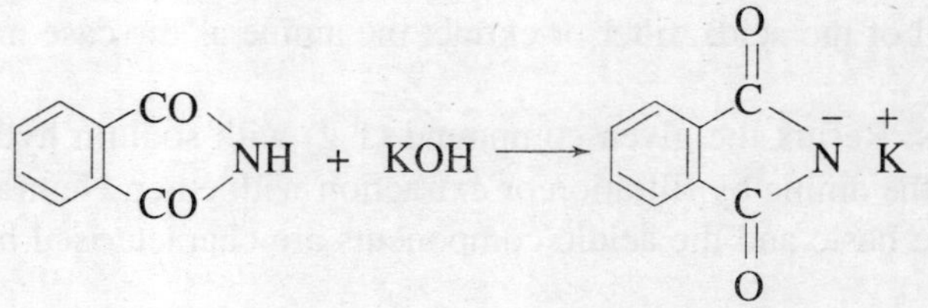


**(iv) Biuret test.** This test is performed to confirm the presence of urea.

On heating, urea forms biuret, which when treated with $Cu^{+2}$ in the presence of a base, forms a purple coloured complex.

$$2NH_2CONH_2 \xrightarrow[-NH_3]{\text{dry } \Delta} NH_2CONHCONH_2$$

Biuret

$\downarrow CuSO_4$

Purple coloured complex

*Procedure* Heat the given compound (0.2 g) in a dry test tube till the melted compound solidifies and evolution of $NH_3$ ceases, cool, dissolve the solid in dil. NaOH and add a dilute solution of copper sulphate (1-2 drops). Formation of a purple colour indicates the presence of urea.

Amids, nitriles, imides can be distinguished on the basis of their IR spectra.

**(v) Spectroscopy of amides.** *IR*—The carbonyl stretching of amides causes an absorption at 1650–1690 $cm^{-1}$. The lower absorption frequency as compared to other carbonyl groups is due to the reasonance in the —$CONH_2$ group.

Primary amides show two bands in the region of 3150–3500 $cm^{-1}$ due to N—H stretching, secondary amides show one band and tertiary amides show none since they do not have a N—H bond.

N—H bending absorption is observed between 1650–1600 $cm^{-1}$. C—N stretch is observed at 1360 – 1180 $cm^{-1}$.

*NMR*—The-$CON\underline{H}_2$ protons are shifted downfield as compared to –$N\underline{H}_2$ of amines due to an electron withdrawing carbonyl group. The position of the broad peak due to —$CON\underline{H}_2$ and-$CON\underline{H}R$ is variable, depending on concentration, appearing between 5.0–8.0 $\delta$. The protons attached to the $\alpha$-carbon atom absorb between 2.0–2.8 $\delta$.

**(vi) Spectroscopy of nitriles.** *IR*—A strong band between 2200–2270 $cm^{-1}$ is due to C≡N of the nitrile.

*NMR*—The protons attached to the carbon atom adjacent to the C≡N group absorb between 2.0–3.0 $\delta$ (e.g. $C\underline{H}_3CN$ absorbs at $\delta$ 2.00).

## (d) Anilides and substituted amides

Anilides and substituted amides are hydrolysed in presence of an acid or a base to produce an acid and an amine.

$$RNHCOR' + H_2O \xrightarrow[\text{or } OH^-]{H^+} RNH_2 + R'COOH$$

**(i) Acid hydrolysis.** Reflux the given compound (1 g) with sulphuric acid (50 %, 10 ml) for about 30 minutes. The acid obtained by hydrolysis will separate as a solid (filter) or an oil (extract with ether). Work up the ether extract and convert the acid to a derivative. Neutralise the acidic solution obtained above (after the removal of the acid), filter or extract the amine as the case may be and convert it into a crystalline derivative.

**(ii) Alkaline hydrolysis**. Reflux the given compound (1 g) with sodium hydroxide (20 ml, 15%) for about one hour. Isolate the amine by filtration or extraction with ether. Neutralise the alkaline solution and isolate the acid. The basic and the acidic components are characterised by preparation of suitable derivatives.

## (e) Amino acids

Amino acids contain both—$NH_2$ and—COOH groups and show general reactions of amines and carboxylic acids. $\alpha$-Amino acids belonging to the configuration L-series occur in proteins and peptides, found in plants and animals. At a particular pH, each amino acid exists as a zwitter ion.

$$\overset{+}{N}H_3\text{—}\underset{\underset{R}{|}}{CH}\text{—}COOH \underset{+H^+}{\overset{-H^+}{\rightleftharpoons}} \overset{+}{N}H_3\text{—}\underset{\underset{R}{|}}{CH}\text{—}CO\overset{-}{O} \underset{H^+}{\overset{-H^+}{\rightleftharpoons}} NH_2\text{—}\underset{\underset{R}{|}}{CH}\text{—}COO^-$$

The following tests may be performed for their identification.

**(i) Ninhydrin test**. It is widely used for the identification of amino acids, proteins and peptides. Heating a solution of amino acid with a solution of ninhydrin results in the development of a blue-violet colour. Proline, hydroxyproline, o-, m- and p- amino benzoic acids do not give a positive test.

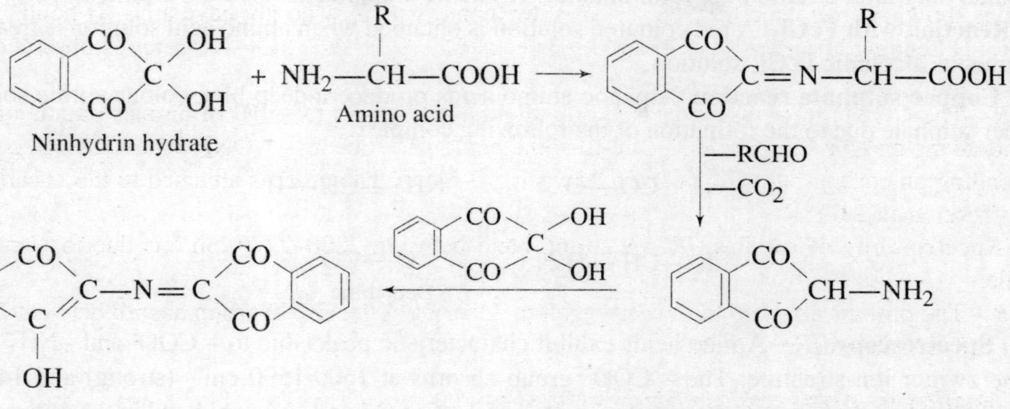


Blue

*Procedure* Heat a small amount of the aqueous solution of the organic compound (20 mg in 1 ml water) with ninhydrin solution (1 ml). The appearance of a blue-violet colour indicates the presence of an $\alpha$- or a $\beta$-amino acid.

**(ii) Reaction with Isatin**. Isatin reacts with amino acids to produce a range of colours, e.g. proline produces an intense blue, valine a red and alanine a pink-red colour with isatin. The structure of the product with proline is shown below.

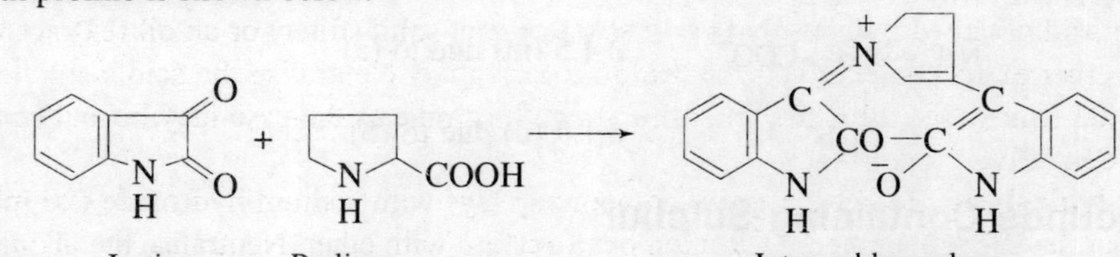

Isatin Proline Intense blue colour

*Procedure* To a solution of isatin in alcohol (1%, 1 ml) add an aqueous solution of the organic compound (15-20 mg). Appearance of an intense blue, red or pink-red colour indicates the presence of an amino acid.

**(iii) Reaction with nitrous acid.** Nitrogen is liberated when amino acids are treated with cold aqueous nitrous acid (page 35). Aromatic amino acids form diazonium salts which couple with $\beta$-naphthol (page 36). Some aromatic amino acids like anthranilic acid, form the diazonium salts which are unstable and undergo decomposition liberating nitrogen.

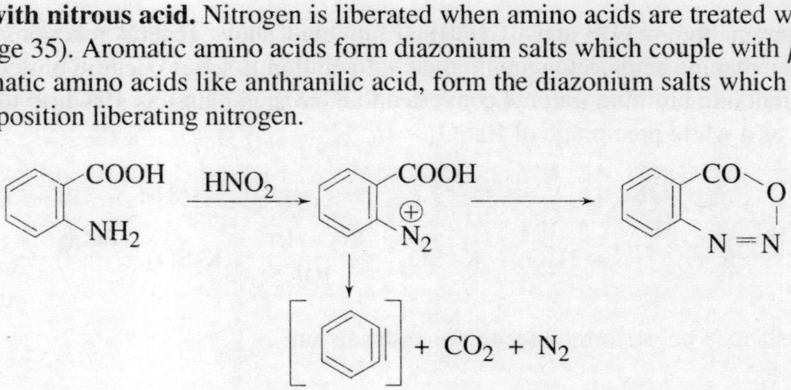

**(iv) Carbylamine reaction.** Isocyanides are obtained from amino acids (page 36) but the odour is not readily detected because an acidic isocyanide is formed which forms a non volatile salt.
**(v) Reaction with $NaHCO_3$.** Aliphatic amino acids react with $NaHCO_3$ solution (page 19) but $CO_2$ is liberated only after a lapse of several minutes. Aromatic amino acids give $CO_2$ immediately.
**(vi) Reaction with $FeCl_3$.** A red coloured solution is obtained when amino acid solution is treated with an aqueous alcoholic $FeCl_3$ solution.
**(vii) Copper sulphate reaction.** Alphatic amino acids produce a deep blue colour with a solution of copper sulphate due to the formation of the following complex.

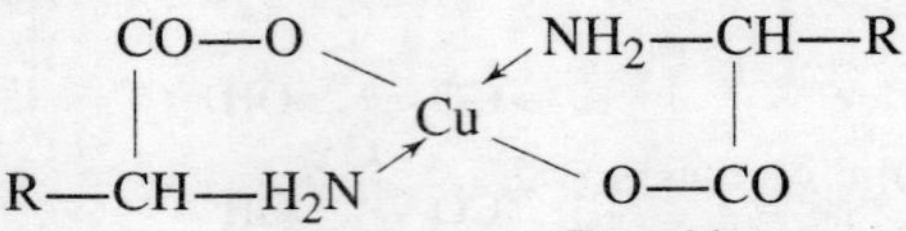


Deep blue

**(viii) Spectroscopy.** *IR*—Amino acids exhibit characteristic peaks due to – $COO^-$ and $-NH_3^+$ because of the zwitter ion structure. The – $COO^-$ group absorbs at 1600-1590 $cm^{-1}$ (strong) and 1400 $cm^{-1}$ (weak) due to symmetric and asymmetric stretch of the C=O bond. A board band in the range of 3100–2600 $cm^{-1}$ (strong) is due to the $-^+NH_3$ stretching. Absorptions at 1660–1600 $cm^{-1}$ (weak) and 1400 $cm^{-1}$ (strong) are due to the asymmetric and symmetric bending of $-NH_3^+$.
*NMR*—A broad absorption due to – $\overset{+}{N}H_3$ of the amino acid zwitter ion occurs between 7.4–7.5 $\delta$. The protons attached to the $\alpha$-carbon atom of the $\alpha$-amino acids absorb between 4.2–4.6 $\delta$.

For example $\overset{+}{N}H_3$—$C\underline{H}_2$—$COO^-$ $\delta$ 4.2 (q)

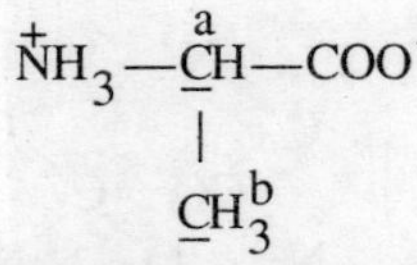


$\delta$ 4.5 (m) due to (a)

$\delta$ 1.9 (d) due to (b)

## 2.4 Compounds Containing Sulphur

Compounds containing sulphur as an extra element may have functional groups like $-SO_3H$, –SH, R—S—R, R—S—S—R.

### *(a) Sulphonic acids*

Compounds containing the – $SO_3H$ group are called sulphonic acids. These are acidic in nature and are generally water soluble. Sulphonic acids on fusion with KOH form $K_2SO_3$, which on oxidation with a mild oxidising agent like bromine water is converted to $K_2SO_4$. Addition of $BaCl_2$ to the clear solution thus obtained gives a white precipitate of $BaSO_4$.

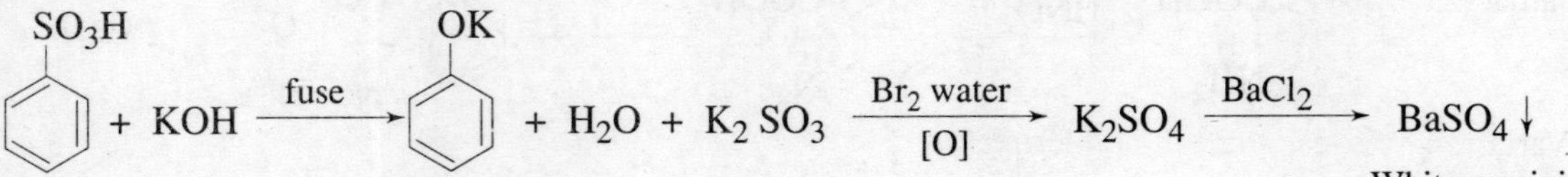


White precipitate

*Procedure* Fuse the organic compound (0.5 g) with KOH (1 g) in a nickel crucible. Extract the fused mass (after cooling) with water (2–3 ml). Acidify the aqueous solution with dil. HCl, add a few drops of bromine water and shake well. To this clear solution add a solution of $BaCl_2$ (1 ml). Formation of a white precipitate indicates the presence of a sulphonic acid.

### (b) Thiols and Mercaptans

Presence of the –SH group in the given organic compound can be confirmed by the following tests.

**(i) Reaction with lead acetate.** Thiols react with lead acetate to give a yellow precipitate of lead mercaptide.

$$2\ RSH + Pb(OCOCH_3)_2 \longrightarrow 2\ CH_3COOH + Pb(SR)_2 \downarrow$$

*Procedure* Add the given organic compound (0.1 g) to lead acetate solution (1 ml). Formation of a yellow precipitate indicates the presence of the—SH group.

**(ii) Reaction with nitrous acid.** Mercaptans react with nitrous acid (or better with RONO + HCl) to produce a red precipitate.

$$RCH_2SH + HNO_2 \longrightarrow \underset{\text{Red precipitate}}{RCH_2SNO} + H_2O$$

$$RCH_2SH + R'ONO \longrightarrow \underset{\text{Red precipitate}}{RCH_2\ SNO} + R'OH$$

$R_3C$—SH gives a pale yellow-green colouration.

*Procedure* Dissolve the given compound (20 mg) in ether (1-2 ml) and add HCl (6N, 1-2 drops) followed by ethyl or n-butyl nitrite (2-3 drops). Shake Well. A red colour indicates $RCH_2SH$ or $R_2CHSH$, a yellow-green colour in the ether layer indicates the presence of $R_3CSH$.

**(iii) Reaction with sodium nitroprusside.** Thiols give a purple colouration on reaction with sodium nitroprusside.

*Procedure* Dissolve the given organic compound (0.1 g) in ammonium hydroxide and then add sodium nitroprusside (2-3 drops). A purple colouration indicates the presence of a thiol.

**(iv) Spectroscopy**. *IR*—A weak band in the region of 2600–2550 $cm^{-1}$ is due to S-H stretching. Since most of the other organic compounds do not show any absorption in this region, the S-H stretch, though a weak band, may be seen in the IR spectrum of mercaptans and thiophenols.

*NMR*–The protons attached to the $\alpha$-carbon atom of a mercaptan absorb in the following region.

| | |
|---|---|
| $C\underline{H}_3$-SH | 2.1 $\delta$ |
| $RC\underline{H}_2$-SH | 2.6 $\delta$ |
| $R_2C\underline{H}$-SH | 3.1 $\delta$ |

The hydrogen attached to –SH does not undergo rapid exchange and hence is coupled to protons on the adjacent carbon atoms. $D_2O$ exchange is however easy.

| | |
|---|---|
| Ar S-$\underline{H}$ | 2.8-3.6 $\delta$ |
| R S-$\underline{H}$ | 1.2-1.6 $\delta$ |

Variation in concentration and temperature do not result in a large shift.

## 2.5 Compounds Containing Nitrogen and Sulphur

### (a) Sulphonamides

**(i) Reaction with NaOH**. Like simple amides, the sulphonamides give $NH_3$ on fusion with an alkali. Acidification of the resultant solution liberates $SO_2$. Formation of $SO_2$ distinguishes $SO_2NH_2$ from $CONH_2$.

$$RSO_2NH_2 + NaOH \longrightarrow RSO_3H + NH_3$$

$$\downarrow NaOH$$

$$Na_2SO_3 \xrightarrow{HCl} SO_2 \uparrow$$

*Procedure* Fuse the organic compound (0.5 g) with KOH (1 g) in a nickel crucible. Ammonia is evolved (test by usual methods). Extract the fused mass (after cooling) with water (5 ml) and divide the solution in two parts. Acidity and part with dil. HCl, $SO_2$ is evolved (test by usual methods). To the second portion add bromine water (1-2 ml) after acidification. To the clear solution add barium chloride solution (1 ml). Formation of a white precipitate indicates the presence of a sulphonamide.

**(ii) Spectroscopy.** *IR*—Sulphonamides show strong bands between 1370–1335 $cm^{-1}$ and 1170–1155 $cm^{-1}$.

Primary and secondary sulphonamide also show bands due to N-H stretching. Two bands at 3390-3330 $cm^{-1}$ and 3300-3240 $cm^{-1}$ for N-H of the $-SO_2NH_2$ group and an absorption at 3265 $cm^{-1}$ is due to N–H of the $-SO_2$ NHR group.

# 3

# Preparation and Recrystallisation of Derivatives

After identifying the functional group present in the unknown organic compound, refer to the tables on page 113 to 266. For example, if the compound has been identified as a phenol, see the table which gives a list of phenols. The compounds are arranged in order of increasing melting/boiling points. There may be a few phenols having nearly the same m.p./b.p. as the unknown. For example, if the m.p. of the unknown is found to be 109–110°, it may be one of the following phenols:

| *Compound* | *Melting point* |
|---|---|
| 1. Orcinol | 107° |

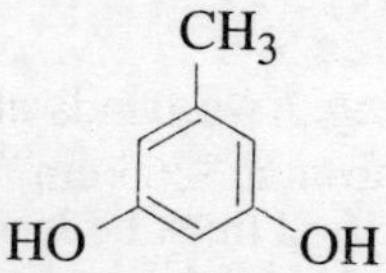

| | |
|---|---|
| 2. p-Hydroxyacetophenone | 109° |

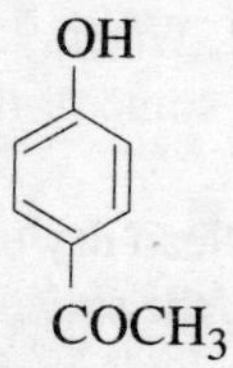

| | |
|---|---|
| 3. Resorcinol | 110° |

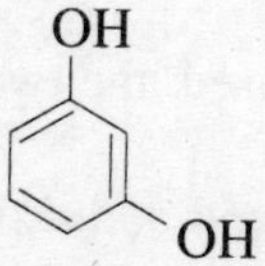

| | |
|---|---|
| 4. Bromohydroquinone | 110° |

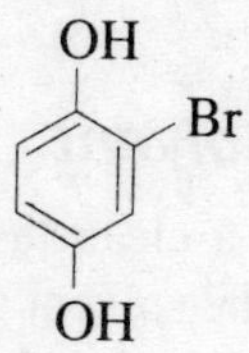

But if the given unknown compound does not contain an extra element, it cannot be bromohydroquinone, if it does not have a keto group, it cannot be p-hydroxyacetophenone. It may be either resorcinol or orcinol. Now prepare two crystalline derivatives, keeping in mind the following points.

1. The derivative should be such that it can be easily and quickly prepared.
2. The melting point of the chosen derivative should be between 50–200°C.
3. Physical and chemical properties of the derivatives should be different from the compound, especially its melting point.

The derivatives should be recrystallised before the determination of their melting points. A brief account of the technique employed for recrystallisation is given below.

## Recrystallisation

Recrystallisation is a very important technique often employed for the purification of solids.It is based on the fact that all organic solids are more soluble in hot than in cold solvents, so that the solid dissolved on heating can be obtained back on cooling.

The following steps should be performed for purification by crystallisation.

(a) Selection of a solvent or a mixture of solvents,
(b) Dissolution of the solute in required amount of the solvent,
(c) Decolorisation (charcolisation) for removal of coloured impurities,
(d) Crystallisation, filtration and drying of the solid.

### *(a) Selection of a Solvent*

The solvent chosen for recrystallisation should be such that the solid to be purified must be insoluble or nearly insoluble in the cold and fairly soluble in the hot solvent. The boiling point of the solvent should be lower than the melting point of the solid so that it can be removed during the drying of the solid.

The rule of like dissolves like should be kept in mind while selecting a solvent. Water, 95% ethanol, methanol, petroleum ether, THF, $CCl_4$, ethyl acetate, acetone and acetic acid are some of the common solvents employed for recrystallisation.

*Procedure* For selection of the solvent, take about 20–30 mg of the solid in a clean dry tube and add about 0.5–1 ml of the solvent, stir well, if the solid is soluble at room temperature, the solvent cannot be used for crystallisation. If the solid is insoluble, heat the test tube in a hot water bath or directly on a flame (caution). If the solid dissolves in the boiling solvent, filter and then cool the solution. If crystals appear on cooling, the solvent is suitable for recrystallisation.

If a pure (single) solvent is found unsuitable, a mixture of solvents may be used, the two solvents should be miscible with each other e.g. water-ethanol or ethyl acetate-benzene.

Dissolve the compound in the hot solvent (A) in which it is freely soluble and add dropwise the solvent (B) in which the compound is insoluble till a turbid solution is obtained. Add dropwise the solvent (A) to get back a clear solution and allow the solution to cool undisturbed for the separation of crystals.

### *(b) Dissolution of the Solute in the Required Amount of Solvent*

After selecting a suitable solvent for crystallisation, take the impure solid in a clean test tube or a conical flask (never use a beaker for recrystallisation). Add a few millilitres of the solvent and heat the solution on a water bath with constant stirring or siwrlling (do not heat inflammable liquids directly on a flame). If the solid does not dissolve, add a few more millilitres of the solvent and heat again (do not add excess of solvent). If near the end of the dissolution process, it appears that an additional amount of the solvent is not dissolving any more of the solid, then this small amount of undissolved solid is likely to be an insoluble impurity. Do not add any more solvent but filter the hot solution through a fluted filter paper and allow the filtrate to stand undisturbed for crystallisation.

### *(c) Decolourisation*

If the resultant solution is highly coloured, add a small amount of charcoal, boil the solution for some more time (the coloured impurities are adsorbed on the surface of charcoal), filter the hot solution through a fluted filter paper and allow the filtrate to cool for the separation of crystals.

### *(d) Crystallization*

Do not cool the solution obtained above rapidly, allow it to slowly cool to room temperature. If no crystals form, cool the solution in an ice-bath. Scratch the sides of the tube with a glass rod, just above the surface of the solution to initiate crystallisation or add a very small amount of solid compound to induce crystal formation.

If an oil separates an cooling (1) either scratch it against the sides of the tube (or flask) to solidify, (2) add a few drops of the solvent, heat, allow to solidify, or (3) add a few crystals of the solute to initiate the crystal formation.

## Filtration, Drying etc.

Filter the crystals by vacuum filtration and collect the mother liquor in a clean dry flask or a filter tube. Cool the mother liquor in an ice bath or concentrate and cool to obtain the second crop of the crystals. Do not mix with the first crop until their purity has been confirmed by melting point determination. Transfer the crystals from the filter funnel on to a filter paper and air dry or dry over hot water bath or in an air oven maintained at a much lower temperature than the melting point of the solid or by keeping in a vacuum desiccator.

## 3.1 Compounds containing Carbon, Hydrogen and Oxygen

### *(a) Derivatives of Carbohydrates*

**(i) Acetate.** All carbohydrates can be converted to acetates. Acetylation should be complete (i.e. all OH groups should be converted to $-OCOCH_3$). The $\alpha$- or the $\beta$-isomer of the acetate results depending on the catalyst employed for acetylation.

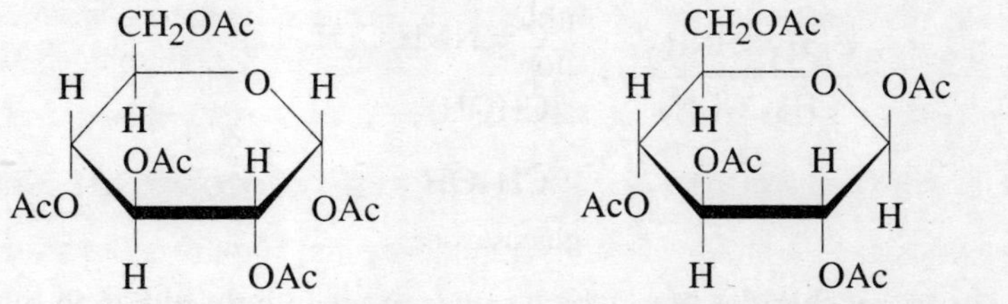


$\alpha$-D-glucose penta-acetate $\quad$ $\beta$ - D-glucose penta-acetate

***β-Acetate*** Fused sodium acetate is used as a catalyst for acetylation.

Take fused sodium acetate (1 g), acetic anhydride (5 ml) and the carbohydrate (1 g) in a dry boiling tube fitted with a condenser. Reflux the reaction mixture (30 min. – 2hrs). Pour the reaction mixture over ice cold water (~ 50 ml) with vigorous stirring to decompose the excess acetic anhydride. Filter the solid, wash with some cold water and recrystallise from ethanol.

***Fused sodium acetate.*** Sodium acetate is available in the laboratory as trihydrate or in the anhydrous form. The trihydrate loses water in two stages but decomposes at a high temperature (above 120°C). The anydrous form also contains moisture.

$$CH_3COONa{\cdot}3H_2O \xrightarrow[120°]{\Delta} CH_3COONa.\ H_2O \xrightarrow{\Delta} CH_3COONa$$

*From trihydrate.* Heat (120°C) the sodium acetate trihydrate in a dry china dish, the solid will melt. As the heating is continued, it solidifies; continue heating till the solid melts again, allow the melted solid to cool and stir while cooling. Use this fused sodium acetate immediately or keep in a dessicator.

From monohydrate or anhydrous sodium acetate. Heat in a dry china dish till the solid melts, allow it to cool to room temperature with continuous stirring. Use it immediately or store in a dessicator.

***α-Acetate.*** Reflux a mixture of acetic anhydride (10 ml), fused zinc chloride (0.5 g) and the carbohydrate (1.5 g) in a water bath for one hour (add the carbohydrate preferably in small lots during refluxing). Pour the reaction mixture into a beaker containing ice cold water (100 ml). Stir well to decompose acetic anhydride. Filter the formed product and recrystallise from methanol or ethanol.

***Fused zinc chloride.*** Zinc chloride is a highly hygroscopic substance. Heat in a dry china dish till the liquid starts emitting white fumes. Stop heating and continuously stir while cooling, use the fused zinc chloride immediately.

***Acetate of di- and tri-saccharides.*** Place sugar (1g), fused sodium acetate (1 g) and acetic anhydride (10 ml) in a round bottomed flask fitted with a condenser. Heat the mixture in a boiling water bath to get a clear solution and then gently heat on a wire gauze for half an hour. Transfer the mixture in a porcelain dish and evaporate several times by adding a little ethanol (to remove excess acetic anhydride) extract the residue with a little hot water (to remove $CH_3COONa$) and crystallise the remaining solid from ethanol.

**Note** Alternatively the procedure adopted for the preparation of the *β*-acetate (given above) is followed.

**(ii) Osazone**. Reducing carbohydrates of the type CHO—CHOH—R and $CH_2OH$—CO—R can be converted to osazones by treatment with an excess of phenylhydrazine.

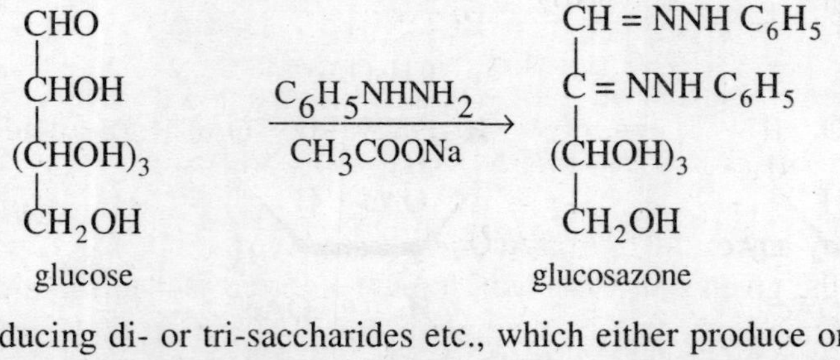


glucose　　　　glucosazone

Nonreducing di- or tri-saccharides etc., which either produce only one type of monosaccharide on hydrolysis or produce a mixture of such monosaccharides which form the same osazone, can also be converted to osazones after hydrolysis. For example sucrose.

***Osazone of reducing carbohydrates.*** Dissolve the carbohydrate (1 g), phenyl hydrazine hydrochloride (2 g) and sodium acetate (3 g) in water (15 ml). Filter if necessary and heat the solution (loosely corked) in a boiling water bath for 30 minutes. Note the time required for the formation of a yellow to orange coloured osazone. Filter and recrystallise from alcohol.

The time required for the formation of osazones of some of the carbohydrates is given in Table 3.1.

**Table 3.1**

| *Carbohydrates* | *Time (in minutes)* |
|---|---|
| Mannose | 0.5 |
| Fructose | 2-3 |
| Glucose | 4-5 |
| Xylose | 8-10 |
| Arabinose | 10 |
| Rhamnose | 10 |
| Galactose | 15-20 |
| Maltose and lactose | 20-25 (on cooling) |
| Sucrose | 30 |

*Alternative method* Take the compound (0.4 g) and dissolve in water (16 ml). In another tube take glacial acetic acid (10–12 drops), water (4 ml) and phenyl hydrazine (8–10 drops), mix well to get a homogeneous solution. Mix the two solutions and heat the mixture in a boiling water bath to get the osazone.

***Osazone of nonreducing carbohydrates***. Some nonreducing disaccharides on hydrolysis given a mixture of two monosaccharides (e.g. glucose and fructose from sucrose) which yield the same osazone. Such disaccharides may be converted to the osazone by the following procedure.

Place sucrose (1 g) in a round bottomed flask, add water (5-6 ml), concentrated hydrochloric acid (2-3 drops) and reflux for half an hour. After the hydrolysis neutralise the solution with solid sodium acetate and prepare the osazone as described above.

## (b) Derivatives of Carboxylic Acids

**(i) Amide, anilide or p-toluidide**. The acid is first converted to acid chloride and the acid chloride is then converted to any of the above mentioned derivatives.

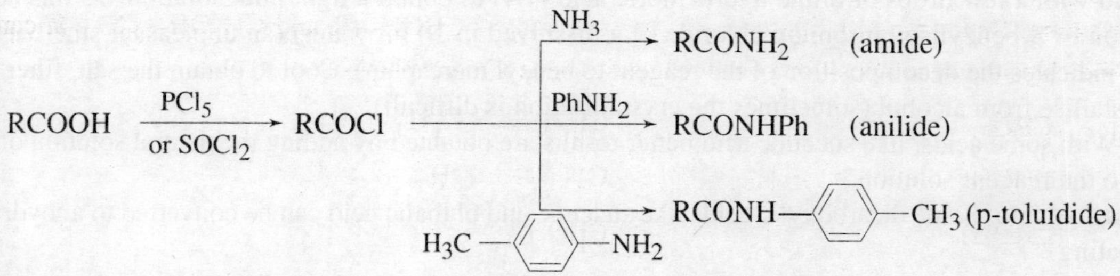


The acid chloride may be prepared by any of the following procedures:

*(a) Using thionyl chloride*: Reflux the given acid (1 g) with thionyl chloride (3-4 ml) in a dry round bottomed flask in a water bath (in a fume cupboard) for 30 minutes. Distill off excess of thionyl chloride and cool the residue.

**Note** 1. Formic, malonic and oxalic acid cannot be converted to their chlorides by the above method.
2. Excess of thionyl chloride cannot be removed by distillation if the boiling point of acid chloride is close to that of thionyl chloride; in such cases use only one molar equivalent of thionyl chloride.
3. Use two drops of dry pyridine for conversion of phthalic acid and p-nitrobenzoic acid to their chlorides to speed up the reaction.

*(b) Using phosphorus pentachloride:* In a dry boiling tube take the acid (1 g), phosphorus pentachloride (4 g) and heat in a boiling water bath (in a fume cupboard) (continuously stir the mixture as it is heated to obtain a liquid). The crude acid chloride thus obtained may be used for further reactions.

***Amide*** Treat the acid chloride cautiously with concentrated ammonia (20 parts, specific gravity 0.88), warm for a few minutes and then cool. If no solid separates on cooling, evaporate to dryness on a water bath and recrystallise the crude amide from water or dilute ethanol.

**Note** Amides of lower aliphatic acids are very soluble in water and hence difficult to isolate.

***Anilide*** Treat the acid chloride with a solution of aniline (1-2 ml) in benzene, warm on a water bath, cool and filter if a solid separates out. If the solid anilide does not separate, wash the benzene layer with water, dilute hydrochloric acid, dilute sodium hydroxide and finally with some water. Dry the remaining benzene solution over anhydrous sodium sulphate or anhydrous calcium chloride, remove the solvent by distillation and recrystallise the residue from benzene or alcohol.

*Alternative procedure* Take the acid (0.5 g) and aniline (1 g) in a dry round bottomed flask, fitted with an air condenser. Heat the mixture in an oil bath for 1–2 hours at 140–160°C. Cool and dissolve in benzene or ether. Wash the solution with dilute hydrochloric acid, sodium hydroxide and finally with water. Dry the organic layer, remove the solvent carefully and recrystallise the residue from ethanol.

***p-Toluidide*:** This can be prepared by using *p*-toluidine instead of aniline.

**(ii) S-Benzylisothiuronium salt**. Acids give crystalline salts when treated with S-benzyl isothiuronium chloride.

$$RCOONa + C_6H_5CH_2SC\begin{smallmatrix} \diagup NH_2 \\ \diagdown\!\!\diagdown \overset{+}{N}H_2Cl^- \end{smallmatrix} \longrightarrow C_6H_5CH_2SC\begin{smallmatrix} \diagup NH_2 \\ \diagdown\!\!\diagdown \overset{+}{N}H_2R\overset{-}{C}OO \end{smallmatrix} + NaCl$$

To a solution (or suspension) of the acid (0.5 g) in water or alcohol, (10 ml), add 1-2 drops of phenolphthalein and neutralise with sodium hydroxide solution (1N). Treat the pink coloured solution so formed with a few drops of dilute hydrochloric acid (1N) to obtain a light pink solution. To this add a solution of S-benzylisothiuronium chloride (2 g dissolved in 10 ml water) (an unpleasant smell at this stage indicates the decomposition of the reagent to benzyl mercaptan). Cool to obtain the salt, filter and recrystallise from alcohol (sometimes the crystallisation is difficult).

**Note** With some acids, like succinic acid better results are obtained by adding the neutral solution of the acid to the reagent solution.

**(iii) Anhydride**. Some dicarboxylic acids like succinic and phthalic acid can be converted to anhydrides by heating

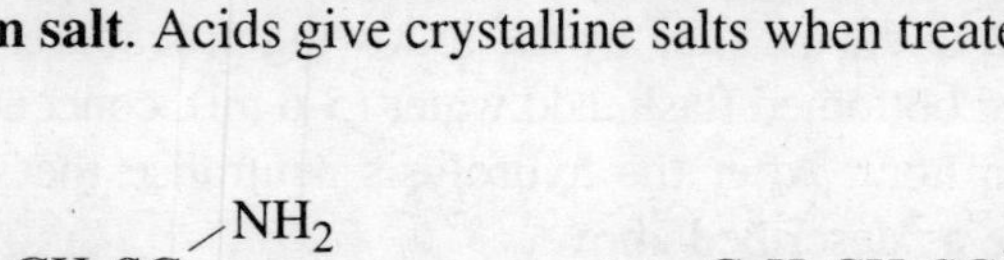


Phthalic acid Phthalic anhydride

Take the dicarboxylic acid (1 g) in a dry china dish. Cover the dish with a filter paper punched with small holes. Keep an inverted funnel on the dish and loosely plug the stem of the funnel with cotton. Heat the dish on a sand bath using a small flame. The anhydride sublimes and deposits on the inner sides of the funnel, from where it is collected.

## *(c) Derivatives of Carbonyl Compounds*

Both aldehydes and ketone contain a $>C{=}O$ functional group and hence form crystalline derivatives with similar reagents.

**(i) 2, 4-Dinitrophenylhydrazone**. Aldehydes and ketones form crystalline 2, 4-dinitrophenylhydrazones when treated with 2, 4-Dinitrophenylhydrazine.

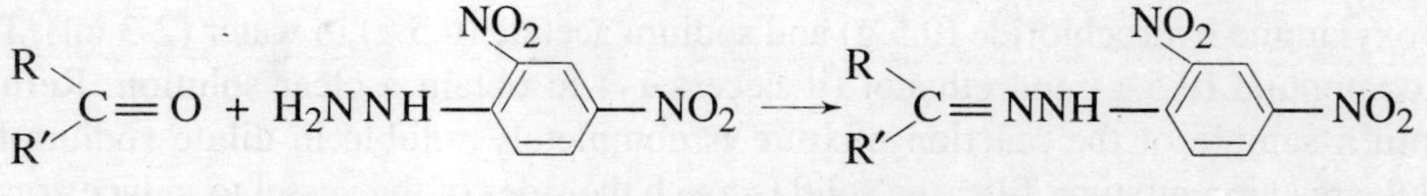

Take 2, 4-dinitrophenylhydrazine (0.2 g) in methanol (10 ml), in a conical flask. To this add concentrated hydrochloric acid (1 ml, dropwise). Filter if necessary, cool and add this to the solution of aldehyde or ketone (0.2 g in 2-3 ml methanol). Warm on a water bath for 1-2 minutes, cool, filter and recrystallise from alcohol.

**Note** 1. The reagent itself may get precipitated at times, m.p. 194°C. The precipitated reagent is soluble in dilute sulphuric acid where as the 2, 4-DNP of the carbonyl compound is insoluble.

2. In case of lower aliphatic ketones like acetone etc., it is better to add the ketone solution dropwise with shaking to a solution of 2, 4-DNP.

**(ii) p-Nitrophenylhydrazone.**

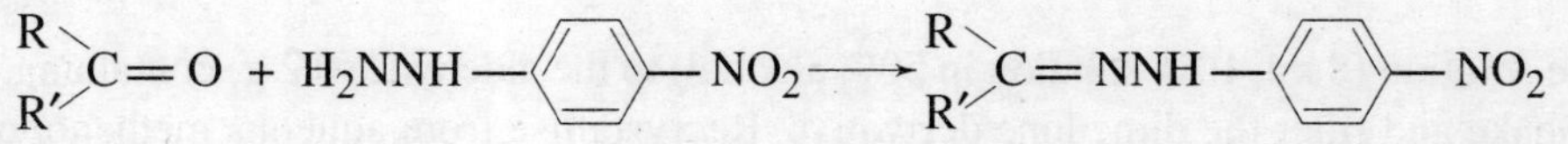

Reflux for 5 minutes, a mixture of p-nitrophenylhydrazine (0.5 g) in alcohol (15 ml), the carbonyl compound (0.5 g) and glacial acetic acid (5 drops). Cool, filter, wash with a little alcohol and recrystallise from alcohol.

**(iii) Phenylhydrazone.**

$$RR'C{=}O + PhNHNH_2 \longrightarrow RR'C{=}NNHPh$$

Dissolve phenylhydrazine (0.4 g) in water (1.5 – 2 ml) and add the compound (0.2 g dissolved in 5 ml ethanol). Boil the mixture for 1 minute, add glacial acetic acid (2 drops) and boil for 5 minutes again. Cool, filter and crystallise from alcohol.

**Note** In case phenylhydrazine hydrochloride is used, sodium acetate (2g) is added to the reaction mixture.

**(iv) Semicarbazone.**

$$RR'C{=}O + NH_2NHCONH_2 \longrightarrow RR'C{=}N{\cdot}NHCONH_2$$

Dissolve semicarbazide hydrochloride (0.5 g) and sodium acetate (1 g) in a minimum amount of water and add the carbonyl compound (0.5 g) (clarify the solution by dropwise addition of ethanol if the carbonyl compound is water insoluble) and shake well. If the semicarbazone does not form on shaking, warm on a water bath for 10 minutes, cool, filter and recrystallise from methanol, benzene or acetic acid.

**Note** On no account should prolonged heating be used in the above preparations as the reagent may undergo self condensation to form hydrazodicarbonamide ($NH_2CONH\text{-}NHCONH_2$), m.p. 246°C.

**(v) Oxime.**

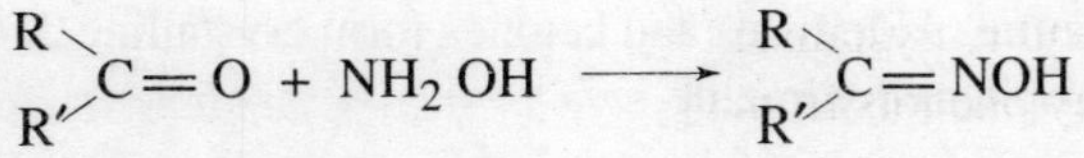

Dissolve hydroxylamine hydrochloride (0.5 g) and sodium acetate (0.5 g) in water (2-3 ml). To this add the carbonyl compound (0.5 g) and ethanol (if necessary) to obtain a clear solution. Reflux for ten minutes or until a sample of the reaction mixture is completely soluble in dilute sodium hydroxide solution. Cool the reaction mixture, filter the solid (scratch the sides of the vessel to induce crystallisation if a solid does not form) and recrystallise from dilute ethyl alcohol.

**(vi) Iodoform.** Iodoform can be prepared from ketones and aldehydes containing $CH_3CO$ group (page 21)

**(vii) Alkylidene dimethone or dimedone**. Aldehydes react with dimedone (5, 5-dimethylcyclohexane-1, 3-dione) to form crystalline derivatives. Ketones do not react

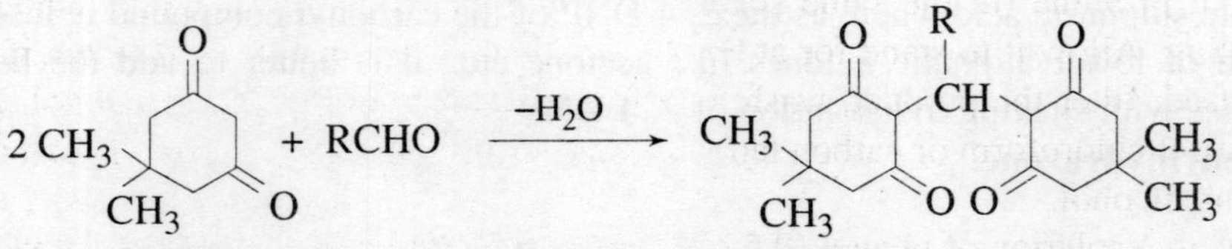

Add dimedone solution (8 ml, 10% solution in 50% alcohol) to the aldehyde (0.2 g) containing piperidine (1-2 drops). Shake and filter the dimedone derivative. Recrystallise from aqueous methanol or ethanol.

**Note** Reflux the reaction mixture in a water bath for 5-10 minutes if the derivative does not separate on shaking.

## *(d) Derivatives of Phenols*

**(i) Benzoate.** Crystalline benzoates can be prepared for phenols (especially for polyhydric phenols) by reaction with benzoyl chloride in sodium hydroxide solution (Schotten-Baumann reaction).

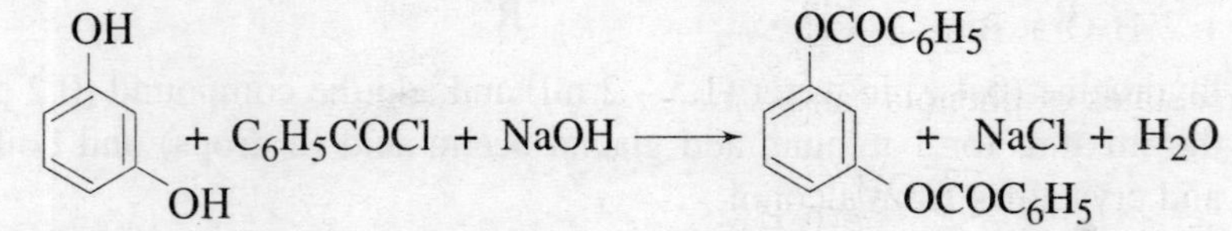

In a conical flask (100 ml) take the phenol (1 g) and freshly prepared sodium hydroxide solution (20 ml, 10%). To this add benzoyl chloride (2 ml) in small quantities (cork the falsk and shake well after each addition). Maintain the alkalinity of the reaction mixture throughout. Shake for some time, cool if the reaction mixture becomes hot. The completion of the reaction is indicated by the disappearance of the smell of benzoylchloride. Filter, wash the solid with water and recrystallise from ethanol.

**(ii) 3, 5-Dinitrobenzoate.**

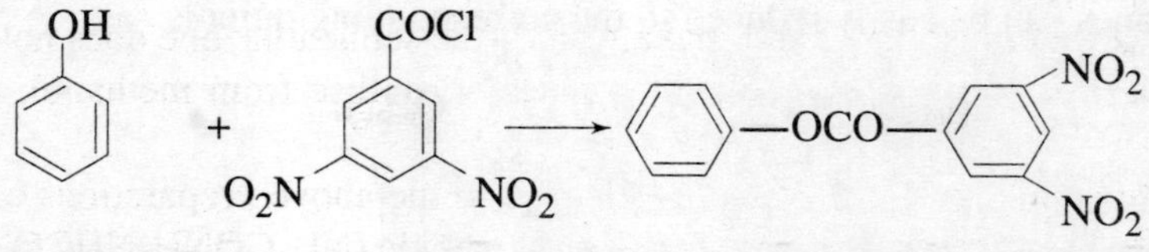


Gently warm 3, 5-dinitrobenzoyl chloride (0.5 g, for prep. of chloride see p. 51) and phenol (1g) (in a fume chamber). Cool, add ice cold water (20 ml) and filter. Wash the solid with excess sodium bicarbonate solution, water and recrystallise from aqueous alcohol.

**(iii) Bromo-derivative**. Many phenols from crystalline bromo derivatives on reaction with bromine in a suitable solvent.

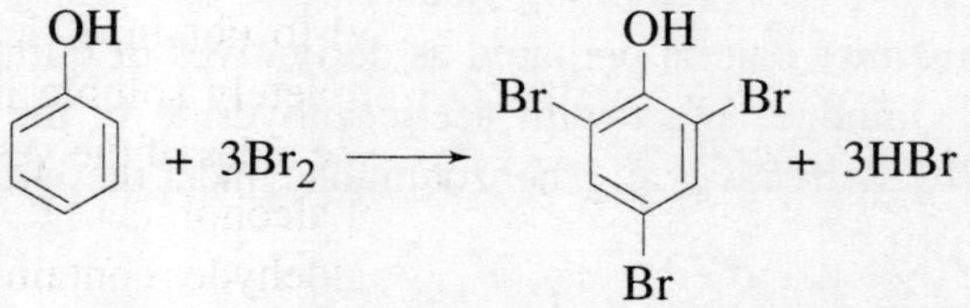


Bromo derivatives of some polyhydroxy phenols (which oxidise easily) are difficult to prepare.

To a solution or a suspension of the phenol in glacial acetic acid, carbon tetrachloride or chloroform, add a solution of bromine (in the same solvent) slowly from a burette until the colour of bromine persists on shaking. Allow it to stand for at least 15 minutes. Pour into ice cold water, if glacial acetic acid has been used, filter the product, wash with water and recrystallise from alcohol. Evaporate the solvent (caution), if chloroform or carbon tetrachloride has been used, wash the residue with water and recrystallise from alcohol.

Alternatively, to a solution of phenol (0.5 g) in water, ethanol or acetone, add a solution of aqueous bromine (10%) dissolved in KBr solution. Shake well, allow the solution to stand for 10-15 minutes. Pour into ice cold water, filter the solid, wash with water and recrystallise from ethyl alcohol.

**Note** Add some dilute sodium bisulphite solution if the solution or the precipitate is yellow coloured, excess bromine reacts with bisulphite.

$$NaHSO_3 + Br_2 + H_2O \longrightarrow NaHSO_4 + 2HBr$$

If the solution is acidic, $H_2SO_3$ is first formed, which reacts with excess bromine as follows

$$H_2SO_3 + Br_2 + H_2O \longrightarrow H_2SO_4 + 2HBr$$

$$\text{or } SO_2 + 2\ H_2O + Br_2 \longrightarrow H_2SO_4 + 2HBr$$

**(iv) Acetate**. The acetates of phenols can be prepared as given below:

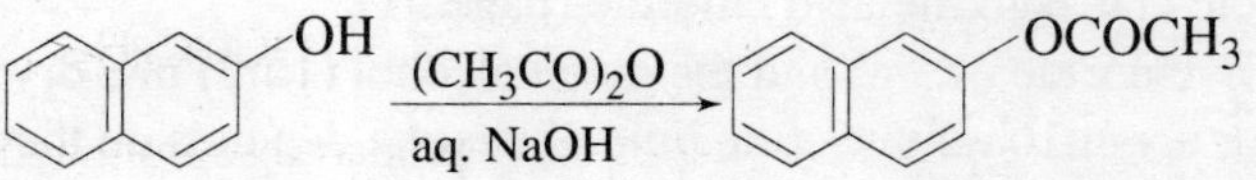


Dissolve the phenol (1 g) in dil. NaOH (10 ml, 10%), add some crushed ice to the solution and then add acetic anhydride (2-3 ml) and shake the mixture vigorously for a few minutes. The solid acetate separates out. Filter, wash with water and crystallise from hot water or dilute alcohol.

## (e) Derivatives of Quinones

**(i) Reduction**. Quinones can be easily reduced to the corresponding quinols.

Take the quinone (0.5 g), zinc dust (0.2 g), dilute hydrochloric acid (5-6 ml) in a conical flask and shake the reaction mixture for 10–15 minutes (till the dark colour of the quinone disappears). Neutralise the solution with $NaHCO_3$ or $Na_2CO_3$ solution. Extract the reduced product with ether, wash the ether layer with water, dry it and remove ether (caution). Recrystallise the quinol from aqueous alcohol.

**(ii) Oximes and semicarbazones** can be prepared as derivatives of quinones (page 53-54).

**(iii) Thiele's acetylation**. Quinones react with acetic anhydride in the presence of concentrated sulphuric acid to produce acetoxy derivatives, e.g. benzoquinone under these reaction conditions produces 1, 2,4-triacetoxybenzene.

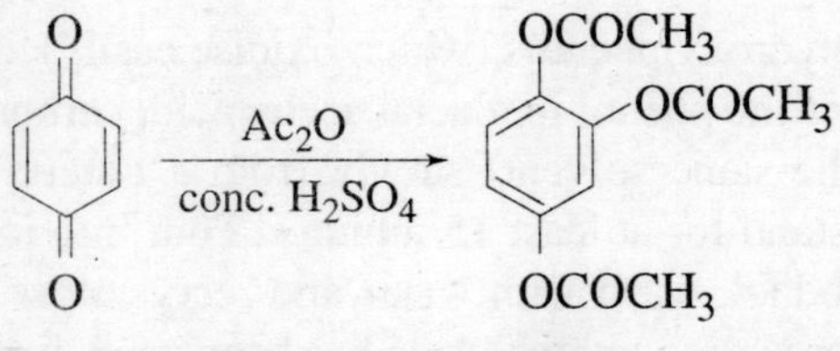


To a mixture of the quinone (2 g) and acetic anhydride (10–15 ml), add conc. $H_2SO_4$ (0.3 ml) dropwise with shaking, maintaining the temperature of the reaction mixture below 50°C. Allow the reaction mixture to stand for 15–20 minutes. Dilute with water (20–25 ml). Filter the product and recrystallise from ethanol.

## (f) Derivatives of Alcohols

**(i) 3, 5-Dinitrobenzoate**. Alcohols can be converted to 3, 5-dinitrobenzoates by reaction with 3, 5-dinitrobenzoyl chloride.

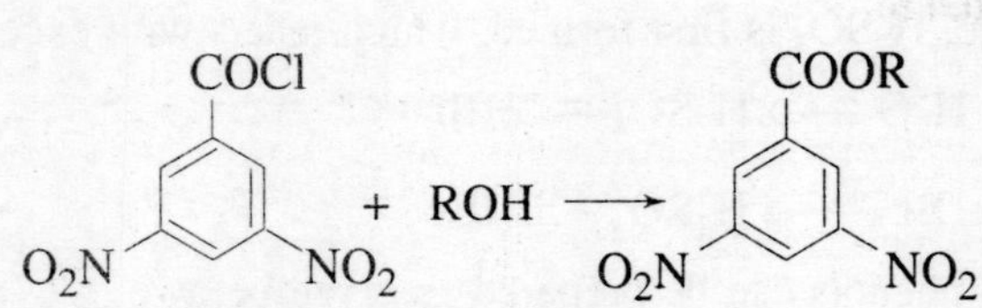


Convert 3, 5-dinitrobenzoic acid into the acid chloride (page 51).

Take 3, 5-dinitrobenzoylchloride (0.5 g) and the given alcohol (1 ml) in a dry boiling tube. Heat the mixture in a water bath for about 10 minutes (the time of heating depends on the nature of the alcohol). After the completion of the reaction, pour the reaction mixture into ice cold water (100 ml). Filter the formed solid, wash it well with sodium bicarbonate solution to remove any acid present. Wash the remaining solid with water and recrystallise from alcohol.

**(ii) p-Nitrobenzoate.**

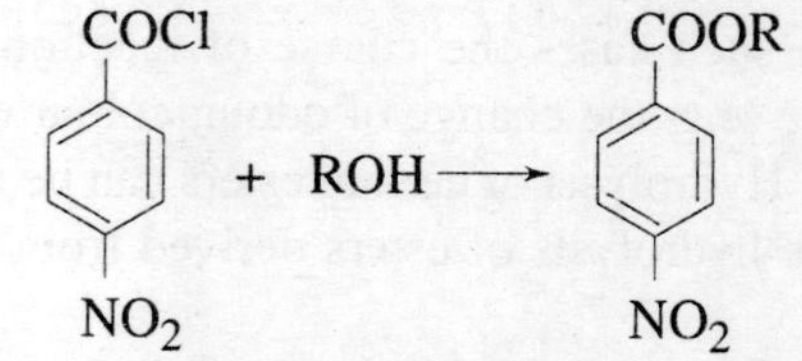


Follow the procedure used for the preparation of 3, 5-dinitrobenzoate.

**(iii) Oxidation of alcohol to the carbonyl compound**. Primary and secondary alcohols can be oxidised to the aldehydes or ketones and crystalline derivatives can then be obtained from the carbonyl compounds. Add the alcohol (2 ml) or its concentrated aqueous solution to an ice cold mixture of concentrated sulphuric acid and aqueous solution of potassium dichromate (1 ml conc. $H_2SO_4$ + 5 ml of concentrated aqueous solution of potassium dichromate). After 5–10 minutes, dilute the solution with water (10 ml). Distil and collect the first few millilitres of the distillate. Perform 2, 4-DNP test with a few drops of the distillate and convert the rest of the distillate into a suitable derivative (2, 4-DNP, oxime etc.).

**(iv) Oxidation of aromatic alcohols**. Aromatic alcohols like benzylalcohol can be oxidised to the corresponding acids.

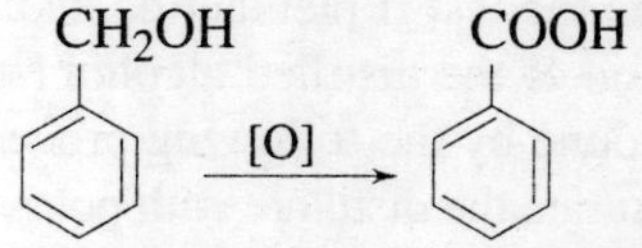


Follow the procedure described for oxidation of the side chain (page 60).

**(v) Iodoform.** Some alcohols containing the $CH_3CHOH$— group can be converted to iodoform, see page 21.

**(vi) Benzoate**. Benzoates of alcohols can be prepared by reaction with benzoyl chloride in presence of sodium hydroxide (p. 54)

Alternatively, Reflux the alcohol (0.5 g) with benzoyl chloride (2 ml) and pyridine (3 ml) for half an hour. Add a solution of sodium bicarbonate (5%, 20 ml) to the cold reaction mixture. Filter the solid benzoate, wash with water and recrystallise from methanol.

## *(g) Derivatives of Esters*

An ester may be derived from an alcohol and an acid or a phenol and an acid and hence, on hydrolysis it may give a phenol and an acid or an alcohol and an acid.

$$C_6H_5COOC_6H_5 \xrightarrow{NaOH} C_6H_5COONa + C_6H_5OH$$

$$C_6H_5COOC_2H_5 \xrightarrow{NaOH} C_6H_5COONa + C_2H_5OH$$

The hydrolysis of an ester can be carried out in an alkaline or an acidic medium. Some esters do not undergo hydrolysis easily. An alternative method is given below for such esters. It is advisable to test the stability of an ester towards alkali on a very small scale before embarking on the hydrolysis of the ester for derivative preparation.

**(i) Hydrolysis with aqueous alkali.** Reflux the ester (1 ml) with freshly prepared sodium hydroxide solution (15%, 15 ml) in a round bottomed flask until the hydrolysis is complete. The time required may vary from half an hour to two hours. If the formed alcohol is water soluble, the completion of hydrolysis is indicated by the disappearance of the insoluble layer of ester. It is difficult to decide in those cases where the alcohol is insoluble. In such cases the course of reaction may be followed by a careful examination of the reaction mixture (e.g. the change of odour, colour or formation of a solid etc.).

***Hydrolysis with methanolic alkali.*** Hydrolysis of certain esters can be done fast with methanolic alkali. This reagent cannot be used for the hydrolysis of esters derived from the alcohols having low boiling points.

***Hydrolysis in diethylene glycol*** Alkaline hydrolysis of almost all the esters can be completed within 10 minutes using diethylene glycol as the solvent. The reagent cannot be used in those cases where the alcohol derived from the ester has a boiling point close to 245°C (which is the boiling point of diethylene glycol). The reagent also does not permit the isolation of acid when it is water soluble, non volatile or sparingly soluble in ether.

In a round bottom flask gently warm potassium hydroxide (2 pellets) with diethylene glycol (2 ml) and water (5 drops) until the alkali has dissolved. Cool the solution, add the ester (1 g) and reflux gently for 5–10 minutes until the hydrolysis is complete.

***Examination for alcohol.*** (1) Cool the hydrolysate obtained above and dilute with water (do not add water if diethylene glycol had been used for hydrolysis. (2) Distil off a small portion and test it for alcohol and keep the residue (A) for identification of acid (perform ceric ammonium nitrate test, for details see page 26. But do not perform the test if methanolic alkali has been used for hydrolysis). (3) Perform the iodoform test with a portion of the distilled alcohol (see page 21).

The isolation of the alcohol can be done by the following procedure.

(1) If the alcohol is water soluble, saturate the distillate with potassium carbonate, remove the separated layer of alcohol, dry it over anhydrous potassium carbonate.

(2) If the alcohol is insoluble or sparingly soluble, extract it with ether, dry the ethereal solution over anhydrous potassium carbonate, and remove the solvent (carefully).

Prepare 3, 5-dinitrobenzoate of the isolated alcohol (see page 56).

***Examination for phenol.*** Add water (10 ml) to the hydrolysate if diethylene glycol method was used.

Saturate the solution with $CO_2$, extract the phenol with ether, wash the extract with a little water and dry. Remove the solvent and prepare the derivative of the phenol (see page 54).

***Examination of the acid.*** The residue (A) after removal of alcohol or phenol contains the sodium salt of the acid derived from the ester. Acidify it with an excess of dilute sulphuric acid and proceed as follows:

(1) Filter, if a solid is formed, recrystallise, determine its melting point and prepare a suitable derivative (see page 52).

(2) Extract with ether if a water insoluble oil separates. Dry the ether solution, evaporate the solvent and prepare a derivative of the remaining acid.

(3) If the formed acid is soluble, neutralise a portion of the alkaline solution and prepare S-benzyl isothiuronium salt of the acid (see page 51).

**(ii) Acid hydrolysis of the ester in presence of 3, 5-dinitrobenzoic acid to obtain 3, 5-dinitrobenzoate of alcohol component.**

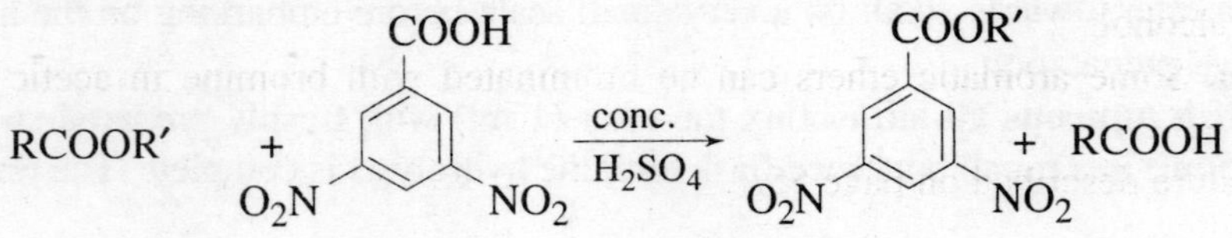

Mix the ester (1 ml) with 3, 5-dinitrobenzoic acid (1 g) and conc. $H_2SO_4$ (5 drops). Heat the mixture gently if the ester has a low boiling point (below 150°) or heat in an oil bath or a sand bath for 15–30 minutes. Stir the mixture while heating. Allow the reaction mixture to cool, pour into ice cold water with stirring and filter the solid. Wash the solid with an excess of sodium bicarbonate or sodium carbonate solution to remove the free acid, filter the crude 3, 5-dinitrobenzoate and recrystallise from alcohol.

**Note** 1. In case the ester is difficult to hydrolyse, it may be identified with the help of NMR.
2. Use a water condenser during heating for lower boiling ester.

## *(h) Derivatives of Lactones*

**(i) Wet methylation**. Ring opening of lactones, e.g., coumarin, can be effected by wet methylation

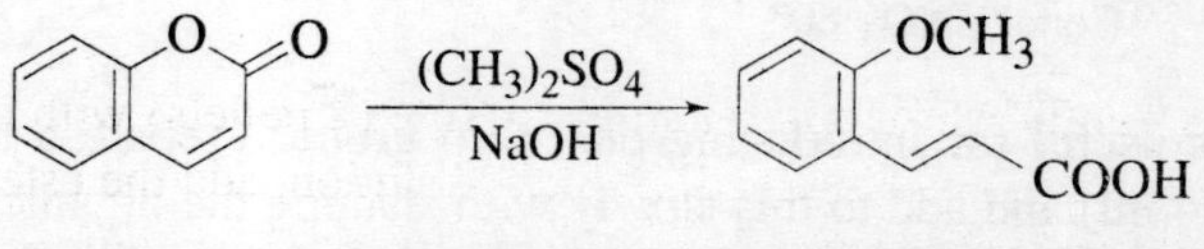


Take the lactone (e.g. coumarin) (1 g) in methanolic potassium hydroxide (1 g KOH dissolved in 10 ml methanol). Add dimethylsulphate (1.5 ml), shake or stir the reaction mixture ocassionally for 20 minutes and heat on a water bath for 5 minutes, cool and acidify with dil. HCl, filter and recrystallise from alcohol.

## *(i) Derivatives of Anhydrides*

Anhydrides may be hydrolysed by water or better by using a dilute sodium hydroxide solution to the acids.

$$(RCO)_2O + NaOH \longrightarrow RCOONa \xrightarrow{H^+} RCOOH.$$

Follow the procedure discussed on page 57 (Hydrolysis with aqueous alkali).

If the formed acid is a solid, determine its melting point. If it is a liquid or soluble in water, extract it with ether, remove ether and convert it into a solid derivative. (see page 51).

## *(j) Derivatives of Ethers*

**(1) 3, 5-Dinitrobenzoate**. Symmetrical aliphatic ethers are converted to 3, 5-dinitrobenzoates.

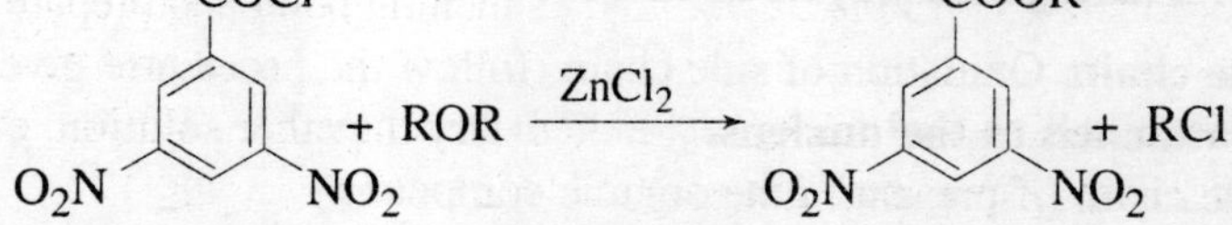


Reflux a mixture of the ether (1 ml), fused zinc chloride (1 g) (page 50) and 3, 5-dinitrobenzoyl chloride (0.5 g page 51) for 1 hour. Allow the reaction mixture to cool, add a dilute solution of sodium carbonate (10 ml) and heat again to 100°C. Cool and filter the solid, wash it well with a solution of sodium carbonate. Dry the solid, extract with hot carbon tetrachloride, evaporate the solvent and recrystallise the residue from alcohol.

**(ii) Bromination.** Some aromatic ethers can be brominated with bromine in acetic acid or carbon tetrachloride.

Follow the procedure described on page 55.

**(iii) Picrates**. Pictrates can be prepared from aromatic ethers following the method described on page 64.

**(iv) Nitration.** Aromatic ethers can be nitrated with (a) fuming $HNO_3$ and glacial acetic acid; (b) conc. $HNO_3$ and conc. $H_2SO_4$. The time required for nitration depends on the nature of the ether (see below).

## *(k) Derivatives of Hydrocarbons*

**(i) Nitration.**

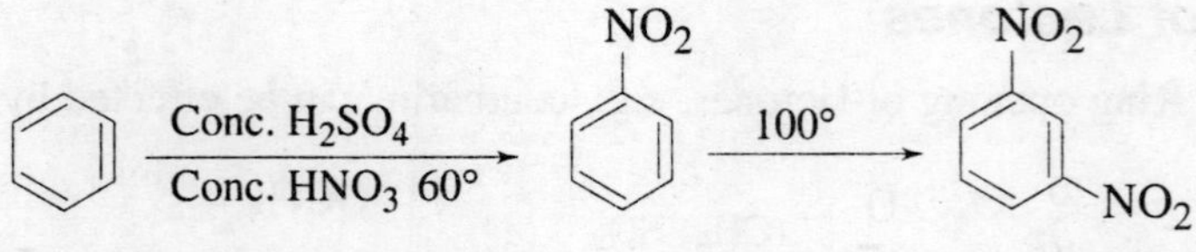


The following method is useful for introducing one nitro group: Prepare a mixture of conc. $H_2SO_4$ (4 ml) and conc. $HNO_3$ (4 ml) and add to this slowly with shaking the organic compound (1 ml). Heat in a hot water bath for 15–20 minutes. Pour the reaction mixture into ice cold water. Filter and recrystallise from dilute alcohol or petroleum ether. In case solid does not separates, isolate it by ether extraction.

***For polynitration.*** Replace concentrated nitric acid by fuming nitric acid and heat the reaction mixture at 100°C for 10–15 minutes.

**Note** The time of heating and the nature of nitration mixture may be chosen according to the nature of the compound.

**(ii) Picrate.** Polynuclear hydrocarbons can be characterised by picrate formation (for details see page 64).

**(iii) Oxidation of side chain.** To a mixture of hydrocarbon (1 g) and water (30 ml), add sodium carbonate (1 g) to make the solution alkaline. Add potassium permanganate (4 g) in small lots with constant shaking. Reflux for 30 minutes, or until the reaction is complete (as indicated by the disappearance of an oil or smell of the starting compound, in some cases, 5-6 hours may be required for oxidation). Cool and acidify the reaction mixture with dilute sulphuric acid, add a solution of sodium sulphite until the precipitated manganese dioxide has dissolved. Filter the acid, or extract with ether if it is water soluble, Recrystallise the solid acid from hot water or alcohol and determine its melting point.

## 3.2 Compounds Containing Halogens

### *(a) Derivatives of Halogeno Hydrocarbons*

**(i) Halogen in the side chain.** Oxidation of side chain (follow the procedure given above)

**(ii) Halogen directly attached to the nucleus.**

(1) Oxidation of side chain (if present in the organic compound)

(2) ***Nitration.*** Either proceed as for hydrocarbons (see above) or apply the following method.

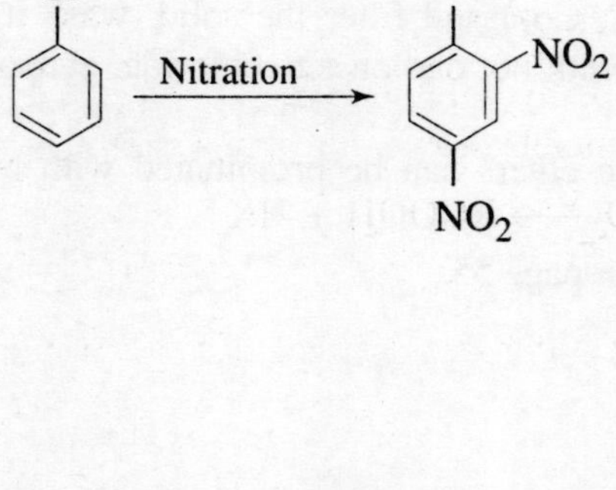

Dissolve the compound (0.3 g) in glacial acetic acid (4 ml) and add dropwise fuming $HNO_3$ (2 ml) to the cold solution. Heat slowly to boiling and allow it to stand for 15 minutes. Pour into ice cold water with stirring (scratch the side of beaker to induce crystallisation if necessary). Filter the solid, wash it with water and recrystallise from alcohol.

**(iii) Alkyl-β-naphthyl ether.** These are formed by reacting an alkyl halide with the sodium or potassium salt of β-naphthol.

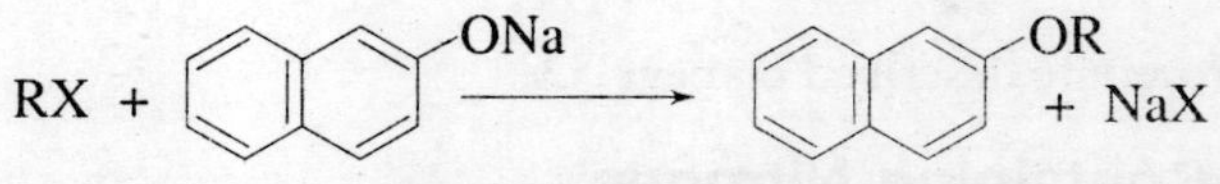


Dissolve β-naphthol (1 g) in alcoholic sodium hydroxide solution (0.3 g of NaOH in 10 ml alcohol). Add the alkyl halide (1 g) to this and reflux for 20–30 minutes. Cool and crystallise the formed alkyl-β-naphthyl ether from alcohol.

***Picrate of alkyl-β-naphthyl ether.*** If the alkyl-β-naphthyl ether is a low melting solid or does not separate out, it may be converted to the picrate by the following procedure. Dissolve the alkyl-β-naphthyl ether in a minimum amount of hot alcohol and add picric acid solution, heat to get a clear solution, cool to obtain the picrate. Crystallise from rectified spirit.

**Note** Some picrates cannot be crystallised as they decompose on heating.

**(iv) S-Alkyl isothiuronium picrate.** Alkyl halides (bromides and iodides) (chlorides react slowly) react with thiourea to form salts which form crystalline picrates when treated with picric acid.

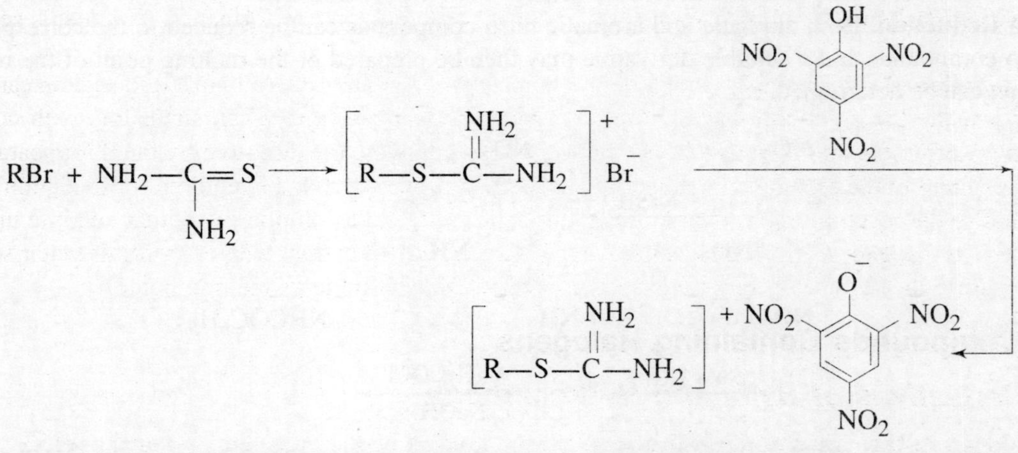


Reflux a mixture of alkyl halide (0.5 g), powdered thiourea (0.5 g) and ethanol (5–6 ml 95%) for 15–20 minutes (alkyl chlorides required 2–3 hours of refluxing). Add picric acid (0.5 g) and heat in a water bath to obtain a clear solution, cool the resultant solution and crystallise the formed picrate from alcohol.

**Note** Add a few drops of water to obtain the solid picrate if it does not separate on cooling.

## *(b) Derivatives of Acid Chlorides*

**(i) Hydrolysis.** Acid halides can be hydrolysed to the corresponding acids. The formed acids may be characterised by conversion to a suitable derivative or by determination of the melting point of the acid.

$$RCOX + H_2O \longrightarrow RCOOH + HX$$

Add the acid halide (1 g) to sodium hydroxide (10 ml, 5–10%) and reflux the solution for 10 minutes. Cool and acidify with dil. HCl. Filter the solid, crystallise, determine its melting point.

**(ii) Amide, Anilide and p-Toluidide.**

$$RCOCl + XNH_2 \longrightarrow RCONHX$$

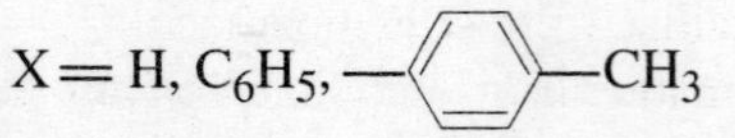


*Procedure*-Follow the procedure described on page 52.

## 3.3 Compounds Containing Nitrogen

### *(a) Derivatives of Nitro compounds*

**(i) Nitration.** simple aromatic nitro-compounds like nitrobenzene can be nitrated to obtain a solid polynitrosubstituted compound.

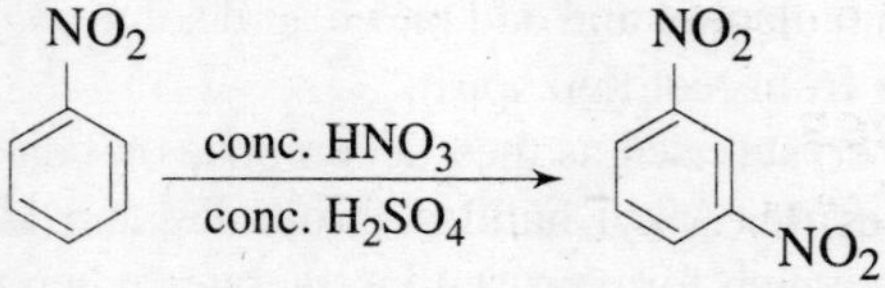


*Procedure* Follow the procedure described on page 60, for polynitration.

**(ii) Reduction.** Both aliphatic and aromatic nitro-compounds can be reduced to the corresponding amino compounds and a suitable derivative may then be prepared or the melting point of the reduced product can be determined, e.g.

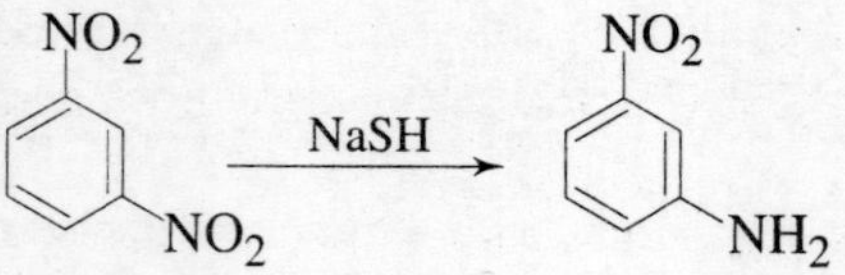


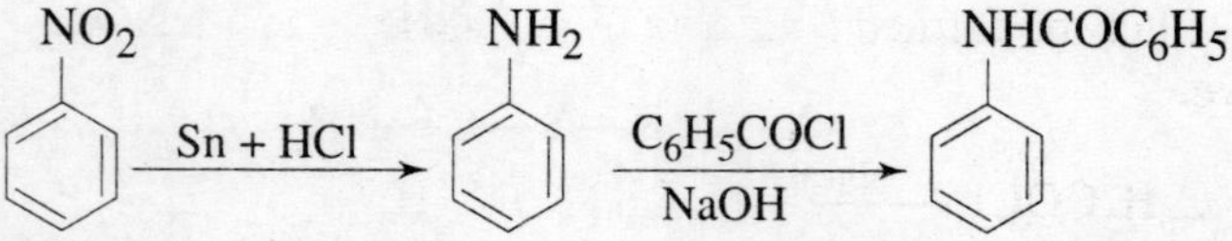


***Reduction using Sn/HCl.*** Reflux the nitro compound (1 g) with tin (2.0-2.5 g) and conc. HCl (20 ml) (add hydrochloric acid in small lots during refluxing) for 10–15 minutes. Cool the solution, dilute with water (10–15 ml) and add enough solution of sodium hydroxide (20%) to dissolve tin hydroxide completely. Extract the alkaline solution with ether, wash the ether extract with water, dry it over potassium hydroxide pellets. Filter the solution. Remove ether by distillation and prepare a suitable derivative of the amine.

**Note** Add some alcohol during refluxing if the nitro compound is highly insoluble in the medium.

***Partial Reduction using NaSH.*** Dissolve sodium sulphide (2 g) in water (5-6 ml), add $NaHCO_3$ (0.7 g) in small portions with the stirring. When the added $NaHCO_3$ has dissolved, add methanol (5-6 ml) and cool the mixture below 20°C. Filter off the precipitated sodium carbonate at the pump, wash the precipitate with a small amount of methanol (2-3 ml) and use the filtrate which contains sodium hydrogen sulphide for further reaction.

Dissolve m-dinitrobenzene (0.75 g) in hot methanol (6 ml) and add the above solution containing sodium hydrogen sulphide to it with shaking. Reflux the mixture for 20 minutes, allow the reaction mixture to cool and distil off most of methanol. Pour the residue into ice cold water (20–30 ml). Filter the reduced compound and recrystallise from dilute methanol.

**(iii) Oxidation of side chain.** Aromatic nitro compound containing a side chain may be oxidised to the carboxylic acid.

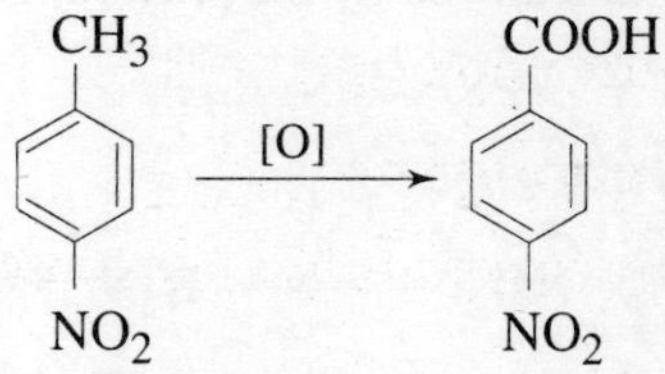


Follow the procedure described on page 60.

## *(b) Derivatives of Amines*

**Primary and secondary amines.** These can form the following derivatives

**(i) Acetyl derivative.**

$$C_6H_5NH_2 + (CH_3CO)_2O \longrightarrow C_6H_5NHCOCH_3 + CH_3COOH$$

(1) Treat the amine (0.5 g) in water (3-4 ml) with acetic anhydride (1 ml) and stir well for 5 minutes. Cool if necessary, then warm to decompose excess acetic anhydride. Cool, filter the solid and recrystallise from water or aqueous alcohol.

(2) Dissolve the amine (0.5 g) in a minimum amount of ice cold dilute hydrochloric acid. Add acetic anhydride (1 ml) and just sufficient sodium acetate to neutralise the hydrochloric acid. Filter the formed solid wash and recrystallise as above.

(3) Heat a mixture of the amine (0.5 g), acetic anhydride (1 ml) and concentrated sulphuric acid (1 drop) for 5 minutes. Cool and pour the mixture over water (5–10 ml). Warm to decompose excess of acetic anhydride, cool, filter the formed solid and recrystallise.

**(ii) Benzoyl derivative.**

$$RNH_2 + C_6H_5COCl \xrightarrow{NaOH} RNHCOC_6H_5$$

Follow the procedure discussed for benzoylation of phenols (see page 54). In the case of solid amine, the rate of benzoylation is increased by the addition of acetone (2-3 ml).

The benzoyl derivative of some amines (e.g. amino acids) are soluble in sodium hydroxide solution. In such cases the solid benzoyl derivative along with some benzoic acid can be obtained by careful acidification of the reaction mixture with dilute sulphuric acid. Pure benzoyl derivative (free from benzoic acid) can then be obtained by extracting the benzoic acid with benzene, light petrol or carbon tetrachloride.

**(iii) p-Toluene sulphonyl derivative (p-tosyl derivative).**

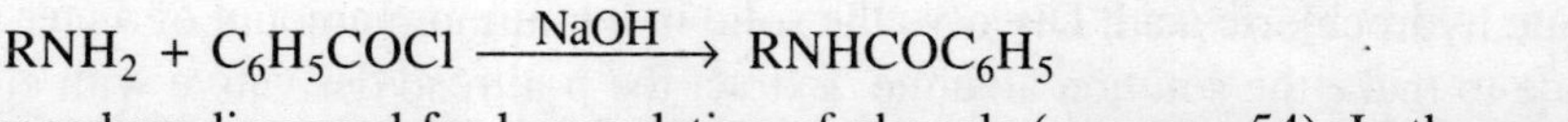

Dissolve p-toluene sulphonyl chloride (1.5 g) and amine (1 g) in a minimum amount of acetone. Treat the solution with dil. NaOH (2N, 25 ml) and shake the reaction mixture for 10 minutes. The reaction mixture should remain alkaline throughout. Cool the reaction mixture in case it becomes hot. The tosyl derivative of secondary amine separates spontaneously. The tosyl derivative of primary amine is soluble in alkali and can be obtained by acidification. Filter, wash with cold water and recrytallise from alcohol.

Alternatively, prepare the tosyl derivatives (especially for nitro anilines) by refluxing a mixture of the amine (1 g), p-toluene sulphonyl chloride (1.5 g) and pyridine (5-6 ml) for 20-25 minutes. Pour the reaction mixture in cold water, filter the solid and recrystallise from alcohol.

**Tertiary amines.** These are easily characterised by the preparation of one of the following derivatives.

**(i) Picrates.**

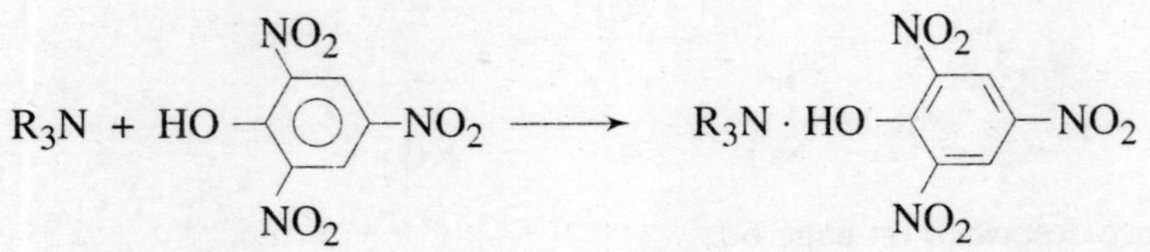


Dissolve the given amine (0.5 g) in a minimum quantity of ethyl alcohol or benzene. Add a saturated solution (2-3 ml) of picric acid in the same solvent. Picrate separates out on mixing. In case no solid forms on mixing, heat the mixture on a water bath and allow it to cool slowly. Filter the product, and recrystallise from dilute alcohol (if necessary). Benzene, acetone, glacial acetic acid or ether (with caution) may be used as alternative solvents.

**(ii) p-Nitroso derivative.**

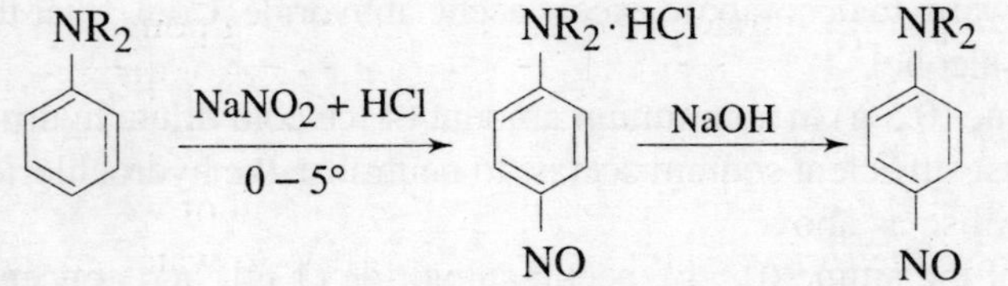


Dissolve the tertiary amine (1.0 g) in dilute hydrochloric acid (1:1, 10 ml), cool the solution to 0-5°C and add slowly with stirring, a cold solution of sodium nitrite (0.7 g) in water (4 ml) keeping the temperature below 5°C during addition. Filter the precipitated hydrochloride of p-nitroso derivative, wash it with dilute hydrochloric acid. Dissolve the solid in a minimum amount of water and add dilute sodium hydroxide to make the solution alkaline. Extract the p-nitroso derivative with ether, evaporate ether (with caution) and recrystallise from light petrol or benzene.

**(iii) Methiodide.**

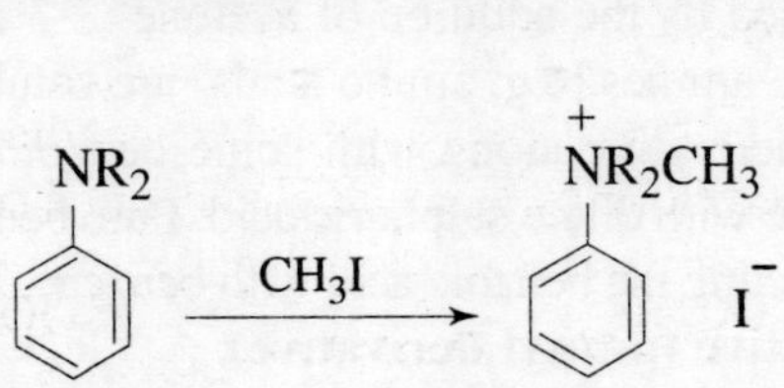


Treat the dry amine (0.5 g) with methyl iodide (2 ml) and allow the mixture to stand for 5 minutes. Heat on a water bath for 5 minutes, cool in ice and initiate crystallisation by scratching the sides of the tube if necessary. Collect the product and recrystallise from dry acetone or absolute alcohol or ethyl acetate.

**(iv) Ethiodides.** These can be prepared from some tertiary amines following the procedure used for preparation of methiodides.

## (c) Derivatives of Amides, Imides and Nitriles

**(i) Alkaline hydrolysis.** Amides on alkaline hydrolysis give the sodium salt of the acid.

$$RCONH_2 \xrightarrow{NaOH} RCOONa \xrightarrow{H^+} RCOOH$$

Place the amide (1g) and sodium hydroxide (20 ml, 10%) in a round bottomed flask. Reflux the mixture gently for about half an hour until the hydrolysis is complete. Cool, acidify the solution with dilute sulphuric acid. Filter, if a solid acid is formed, recrystallise, determine its melting point and convert to a suitable derivative (see page 51).

If no solid is obtained on acidification, the formed acid is water soluble, characterise it by the following procedure.

(1) If the acid is low boiling, water soluble, distil a portion of the acidified solution and use the distillate for preparing S-benzylisothiuronium salt of the acid.

(2) Or extract the acid with ether, dry the ether solution, evaporate (caution) the solvent and use the residue for preparation of a suitable derivative of the acid.

If the given amide is urea (gives a positive biuret test), prepare the following derivatives.

***Urea oxalate***. Dissolve urea (0.5 g) in water (4 ml) and add a solution of oxalic acid (0.6 g in 8 ml water). Mix well, filter the crystalline urea oxalate.

***Urea nitrate.*** Dissolve urea (0.5 g) in water (4 ml) and add concentreated nitric acid (1 ml) with stirring. Filter the crystalline urea nitrate and determine its melting point.

**(ii) Derivatives of nitriles.** Nitriles can be hydrolysed to the acids with an acid or a base.

$$RCN + 2H_2O \longrightarrow RCOOH + NH_3$$

Reflux the nitrile (1 g) with $H_2SO_4$ (5 ml of conc. $H_2SO_4$ + 3-4 ml of water) for half an hour or till the reaction is complete. Filter the formed acid and recrystallise from water or alcohol or extract the acid with ether. If it is soluble, obtain the acid by evaporation of ether (caution) and convert it into a suitable derivative.

**Note** Nitriles can also be hydrolysed with concentrated alkali. The acid can be obtained by neutralisation of the reaction solution with dilute sulphuric acid.

**(iii) Derivatives of imides**. Imides can be hydrolysed to the acid with dilute alkali. Follow the procedure described above for amides.

## (d) Derivatives of Anilides

**(i) Acid hydrolysis.** Anilides may be hydrolysed to the corresponding acid and amine and each of the component may then be characterised either by its melting point or by converting it into a solid derivative.

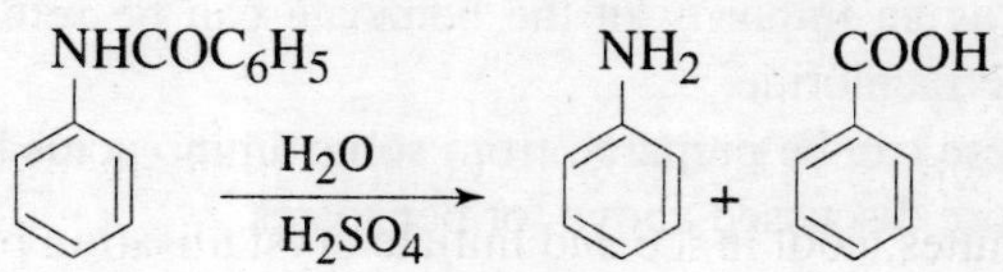

Take the anilide (2 g) and sulphuric acid (20 ml, 70%) and reflux the mixture for 30–35 minutes, wash the condenser with water (10–15 ml) and cool the solution to room temperature. If the acid formed by hydrolysis of the anilide is water insoluble, it separates from the reaction mixture. It may be filtered, recrystallised and its melting point determined. If the acid formed does not separate, extract the solution with ether, remove ether and prepare a crystalline derivative of the acid.

Neutralise the aqueous acidic solution with sodium hydroxide. If the amine separates as a solid, filter and determine its melting point. If the amine does not separate as a solid or is a liquid, extract it with ether, wash the ether extract with water, remove ether by distillation and prepare a derivative of the amine.

**(ii) Bromo derivative.** Some of the anilides (e.g. acetanilide) may be brominated if the para position is free.

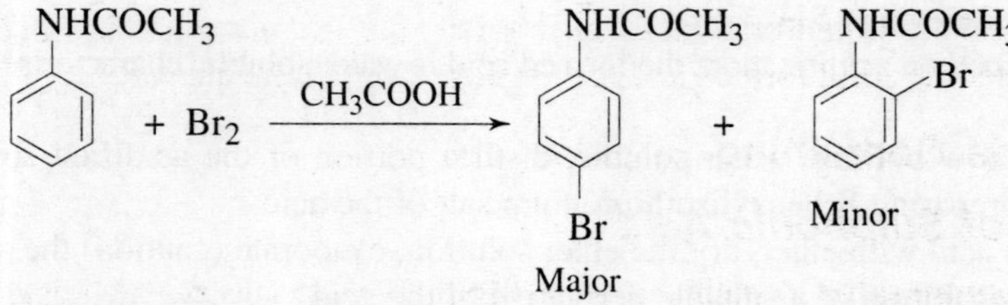


Follow the procedure described on page 55. A minor amount of o-bromoacetanilide is also formed and is removed during crystallisation.

## *(e) Derivatives of Amino Acids*

**(i) Benzoate**. Most of the amino acids can be converted into crystalline benzoyl derivatives by following the Schotten-Baumann method.

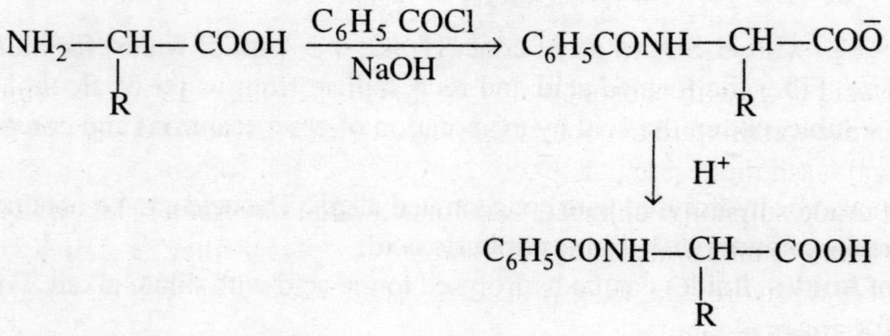


Dissolve the amino acid (0.5 g) in sodium hydroxide solution (10 ml, 10%). Add benzoyl chloride (1 ml) and shake the mixture vigorously for 10-15 minutes. Maintain the alkalinity of the reaction mixture throughout. When the smell of benzoyl chloride has disappeared, acidify the solution with dilute hydrochloric acid to congo red. Filter the solid formed, wash with water and recrystallise from dilute alcohol.

**Note** Benzoic acid, present as an impurity in the benzoate can be removed by washing the crude benzoate with carbon tetrachloride.

**(ii) 3, 5-Dinitrobenzoate.** These can be prepared from some amino acids by using 3, 5-dinitrobenzoyl chloride following the procedure discussed above for benzoates.

**(iii) 2, 4-Dinitrophenyl derivatives.** Amino acids react with 2, 4-dinitrofluorobenzene to from 2, 4-dinitrophenyl derivatives.

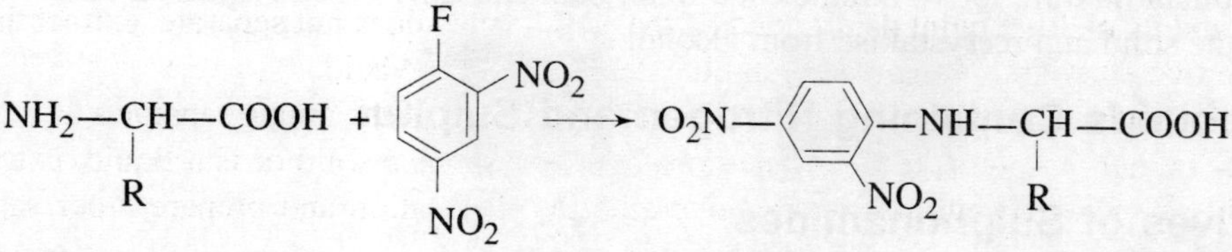

To a solution of the amino acid (0.25 g) in sodium bicarbonate (0.5 g in 5 ml water), add an alcoholic solution of 2, 4-dinitrofluorobenzene (0.5 g in 4 ml ethanol). Shake the reaction mixture for 5 minutes and allow it to stand for one hour with occasional shaking. Add some sodium chloride and extract with ether to remove unreacted 2, 4-dinitrofluorobenzene. Acidify the aqueous solution with dilute hydrochloric acid to congo red, filter the solid formed and recrystallise from dilute alcohol.

## 3.4 Compounds Containing Sulphur

### *(a) Derivatives of Sulphonic Acids*

**(i) S-Benzylisothiuronium salt.** Crystalline S-benzylisothiuronium salts can be prepared from sulphonic acids.
*Procedure* As described on page 52.
**(ii) Sulphonamides** These can be prepared by first converting the acid to its chloride and then reacting it with ammonia.

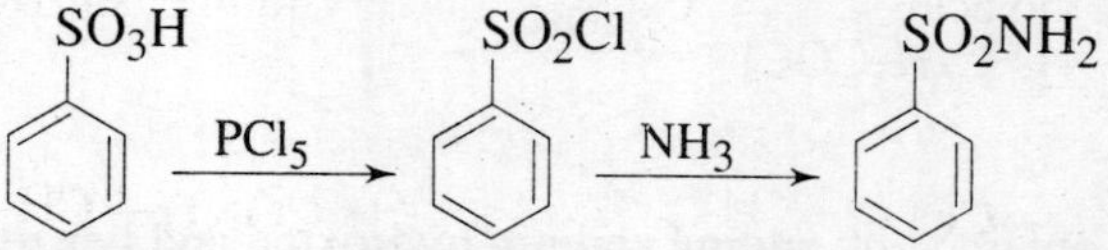

Heat the sulphonic acid (1 g) with phosphorus pentachloride (2 g) at ~ 150°C in a fume chamber for 30 minutes. Allow the reaction mixture to cool and pour it carefully into cold water (20-30ml). Remove the aqueous solution and wash the residual sulphonyl chloride with some cold water. Remove the aqueous solution and to the crude sulphonyl chloride, add concentrated ammonia solution rapidly with stirring. Warm the reaction mixture on a water bath for 5 minutes, cool and filter the sulphonamide. Dissolve it in dilute aqueous sodium hydroxide, filter if necessary and acidify the filtrate. Filter the reprecipitated solid and recrystallise from alcohol.
**Note** Hydroxy and aminosulphonic acids cannot be converted to sulphonamide.

### *(b) Derivatives of Thiols*

**(i) 2, 4-Dinitrophenyl thioether**

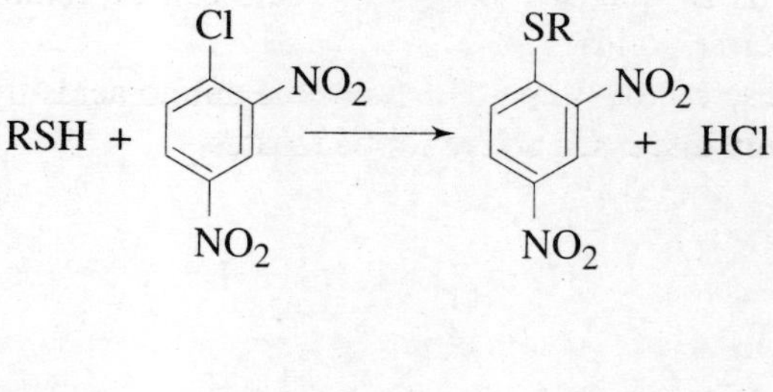

Treat a solution of the mercaptan or thiophenol (1 g) in alcohol (30 ml) with a solution of sodium hydroxide (0.4 g) in water (3 ml). Add 2, 4-dinitrochlorobenzene (2 g dissolved in 10 ml of alcohol). Reflux the resultant mixture for 10 minutes in a water bath and filter the hot solution. Allow the filtrate to cool, filter the solid and recrystallise from alcohol.

## 3.5 Compounds Containing Nitrogen and Sulphur

### (a) Derivatives of Sulphonamides

**(i) Acetyl derivative.**

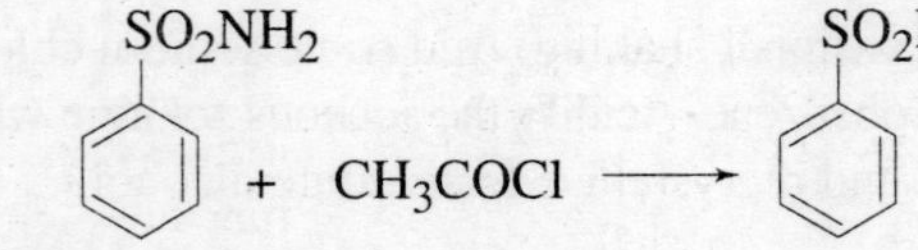


Reflux, in a fume chamber, the sulphonamide (1 g) with acetyl chloride (2.5 ml) on a water bath for 30 minutes (Add glacial acetic acid (~ 2.5 ml) if the reaction mixture is not homogeneous). Distil off the excess of acetyl chloride (in a fume chamber) and pour the cold reaction mixture into water. Filter the product, wash it with water and recrystallise from aqueous alcohol.

**(ii) Benzoyl derivative.** Crystalline benzoyl derivative can be prepared from sulphonamides by the Schotten-Baumann method.

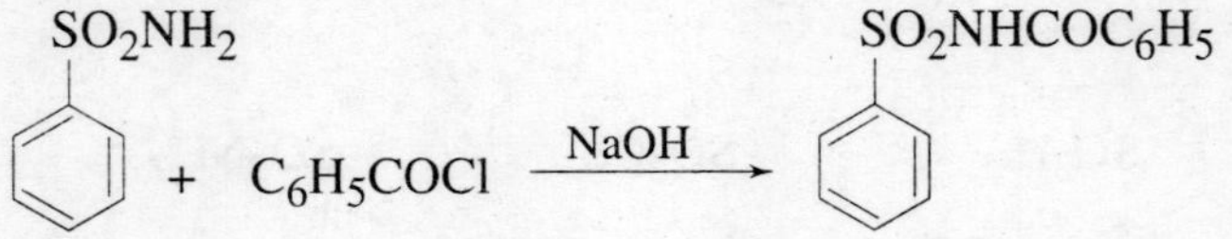


*Procedure* Refer to page 54.

**Note** Better results can be obtained by adding sodium hydroxide and benzoyl chloride in small lots

# 4

# Application of Spectroscopy to the Identification of Organic Compounds

## 4.1 The Infrared Spectrum

Infrared spectroscopy is an important technique for the identification of functional groups in the organic compounds. It is equally important in providing information regarding functional groups which are absent. Infrared spectroscopy does not provide complete information about the structure of the compound. However it is used for establishing the identity (superimposable IR) or non-identity (non-superimposable IR) of the compounds.

The covalent bonds in organic compounds are constantly undergoing vibrations (stretching and bending motions). The frequencies of various vibrations of molecules correspond to those of IR radiations. Only those vibrational changes that result in a change in the dipole moment result in an absorption in the IR region and are said to be IR active.

Following modes of vibrations occur in molecules

**(i) Stretching**. This may be symmetric or asymmetric

(a) Symmetric stretching (b) Asymmetric stretching

**Fig. 4.1**

For example, the two C=O bonds present in the molecule of carbon dioxide can show the (a) and (b) type of stretch shown in the figure 4.1. The symmetric stretching in the case of $CO_2$ does not result in a change in the dipole moment and hence is IR inactive. The asymmetric stretch on the other hand is IR active.

**(ii) Bending.** This may be in plane or out of plane bending. Scissoring, rocking, twisting etc. are some of the important types of bending vibrations

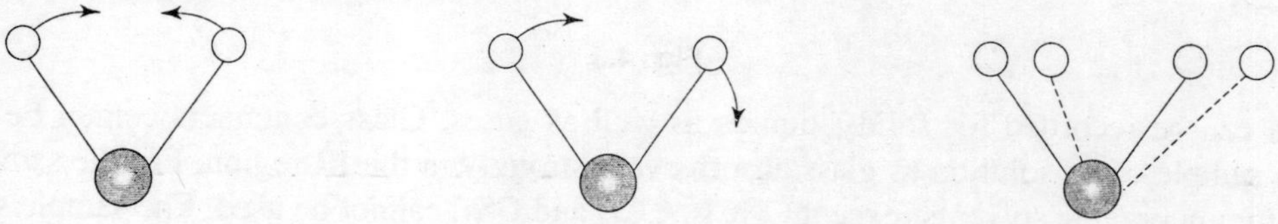

In-plane bending scissoring Rocking Twisting

**Fig. 4.2**

The energy required for causing changes in these modes of vibrations can be provided by the IR radiation. The exact amount of energy required depends on the strength of the bonds and the atoms attached through the bonds. The graph or the plot of energy absorption (% transmittance) vs the wavelength (in microns, $1\mu = 10^4$ Å) or wave number (in $cm^{-1}$) of radiations is known as the infrared spectrum. It shows a large number of adsorptions because of a large number of bonds present and the various types of vibrational changes these bonds can undergo.

The region of interest to an organic chemist is 2.5–15$\mu$ or 4000-650 $cm^{-1}$. Broadly speaking, this region may be divided into the functional group region (2.5 – 7.0 $\mu$), and the finger print region (7–15 $\mu$). The finger print region like the human finger print is a characteristic of each molecule. Careful analysis of the IR spectrum in the functional group region provides a great deal of information concerning the presence or absence of different functional groups. For example, the C═O group absorbs in the region of 1680-1800 $cm^{-1}$. Presence of a strong peak in this region indicates the presence of a C═0 group and the absence of a strong absorption peak in this region indicates the absence of a C═O group and hence the compound cannot be a ketone, aldehyde, carboxylic acid, amide, ester, anhydride, or acid halide etc. An aldehyde can be easily distinguished from a ketone because of the presence of two bands at 2760 and 2850 $cm^{-1}$, which are due to the C—H stretch of the —CHO group and are absent in the spectrum of a ketone.

Similarly, the presence of a broad absorption peak between 3200–3600 $cm^{-1}$ is indicative of the presence of O—H, N—H etc.

## *The IR Spectrophotometer*

A simplified diagram of an IR spectrophotometer is shown in Fig. 4.3.

A continuous beam of IR radiation produced by a heated filament is separated into radiations of different wavelengths by passing through a prism (made of sodium chloride). Rotation of the prism permits the different wavelengths to pass through the focussing slit to produce IR radiation of a single wavelength. The beam is then split into two beams of equal intensity, one passing through the sample and other through the reference. The beams then strikes an IR detector. The detector converts the transmittance into an electric signal that is amplified and sent to the recorder that traces the IR spectrum.

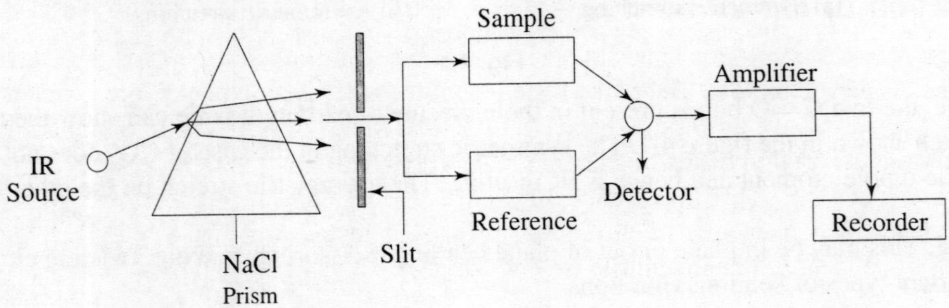


**Fig. 4.3**

IR spectra can be recorded for solids, liquids as well as gases. Glass containers cannot be used for taking the sample or its solution as glass absorbs very strongly in the IR region. For the same reason, quartz, common organic solvents (except $CHCl_3$, $CCl_4$ and $CS_2$) cannot be used. The sample should be free from moisture.

IR spectra may be recorded in one of the following ways:

**(i) KBr pellet**. The solid compound (50 mg) is finely powdered with spectral grade potassium bromide (0.5 g). It is then converted into a pellet by applying high pressure to the mixture. The pellet is placed in the spectrophotometer and IR recorded.

Though this method is generally used for solids, special precaution should be taken to use dry potassium bromide and to exclude moisture during pellet formation. The presence of moisture interferes in the detection of functional groups like O—H, N—H, S—H etc.

**(ii) Solution method**. The dry solid compound is dissolved in anhydrous $CCl_4$, $CHCl_3$ or $CS_2$ (20% solution) and the solution is placed in the sample cell. Pure solvent is taken in the solvent cell and the IR is recorded on a double beam spectrophotometer, which automatically cancels the absorption due to solvent. Small absorption peaks are always observed because of the unequal concentration of solvent in the two cells. These peaks can however be ignored.

**Note** $CS_2$ absorbs at 1500 $cm^{-1}$ and absorption due to C—Cl occurs between 600–800 $cm^{-1}$.

**(iii) Mull method**. A dispersion of the finelly powdered organic compound in Nujol (a mixture of saturated hydrocarbons containing 20–24 carbon atoms) or perfluorokerosene (fluorinated kerosene containing $C_{12}$—$C_{18}$ linear hydrocarbons) can be used instead of a solution of the organic compound to record the spectrum.

**(iv) NaCl windows** The IR spectrum of pure (neat) liquid compound can be recorded using NaCl windows. A small amount of the liquid is placed on a moisture-free NaCl plate (window), it is covered with another NaCl plate and the spectrum recorded. If the compound is a low boiling liquid, some of it tends to evaporate during the time taken to record the spectrum and a uniform concentration of the compound is not maintained throughout. Also a uniform thickness of the sample is not possible, if the spectra of two compounds is to be compared.

Some of the spectrophotometers in use have a fixed distance between NaCl plates and the evaporation of the liquid is prevented by using teflon covers.

**(v) IR of a gaseous compound** When the IR spectrum of a gaseous organic compound is determined, the path length of the cell containing the sample is increased (10–20 cm) so as to have a high concentration of the substance for obtaining a good IR spectrum.

## *Interpretation of the Infrared Spectra*

The position of various absorption bands in the infrared chart is noted and compared with those given in Table 4.1. The diagnostic infrared absorption bands of some important compounds are given in Table 4.2.

Table 4.3 summarises the absorption bands (mostly stretching) of some common functional groups. The correlation of absorption bands of some important bonds is given in Table 4.4.

The infrared spectra of some important compounds, viz. 1-Octene, propylbenzene, benzyl alcohol, phenol, dipropyl ether, phenetole, diethyl ketone, benzaldehyde, acetic acid, ethyl acetate, aniline, benzamide, nitrobenzene, and benzonitrile is given in Figs. 4.1–4.14.

**Table 4.1** Characteristic Infrared Frequencies for Common Compounds

| *Type of compound* | | *Characteristic frequencies* ($cm^{-1}$) | |
|---|---|---|---|
| Alkanes | | 2850–3000 (s) | C—H stretch |
| | | 1480–1440 (s) | $CH_2$ and $CH_3$ bend |
| | | 1370–1380 (s) | |
| | | 2900–2800 (w) | CH stretching |
| Alkenes | RCH=$CH_2$ | 3080–3140 (m) | =C—H stretch |
| | vinyl | 1640–1645 (m) | C=C stretch |
| | $R_2C$=$CH_2$ | 3080–3140 (m) | =C—H stretch |
| | vinylidene | 1645–1650 (m) | C=C stretch |
| | cis RCH=CHR | 1655–1660 (m) | C=C stretch |
| | | 3020–3010 (m) | =C—H stretch |
| | | 675–720 (m) | C—H out of plane bending |
| | trans RCH=CHR | 1670–1675 (m) | C—H stretch |
| | | 3020–3010 (m) | =C—H stretch |
| | | 970–980 (s) | C—H out of plane bending |
| | $R_2C$=CHR (trisubstituted) | 1660–1670 (w) | C=C stretch |
| | $R_2C$=$CH_2$ (tetra substituted) | 1660–1670 (w) | C=C stretch |
| | >C=C—C=$CH_2$ (conjugated) | 1660–1580 (s) | C=C stretch |
| Alkynes | RC≡CH | 3310–3300(s) | C—H stretch |
| | | 2100–2140 (m) | C≡C stretch |
| | | 600–700 (s) | C—H bend |
| | RC≡CR | 2190–2260 (w) | C≡C stretch |
| Benzene and its derivatives | 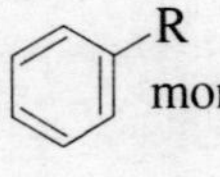 monosubstituted | 3080–3030 (1-5 bands) | C—H stretch |
| | | 1625–1440 (4 bands) | C=C stretch |
| | | 770–690 (2 bands) | |
| | 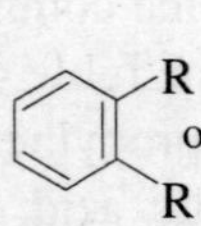 o-disubstituted | 1225-960 (4 bands) | C—H bend |
| | | 770-735 (s) (1 band) | |
| | 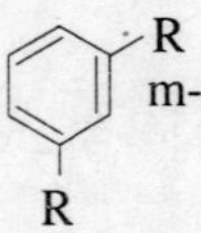 m-disubstituted | 1170-1000 (3 bands) | C—H bend |
| | | 900-680 (3 bands) | |
| | 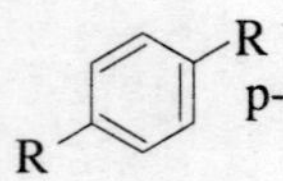 p-disubstituted | 1225-1000 (3 bands) | C—H bend |
| | | 860-800 (s) (1 band) | |
| Alcohols | R—OH | 3600-3200(s) | O—H stretch |

*Contd.*

| Type of compound | | Characteristic frequencies ($cm^{-1}$) | |
|---|---|---|---|
| | (primary) | 1350–1250 (s) | O—H bend |
| | | 1080–1010 | C—O stretch |
| | >CHOH (secondary) | 3600–3200(s) | O—H stretch |
| | | 1350–1250 (s) | O—H bend |
| | | 1120–1030 (s) | C—O stretch |
| | —COH (tertiary) | 3600–3200(s) | O—H stretch |
| | | 1410–1300 (s) | O—H bend |
| | | 1170–1100 (s) | C—O stretch |
| Phenols | 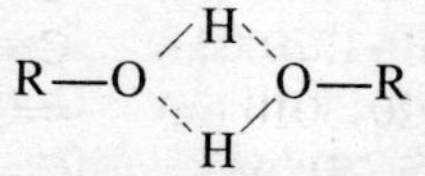 hydrogen bonded intermolecular | 3500–3450 (s) | O—H stretch |
| | 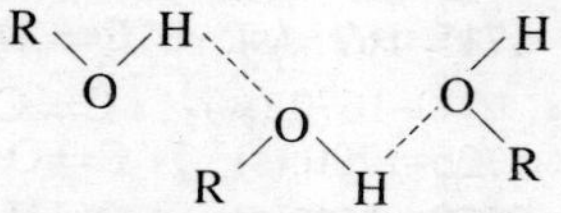 Intermolecular (Chelate) | 3400–3200(s) | O—H stretch |
| | | 3570–3450 (s) | O—H stretch |
| Ethers | —$CH_2$—O—$CH_2$— saturated acyclic | 1150–1060 (s) | C—O stretch |
| | Ar—O—Ar/R aryl and arylalkyl | 1270–1250 (s) | C—O stretch |
| | 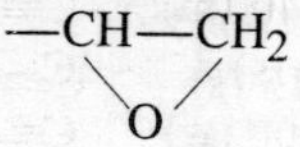 | 980–970 (s) | C—O stretch |
| Ketones | >C=O aliphatic (saturated acyclic) | 1720–1700 (s) | C=O stretch |
| | Ar—CO—R aromatic | 1700–1680 (s) | C=O stretch |
| | Ar—CO—Ar aromatic diaryl | 1670–1660 (s) | C=O stretch |
| | 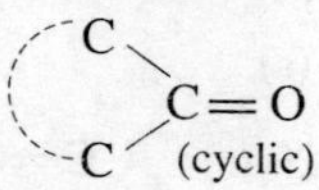 (cyclic) | 1775–1705 (s) | C=O stretch |
| | 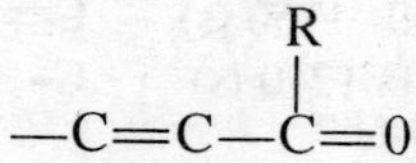 α, β-unsaturated (acyclic) | 1680–1665 (s) | C=O stretch |
| | 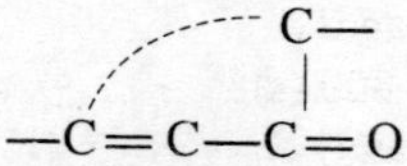 α, β unsaturated (cyclic) | 1725–1665 (s) | C=O stretch |

Contd.

| *Type of compound* | | *Characteristic frequencies* $(cm^{-1})$ | |
|---|---|---|---|
| | —CO—CO—<br>$\alpha$-diketone | 1730–1710 (s) | C=O stretch |
| | —CO—$CH_2$—CO—<br>$\beta$-diketone (enol) | 1640–1535(s)<br>2800–2600 | C=O stretch<br>O—H stretch |
| | —CO—$CH_2$—$CH_2$—CO—<br>$\gamma$- diketone | 1725–1705 | C= O stretch |
| | >C=C=O<br>ketenes | approx. 2150 (s) | C=O stretch |
| Aldehydes | R—CH=O<br>aliphatic saturated | 1740–1720 (s)<br>2870–2710<br>975–780 (w) | C=O stretch<br>C—H stretch<br>C—H bend |
| | R—CH=CH—CH=O<br>$\alpha$, $\beta$ -unsaturated | 1705–1680 (s) | C=O stretch |
| | Ar CHO<br>aromatic | 1715–1695 (s) | C= O stretch |
| Carboxylic acids | R—COOH<br>aliphatic saturated | 1725–1700 (s)<br>3550–3500 (m) | C=O stretch<br>O—H stretch |
| | R—CH=CH—COOH<br>$\alpha$, $\beta$ -unsaturated | 1715–1690 (s) | C=O stretch |
| | ArCOOH<br>aryl | 1700–1680 (s) | C=O stretch |
| | | 1680–1650 (s)<br>(hydrogen bonded) | C=O stretch |
| | X—$CH_2$COOH<br>$\alpha$-halogeno acids | 1740–1715 (s) | C= O stretch |
| Anhydrides | 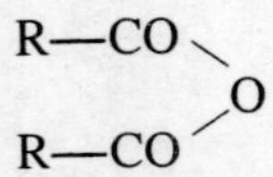<br>acyclic | 1850–1800 (s)<br>1790–1740 (s)<br>1175–1045 (s) | C= O stretch<br>C= O stretch<br>C—O stretch |
| | $(ArCO)_2O$<br>aromatic | 1770–1720 (s) | C= O stretch |
| | 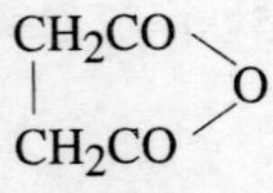<br>cyclic saturated<br>5-membered | 1870–1820 (s)<br>1800– 1750 (s)<br>1310–1210 (s) | C=O stretch<br>C= O stretch<br>C—O stretch |
| | 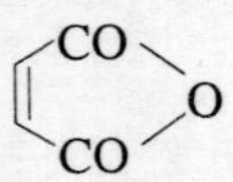 | 1850–1800 (s) | C=O stretch |

*Contd.*

| *Type of compound* | | *Characteristic frequencies* ($cm^{-1}$) | |
|---|---|---|---|
| | cyclic ($\alpha$, $\beta$-unsaturated) | 1830–1780 (s) | C=0 stretch |
| Esters | R—COOR′ saturated acyclic | 1750-1735 (s) | C=O stretch |
| | | 1280-1050 (2 bands) | C—O stretch |
| | —C=C—COOR $\alpha$, $\beta$-unsaturated | 1730-1715 (s) | C=O stretch |
| | —CO—COOR $\alpha$-keto esters | 1755-1740 (s) | C=O stretch |
| | —CO—$CH_2$—COOR $\beta$-keto ester (enolic) | 1655-1635 (s) | C= O stretch |
| | —CO—C—C—COOR ($\gamma$-ketoesters) | 1750-1735 (s) | C=O stretch |
| Lactones | 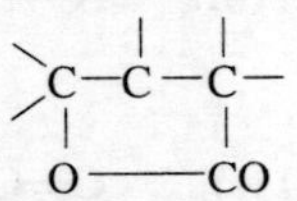 saturated $\gamma$-lactone | 1780-1760 (s) | C=O stretch |
| | 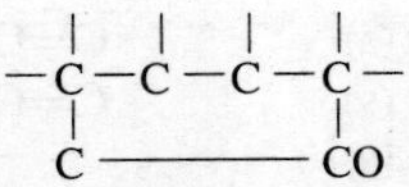 saturated $\delta$-lactone | 1750-1735 (s) | C=O stretch |
| | 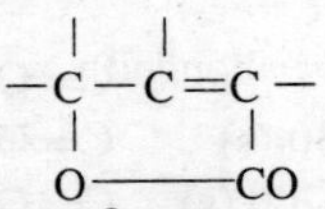 $\alpha$, $\beta$ unsaturated $\gamma$-lactone | 1760-1740 (s) | C=O stretch |
| | 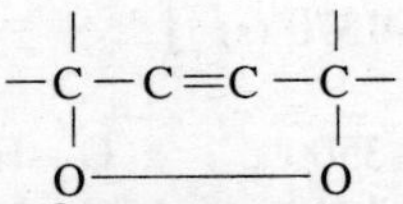 $\beta$, $\gamma$ unsaturated $\delta$-lactone | 1805-1785 (s) | C=O stretch |
| Amines | Ar—$NH_2$ or $RNH_2$ (primary) | 3500-3200 (m) (2 bands) | N—H stretch |
| | | 1340-1200 (s) | C—N stretch (aromatic) |
| | | 1650-1590 (m) | N—H bend |
| |  secondary | 3500-3200 (m) | N—H stretch |
| | | 1360-1310 (s) | C—N stretch (aromatic) |
| | | 1580-1490 (w) | N—H bend |

*Contd.*

| Type of compound | | Characteristic frequencies ($cm^{-1}$) | |
|---|---|---|---|
| | >N— tertiary | 1360–1310 (s) | C—N stretch (aromatic) |
| Imines | =NH | 3400–3000 (m) | N—H stretch |
| Amides | —$CONH_2$ primary free | 3500 (m), 3400 (m) | N—H stretch |
| | | 1690 (s) (dilute solution) | C=O stretch |
| | | 1620-1590 | N—H bend |
| | —$CONH_2$ hydrogen bonded | 3350 (m), 3180 (m) | N—H stretch |
| | —CO—NH— secondary free | 3430(m) | N—H stretch |
| | | 1700–1670 (s) | C=O stretch |
| | | 1550–1510 | N—H bend |
| | hydrogen bonded secondary amide | 3320–3140 (m) | N—H stretch |
| | —CO—N< tert amide | 1670–1630 (s) | C=O stretch |
| Imides | acyclic | 1710 (s) | C=O stretch |
| | cyclic (5-membered) | 1770 (s), 1700 (s) | C=O stretch |
| | cyclic (6-membered) | 1710 (s), 1700 (s) | C=O stretch |
| Ureas | acyclic | 1660 (s) | C=O stretch |
| Lactams | β-lactam | 1760–1730 (s) | C=O stretch |
| | γ-lactam | approx. 1700 (s) | C=O stretch |
| | δ-lactam | approx. 1680 (s) | C=O stretch |
| Nitro | R—$NO_2$ aliphatic & aromatic | 1570–1550 (s), 1380–1370 (s) | N=O stretch |
| Oximes | —C=N—OH | 3650–3500 | O—H stretch |
| | | 1690–1640 | C=N stretch |
| Azo | —N=N— | 1640 | N=N stretch |
| Nitriles | R—CN aliphatic, saturated | 2250 (m) | C≡N stretch |
| | —C=C—CN aliphatic, α, β-unsaturated | 2225 (m) | C≡N stretch |
| | ArCN aromatic | 2240–2220 (m) | C≡N stretch |
| Isocyanides | C—NC | 2220–2070 | C≡N stretch |
| Isocyanates | C—NCO | 2275–2240 | C=N stretch |
| Nitrates | C—O—$NO_2$ | 1655–1610 (s), 1300–1250 (s) | N=O stretch |
| Nitrites | C—O—NO | 1680–1610 (s), 1300–1250 (s) | N=O stretch |

*Contd.*

| *Type of compound* | | *Characteristic frequencies* ($cm^{-1}$) | |
|---|---|---|---|
| Nitroso | C—NO | 1600-1500 (s) | N═O stretch |
| Alkyl halides | R-X | 1400-1000 (s) | C—F stretch |
| | | 800-600 (s) | C—Cl stretch |
| | | 600-500 (s) | C—Br stretch |
| | | 500 (s) | C—I stretch |
| Thiols | C—SH | 2600-2550 (W) | S—H stretch |
| Thioketones | >C═S | 1200-1050 (s) | C═S stretch |

**Table 4.2** Diagonistic infra red frequencies for some compounds

| | | |
|---|---|---|
| Alkenes | | C—H str.; → C—Hstr.; $CH_2$ and $CH_3$ bend |
| Alkenes | $RCH{=}CH_2$ vinyl | C—Hstr.; C=Cstr. |
| | $R_2C{=}CH_2$ vinylidene | C—Hstr.; C=Cstr.; C—H out of plane bend |
| | RCH=CHR cis | C—Hstr.; C=Cstr. |
| | RCH=CHR trans | C—Hstr.; C=Cstr.; C—H bend; C—H out of plane bend |
| | $R_2C{=}CHR$ trisubstituted | C=Cstr. |
| | $R_2C{=}CR_2$ tetrasubstituted | C=Cstr. |
| | $>C{=}C{-}C{=}CH_2$ conjugation | C=Cstr. |
| Alkynes | RC≡CH monosubstituted | ≡C—Hstr.; C≡Cstr. |
| | RC≡CR disubstituted | C≡Cstr. |
| Benzene | monosubstituted | C—H str.; C—Hbend (2 bands) |
| | o-disubstituted | C—H str.; C=Cstr. (4 bands); C—H bend (4 bands) |
| | m-disubstituted | (3 bands); C—Hbend (3 bands) |
| | p-disubstituted | (3 bands) |
| Alcohols | primary | O—H str.; O—H bend; C—O str. |
| | secondary | O—H str.; O—H bend; C—O str. |
| | tertiary | O—H str.; O—H bend; C—O str. |
| Phenols | free OH | O—H str. |
| | intermolecular H-bonded | O—H str. |
| | intramolecular H-bonded | O—H str. |
| Ethers | —$CH_2$—O—$CH_2$— saturated acyclic | C—O str. |
| | Ar—O—Ar(R) alkyl and aralkyl | C—O str. |
| | $CH_2$—$CH_2$ \ O / | C—O str. |

Wavelength scale: 2 3 4 5 6 7 8 9 10 11 12 13 14 15 μ

Wavenumber scale: 3000 2500 2000 1800 1600 1400 1200 1000 900 800 700 $cm^{-1}$

(contd.)

2 3 4 5 6 7 8 9 10 11 12 13 14 15

| Class | Compound type | Absorption bands (left to right) |
|---|---|---|
| Ketones | >C=O aliphatic saturated | C=O str. |
| | Ar—CO—R aromatic aryl | C=O str. |
| | Ar—CO—R aromatic diaryl | C—O str. |
| | —CO—$CH_2$—CO— β–diketone | O—H str.; C=O str. |
| Aldehydes | RCHO aliphatic saturated | C—Hstr.; C=O str.; C—H |
| | ArCHO aromatic | C=O str. |
| | R—CH= CH—CHO α, β–unsaturated | C=O str. |
| Carboxylic acids | R—COOH aliphatic | O—H; C=O str. |
| | Ar—COOH aromatic | C=O str. |
| | RCH=CHCOOH α, β–unsaturated | C=O str. |
| Anhydrides | R—CO—O—CO—C acyclic | C=O str. |
| | Ar—CO—O—CO—Ar aromatic | C=O str. |
| Esters | R—COOR saturated acyclic | C=O str.; C—O str. (2 bands) |
| | RCH—CHCOOR α, β–unsaturated | C=O str. |
| | —CO—COOR α, β–unsaturated | C=O str. |
| Lactones | δ–lactones | C=O str. |
| | α, β–unsaturated δ–lactones | C=O str. |
| Amines | R—$NH_2$ primary | N—H str. 2 bands; N—H bend; C—N str. |
| | > NH secondary | N—H str.; N—H bend; C—N str. |
| | >N tertiary | C—N str. |
| Imines | —C=NH | N—H; C=N str. |
| Amides | RCONH— | N—H; C=O str. and N—H bend |
| Nitro | R—$NO_2$ | N= O str. |
| Oximes | —C= N—OH | O—H str.; C=N str. |

3000 2500 2000 1800 1600 1400 1200 1000 900 800 700 $cm^{-1}$

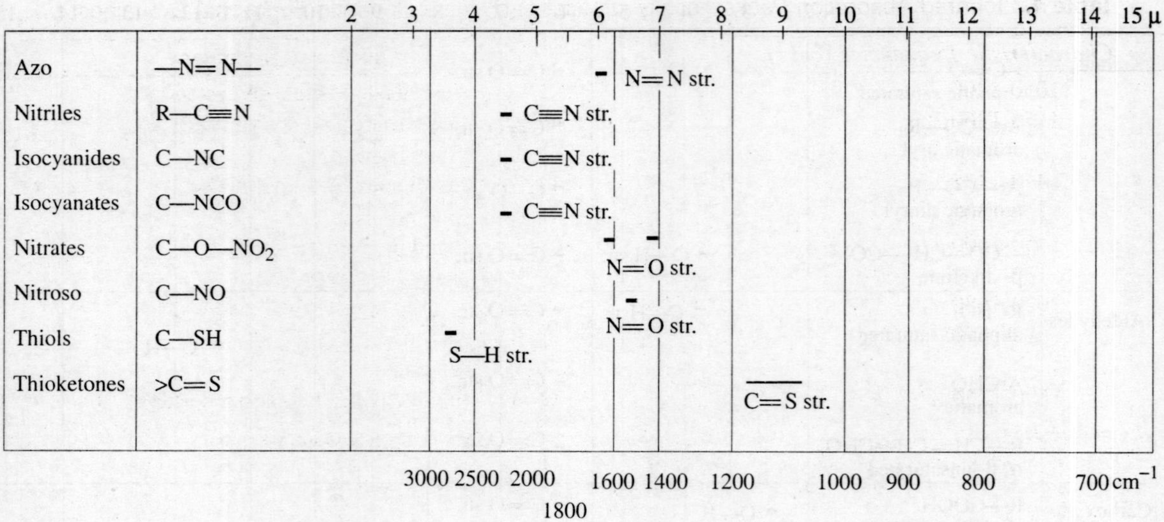
2 3 4 5 6 7 8 9 10 11 12 13 14 15 μ
Azo
—N=N—
N=N str.
Nitriles
R—C≡N
C≡N str.
Isocyanides
C—NC
C≡N str.
Isocyanates
C—NCO
C≡N str.
Nitrates
C—O—$NO_2$
N=O str.
Nitroso
C—NO
N=O str.
Thiols
C—SH
S—H str.
Thioketones
>C=S
C=S str.
3000 2500 2000 1800 1600 1400 1200 1000 900 800 700 $cm^{-1}$

**Table 4.3** Infrared Absorption Peaks (mostly stretching) of some Common Functional Groupings

| *Characteristic frequencies* ($cm^{-1}$) | *Functional groupings* | |
|---|---|---|
| 1050-1400 | C—O | (in ethers, alcohols and esters) |
| 1150-1360 | $SO_2$ | (in sulfonic acid derivatives) (2 bands seen) |
| 1440-1475 | C—H | (in alkanes) bending mode |
| 1340-1500 | $NO_2$ | (2 bands) |
| 1450-1600 | $C \cdots C$ | (band in aromatic ring) usually several peaks |
| 1620-1680 | C=C | (not for symmetrical alkenes) |
| 1630-1690 | C=O | (in amides, O=C—N) (2 bands) |
| 1690-1750 | C=O | (in aldehydes, ketones and esters) |
| 1695-1715 | $\underset{\mid \atop OH}{C}{=}O$ | (in carboxylic acids) |
| 1770-1820 | $\underset{\mid \atop Cl}{C}{=}O$ | (in acid chlorides) |
| 2100-2200 | C≡C | (not for symmetrical alkynes) |
| 2210-2260 | C≡N | (nitrile) |
| 2500 (one peak) | S—H | |
| 2700-2820 | C—H | (of aldehyde group) |
| 2500-3000 | O—H | (in—COOH) |
| 3000-3100 | C—H | (C—is a part of aromatic ring) |
| 3300 | C—H | (C is acetylenic) |
| 3020-3080 | C—H | (C is ethylenic) |
| 2800-3000 | C—H | (C is in alkanes) |
| 3300-3500 | N—H | in amines and amides |
| 3200-3600 | O—H | (in H-bonded ROH and ArOH) |
| 2100 | O—D | |

Bending frequencies of Hydrocarbons ($cm^{-1}$)

| | | | | | |
|---|---|---|---|---|---|
| Alkanes | $CH_3$ 1420-1470, 1375 | $=CH_2$ 1430-1470 | $—CH(CH_3)_2$ doublet of equal intensity at 1375, 1875 and also 1170 | $—C(CH_3)_3$ doublet at 1370 (strong) 1395 (moderate) | |
| Alkenes (out of plant) | $R\ CH{=}CH_2$ 910-920 990-1100 | $R_2\ C{=}CH_2$ 880-900 | $R\ CH{=}CHR$ cis 675-730 trans 965-975 | | |
| Aromatic C—H (out of plane) | | Monosubstituted 690-710 730-770 | Disubstituted ortho 700-770 | meta 690-710 750-810 | para 810-840 |

**Table 4.4** Correlation of absorption bands of some important boundings.

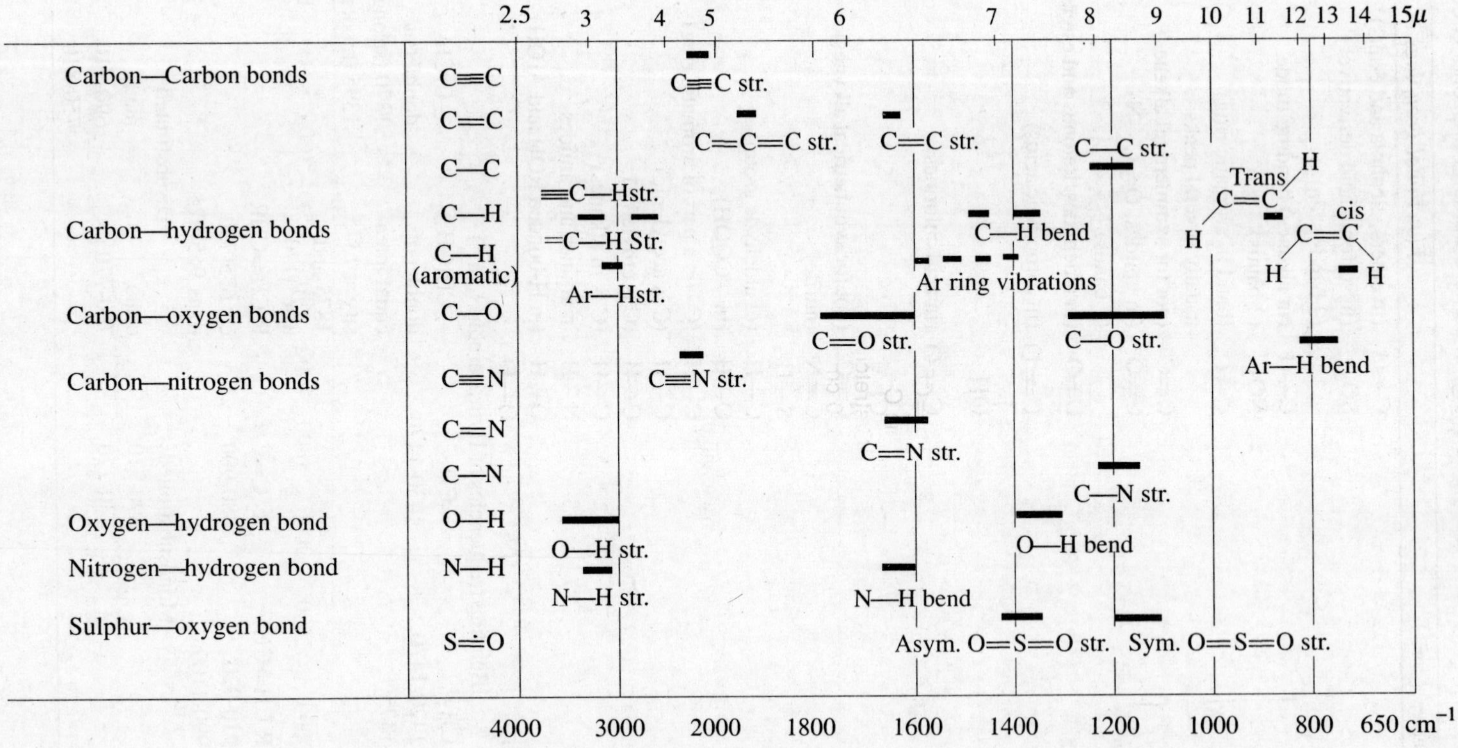

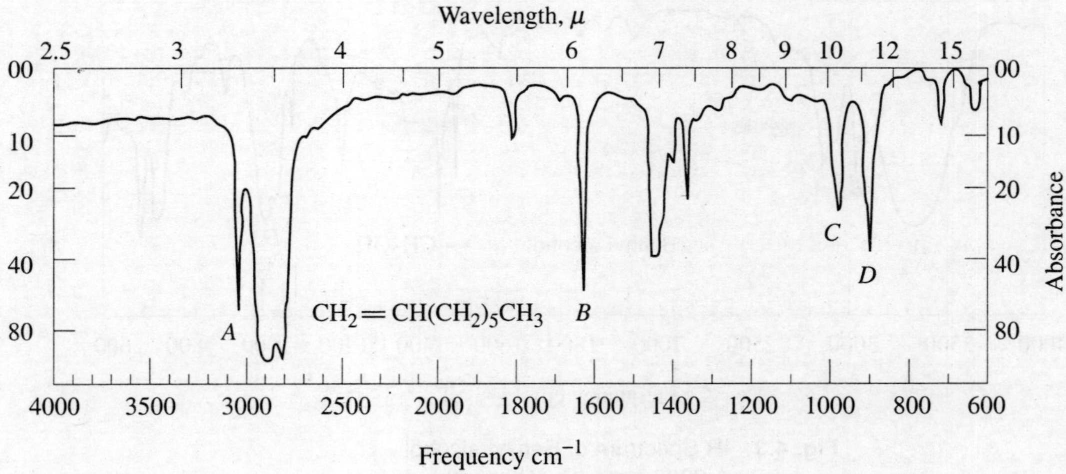


**Fig. 4.1** IR Spectrum of $CH_2{=}CH\ (CH_2)_5\ CH_3$
A 3100 $cm^{-1}$ (olefinic C—H stretch)
B 1630 $cm^{-1}$ (C═C stretch)
C and D. 980 and 900 $cm^{-1}$ (═C—H out of plane bend)

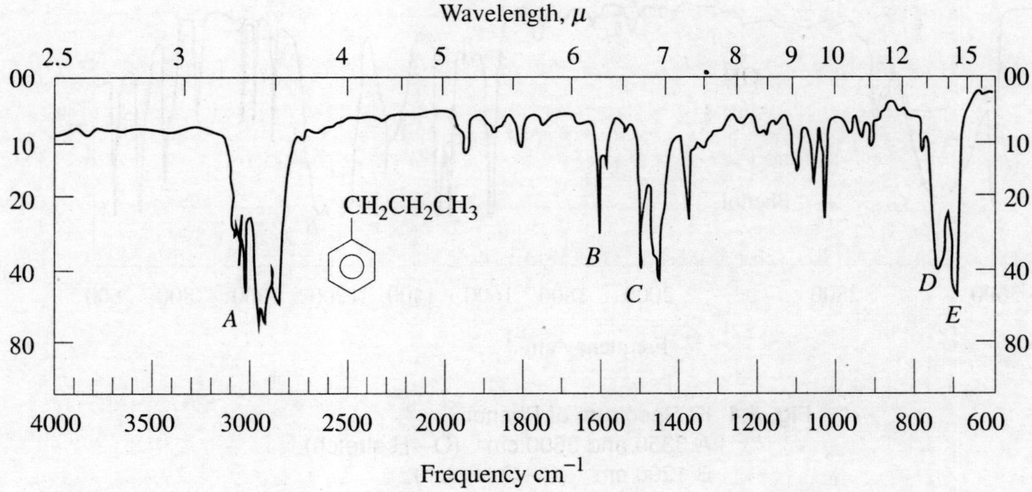


**Fig. 4.2** IR Spectrum of n-propyl benzene
A 3050 $cm^{-1}$ (olefinic C—H stretch)
B and C: (1600 and 1500 $cm^{-1}$ (C═C stretch)
D and E: 750 and 700 $cm^{-1}$ (═C—H out of plane bend)

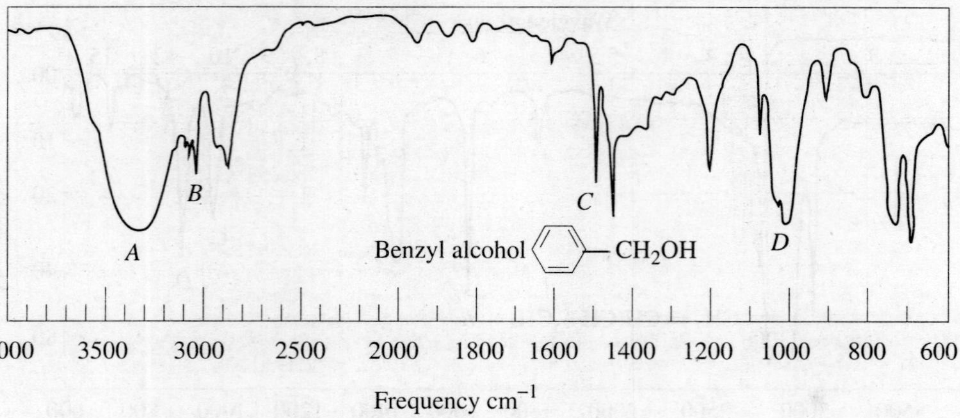


**Fig. 4.3** IR Spectrum of Benzyl alcohol
A 3300 cm$^{-1}$ (O—H stretch);
B 3100 C 1500 cm$^{-1}$ (C=C ring stretch);
D 1017 cm$^{-1}$ (C-O stretch)

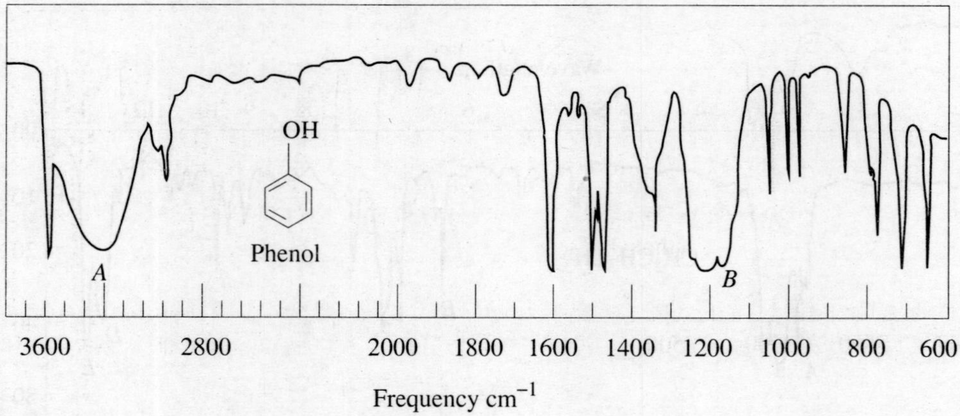


**Fig. 4.4** IR Spectrum of Phenol
A 3350 and 3600 cm$^{-1}$ (O—H stretch)
B 1200 cm$^{-1}$ (C—O stretch)

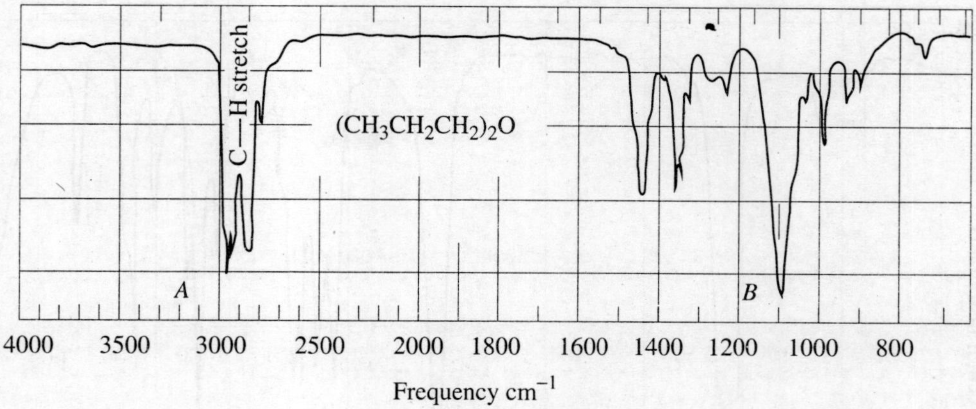


**Fig. 4.5** IR Spectrum of di-n-propyl ether
A 2950 cm$^{-1}$ (C—H stretch);
B 1110 cm$^{-1}$ (C—O stretch)

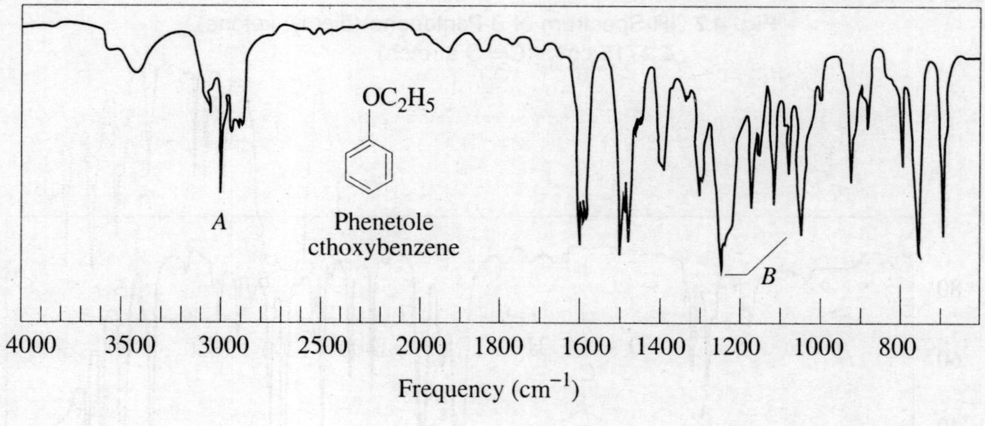


**Fig. 4.6** IR Spectrum of Phenetole
A 2950 cm$^{-1}$ (C—H stretch);
B 1230 and 1050 cm$^{-1}$ (C—O stretch)

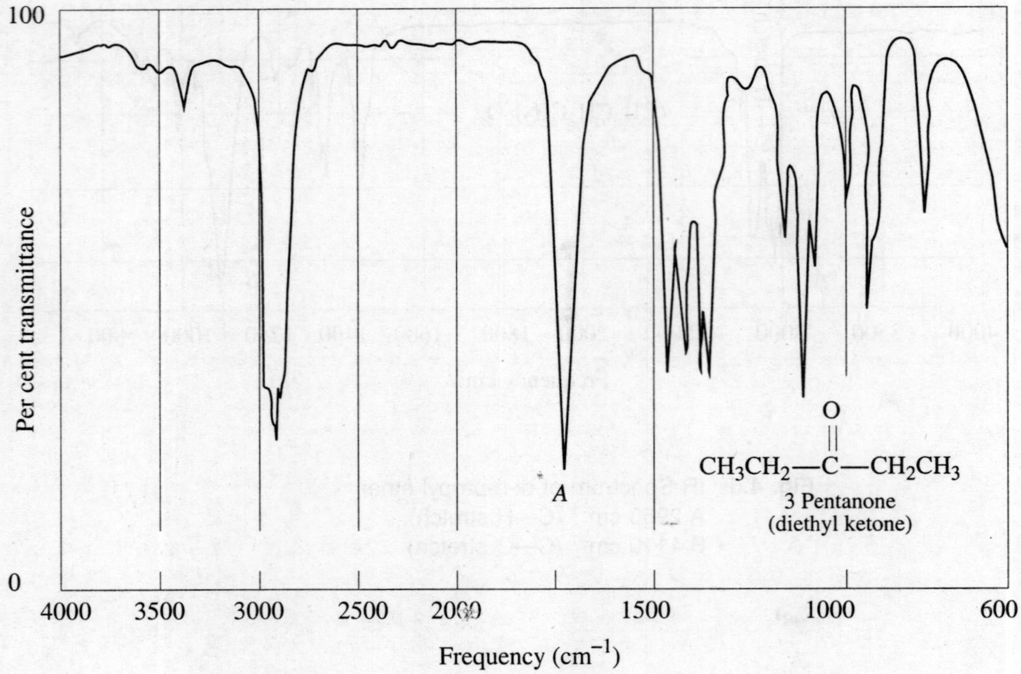


**Fig. 4.7** IR Spectrum of 3-Pentanone (diethyl ketone)
A 1715 cm$^{-1}$ (C═O stretch)

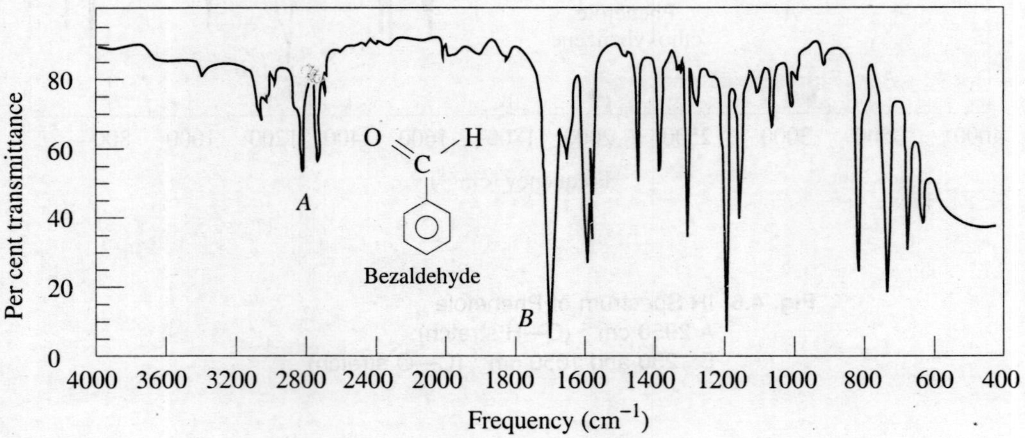


**Fig. 4.8** IR Spectrum of Benzaldehyde
A 2800 and 2780 cm$^{-1}$ (C—H str. of aldehyde)
B 1700 cm$^{-1}$ (C═O stretch)

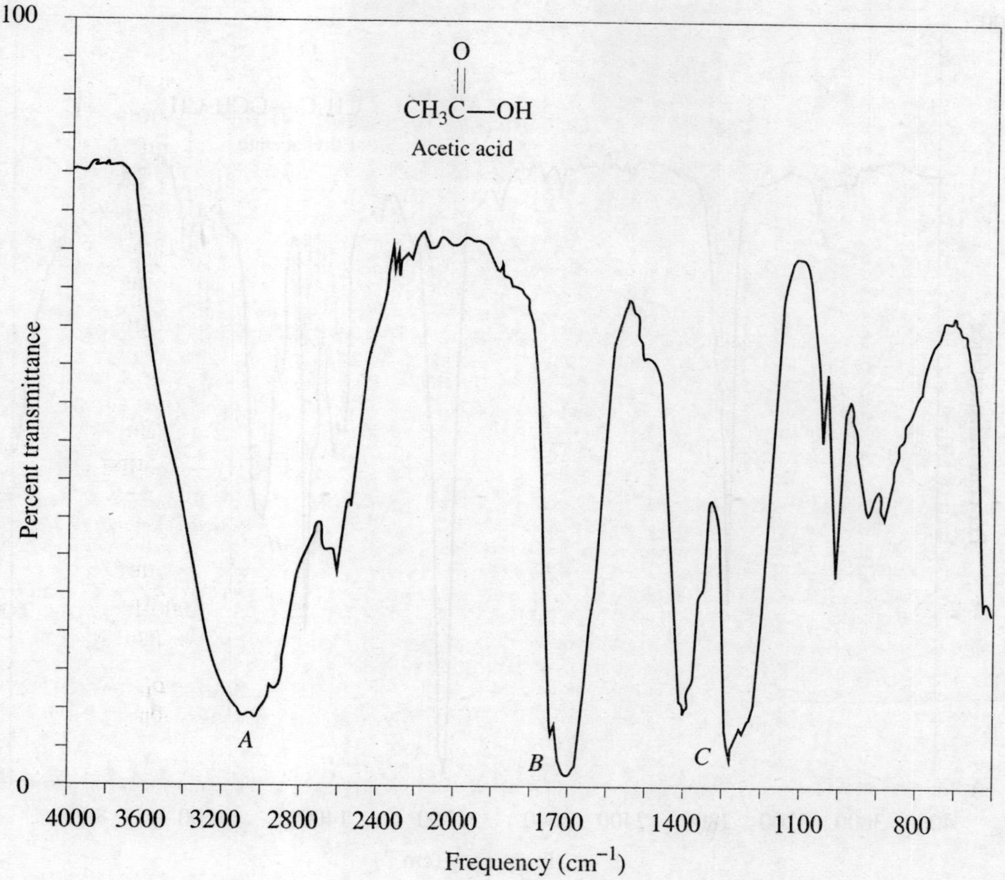


**Fig. 4.9** IR Spectrum of Acetic acid
A 3100 $cm^{-1}$ (O—H stretch);
B 1700 $cm^{-1}$ (C═O stretch)
C 1290 $cm^{-1}$ (C—O stretch)

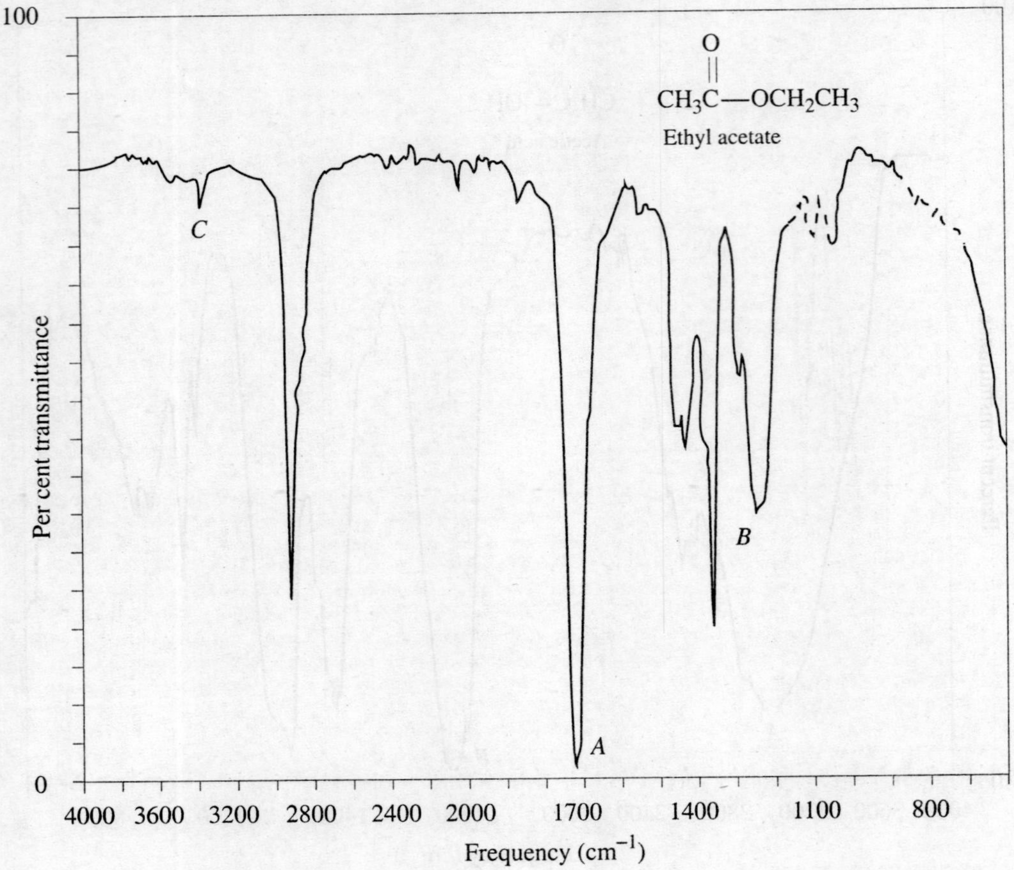


**Fig. 4.10** IR Spectrum of Ethyl acetate
A 1740 cm$^{-1}$ (C═O stretch)
B 1250 cm$^{-1}$ (C—O stretch)
C 3478 cm$^{-1}$ (overtone of C═O stretch)

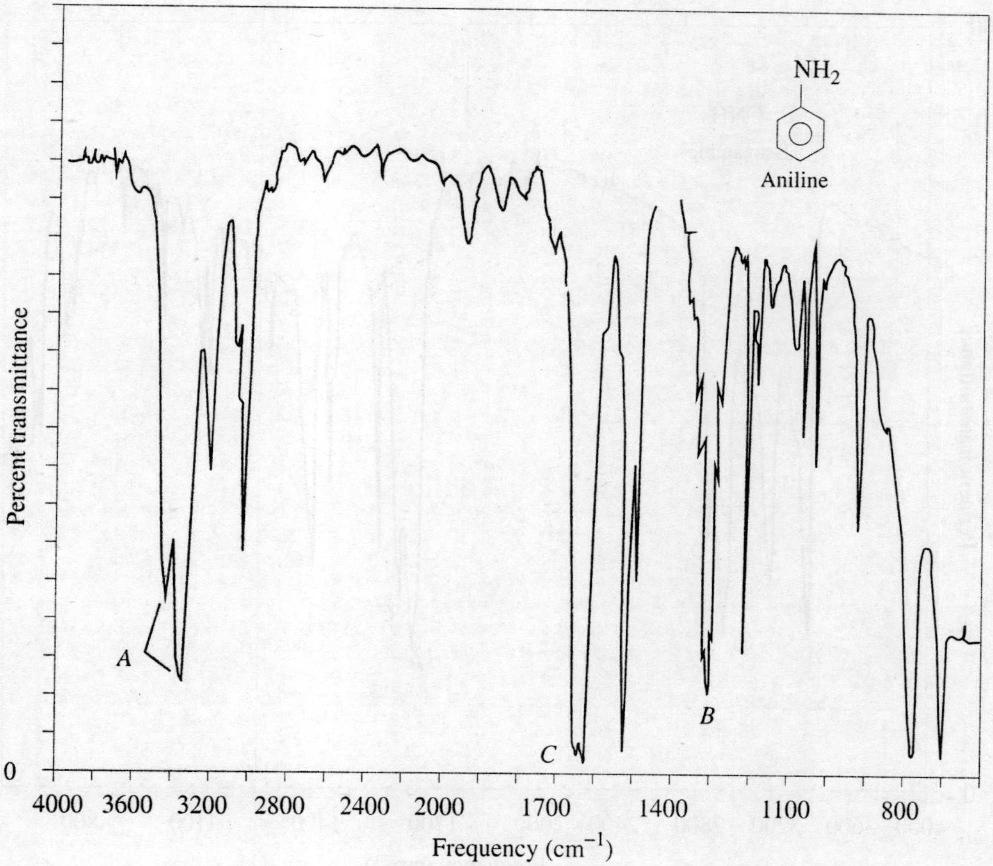


**Fig. 4.11** IR Spectrum of Aniline
A 3440 and 3350 $cm^{-1}$ (N-H stretch)
B 1280 $cm^{-1}$ (C—N stretch)
C 1620 $cm^{-1}$ (N—H bend)

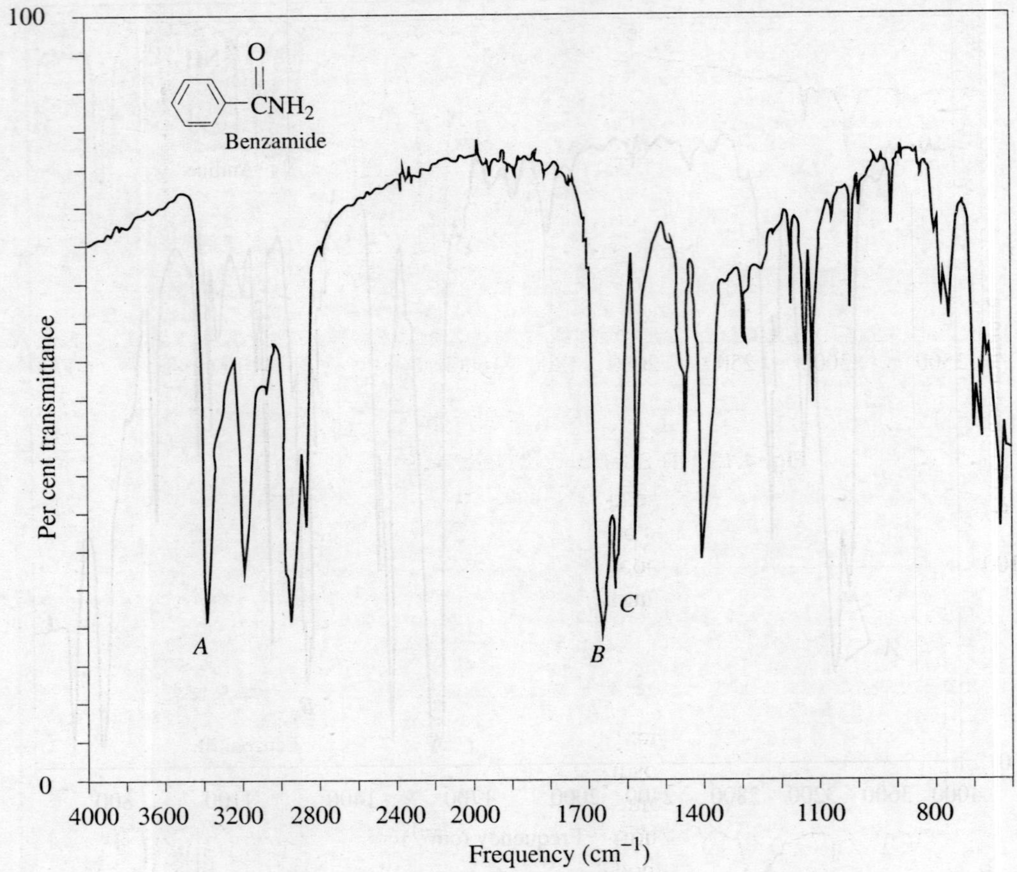


**Fig. 4.12** IR Spectrum of Benzamide
A 3370 $cm^{-1}$ (N—H stretch)
B 1660 $cm^{-1}$ (C═O stretch)
C 1631 $cm^{-1}$ (N—H bend)

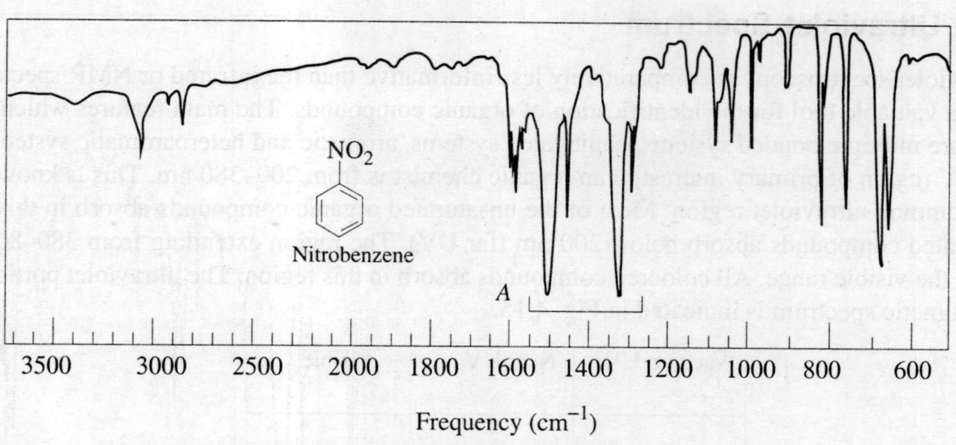


**Fig. 4.13** IR Spectrum of Nitrobenzene
A 1520 $cm^{-1}$ and 1345 $cm^{-1}$ (N=O stretch)

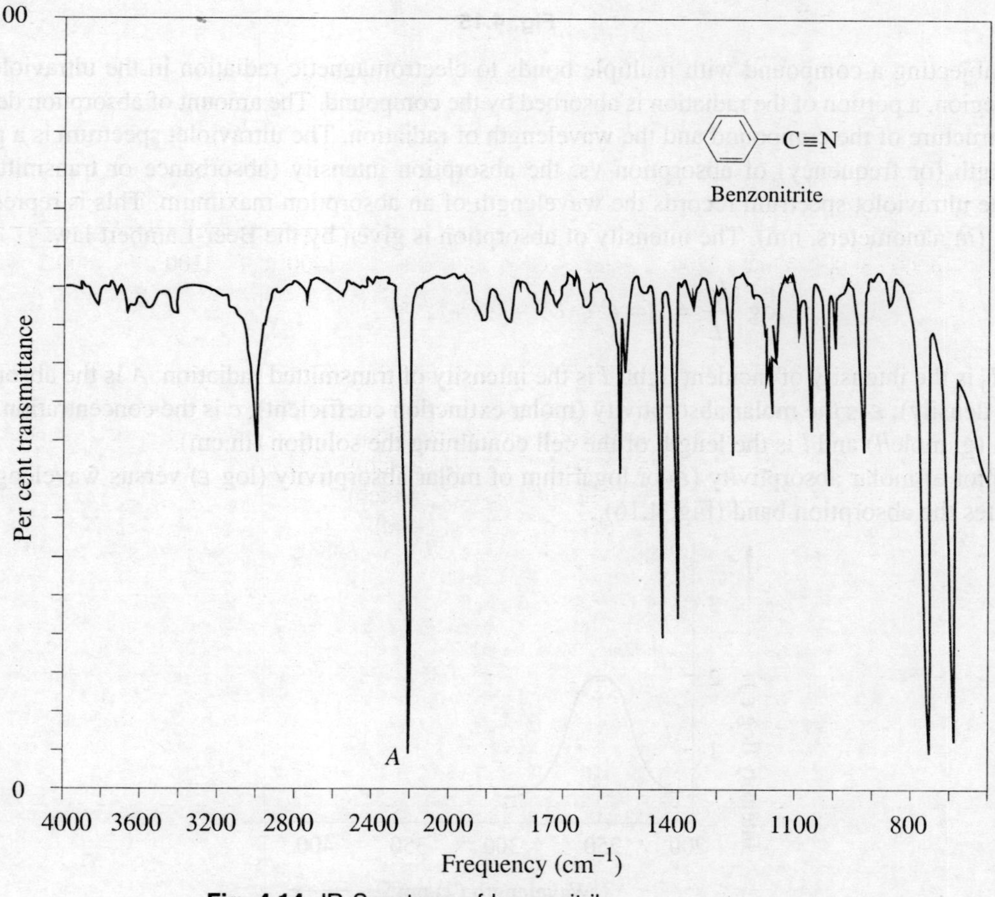


**Fig. 4.14** IR Spectrum of benzonitrile
A 2210 $cm^{-1}$ (C≡N stretch)

## 4.2 The Ultraviolet Spectrum

The ultraviolet spectroscopy is comparatively less informative than the infrared or NMR spectroscopy. Still it is a valuable tool for the identification of organic compounds. The main features which may be detected are multiple bonded systems, conjugated systems, aromatic and heteroaromatic systems.

The UV region of primary interest to an organic chemist is from 200–380 nm. This is known as the near or common ultraviolet region. Most of the unsaturated organic compounds absorb in this region. The saturated compounds absorb below 200 nm (far UV). The region extending from 380–800 nm is known as the visible range. All coloured compounds absorb in this region. The ultraviolet portion of the electromagnetic spectrum is indicated in Fig. 4.15.

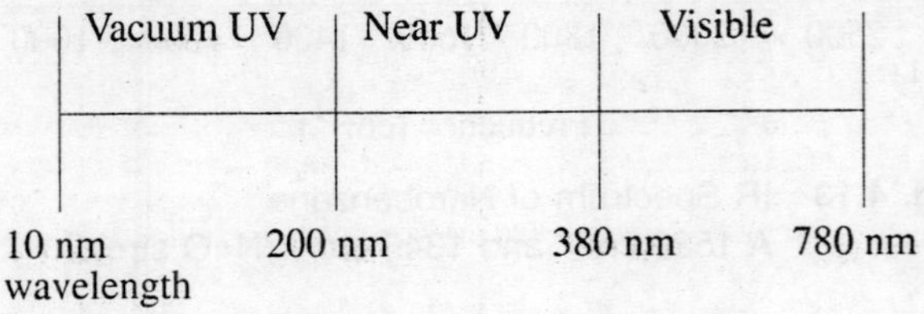


**Fig. 4.15**

On subjecting a compound with multiple bonds to electromagnetic radiation in the ultraviolet and visible region, a portion of the radiation is absorbed by the compound. The amount of absorption depends on the structure of the compound and the wavelength of radiation. The ultraviolet spectrum is a plot of wavelength (or frequency) of absorption vs. the absorption intensity (absorbance or transmittance). Thus, the ultraviolet spectrum records the wavelength of an absorption maximum. This is represented by $\lambda_{max}$ (in nanometers, nm). The intensity of absorption is given by the Beer-Lambert law.

$$\log \frac{I_0}{I} = A = \varepsilon \cdot c \cdot l.$$

where, $I_0$ is the intensity of incident light; $I$ is the intensity of transmitted radiation; $A$ is the absorbance (optical density); $\varepsilon$ is the molar absorptivity (molar extinction coefficient); $c$ is the concentration of the solution (g. mole/$l$) and $l$ is the length of the cell containing the solution (in cm).

The plot of molar absorptivity ($\varepsilon$) or logarithm of molar absorptivity (log $\varepsilon$) versus wavelength ($\lambda$) constitutes the absorption band (Fig. 4.16).

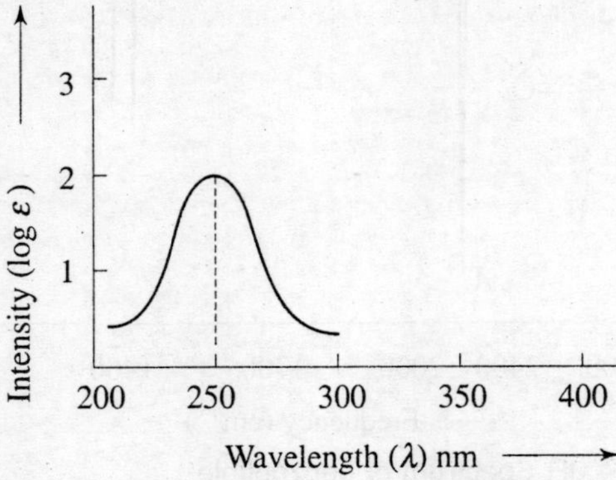


**Fig. 4.16** Representation of a UV absorption band

It will be of interest to explain some terms that are frequently used in the discussion of ultraviolet spectroscopy. A Chromophore is an unsaturated group and is responsible for absorption of radiation, e.g., C=C, C=O and $NO_2$. On the other hand, an auxochrome is a saturated group (e.g., OH, $NH_2$, –SH, Br; etc.) with non-bonded electrons, which when present in a molecule may extend the conjugation and alter the wavelength of absorption.

A bathochromic shift is a shift of absorption to a longer wavelength; this may be due to conjugation, substitution or solvent effect. On the other hand, a hypsochromic shift is the shift of absorption to a shorter wavelength.

Hyperchromic effect is the increase in absorption intensity. On the other hand, Hypsochromic effect is the decrease in absorption intensity.

When electromagnetic radiation in the ultraviolet and visible region falls on a compound, a portion of the radiation is absorbed. This energy produces transition of valence electrons in the molecule from an occupied molecular orbital to the next higher unoccupied orbital, e.g., an antibonding orbital. The relative energies of important orbitals are illustrated in Fig. 4.17.

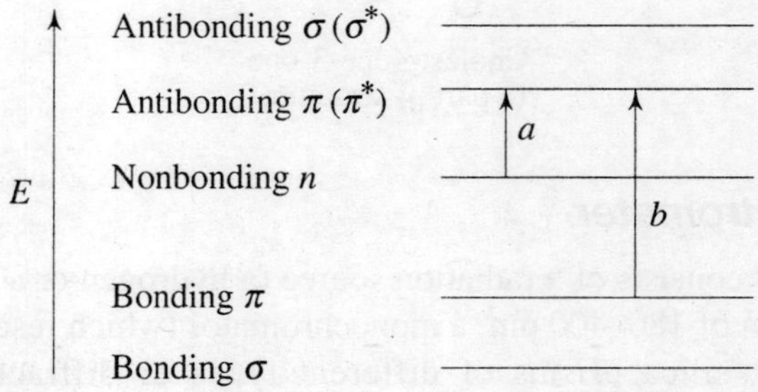


**Fig. 4.17** Designation of Electronic energy levels. The antibonding $\pi$ and $\sigma$ are represented as $\pi^*$ and $\sigma^*$ respectively.

The electronic transitions of concern to an organic chemist are $n \longrightarrow \pi^*$ and $\pi \longrightarrow \pi^*$ (represented as $a$ and $b$ respectively in Fig. 4.17). In $n \longrightarrow \pi^*$ transition, one of the electrons of an unshared pair goes to an unstable (anti bonding) $\pi^*$ orbital. For example,

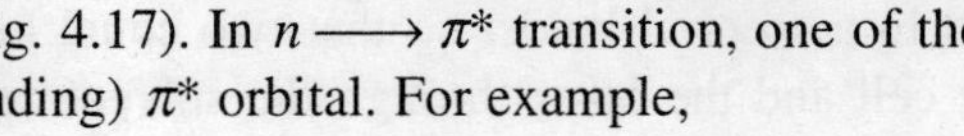


The $n \longrightarrow \pi^*$ transition is also referred to as the K-band. In the $\pi \longrightarrow \pi^*$ transition an electron goes from a stable (bonding) $\pi$ orbital to an unstable (antibonding) $\pi^*$ orbital. For example,

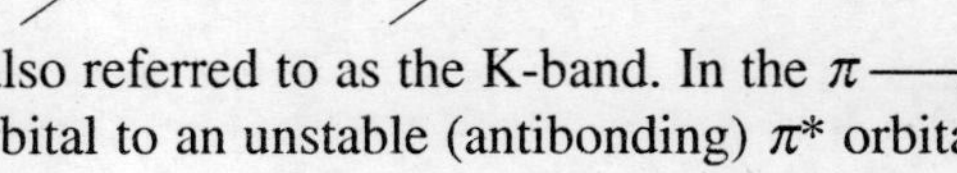


The $\pi \longrightarrow \pi^*$ transition is also referred to as the R-band.

The $n \longrightarrow \pi^*$ transition (Fig. 4.17) requires less energy compared to the $\pi \longrightarrow \pi^*$ transition. The $\pi \longrightarrow \pi^*$ transition can occur in a simple alkene, like ethylene, but the absorption occurs in the far ultraviolet. The conjugation of double bonds lowers the energy required for the electronic transition and the absorption moves to longer wavelength. In case there are enough double bonds in conjugation, absorption will move into the visible region and the compound will be coloured. For example, $\beta$-carotene is a yellow pigment found in carrots and green leaves and is a precursor of vitamin A; it contains eleven carbon-carbon double bonds in conjugation and owes its colour to absorption at the violet end of the visible spectrum ($\lambda_{max}$ 451 nm).

B-Bands (benzenoid bands) are characteristic of the spectra of aromatic and heteroaromatic nucleus. The E-bands like the B-bands are characteristic of aromatic structures.

The ultraviolet spectrum (in contrast to the infrared spectrum) is not used to show the presence of individual functional groups. It is used to show the relationship between functional groups, mainly conjugation between two or more carbon- carbon double (or triple) bonds; between carbon-carbon and carbon-oxygen double bonds; between double bonds and an aromatic ring. The presence of an aromatic ring can also be inferred. It can also reveal the number and location of substituents attached to the carbons of the conjugated system. This is illustrated by the fact that the spectrum of mesityl oxide, $CH_3COCH{=}C(CH_3){-}CH_3$ closely resembles the spectrum of cholest-4-ene-3-one; the absorption results from the conjugated enol portion of the two compounds. Both compounds exhibit $\lambda_{max}$ at 275 nm.

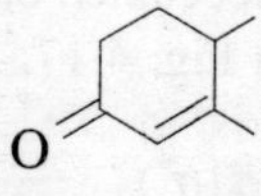


Cholest-4ene-3-one
(a part pf structure)

## *The Ultraviolet Spectrometer*

The ultraviolet spectrometer consists of a radiation source (a hydrogen or a deuterium discharge lamp) giving radiation in the region of 180–400 nm, a monochromator (which resolves the radiation from the source into narrow bands; silica prisms of different types or diffraction gratings are used as monochromators) and a sample cell (made of quartz or fused silica). The spectra are recorded in dilute solutions. The common solvents used are hexane, ethanol and cyclohexane, these are transparent down to about 210 nm. The solution of the substance in a suitable solvent is taken in a cell and placed after the monochromator. A similar cell containing only the pure solvent is also placed along with the sample cell and both cells are mounted in a spectral holder so that two equal beams of ultraviolet light are passed-one through the sample cell and the other through the solvent cell. The transmitted beam is passed through a detector (phototubes or photovoltaic cells); this detects the photons in the transmitted beam of ultraviolet radiation.

## *Interpretation of Spectra*

The ultraviolet spectrum of the organic compound is taken and the absorption band compared with those given in Table 4.5-4.7. Table 4.8, 4.9 and 4.10 give the calculation of the principal band ($\pi \longrightarrow \pi^*$ transition) of substituted benzene derivatives, dienes and enones and dienones respectively.

**Table 4.5** Ultraviolet Absorption Bands for some important Functional Groups

| *Group* | *Structure* | *Example* | *Solvent* | *Absorption Band* | | *Transition* |
|---|---|---|---|---|---|---|
| | | | | $\lambda_{max}$ *(nm)* | $\varepsilon_{max}$ | |
| Alkenes | >C=C< | ethylene | vapour | 165 | 15,000 | $\pi \longrightarrow \pi^*$ |
| Alkynes | —C≡C— | acetylene | vapour | 173 | 6,000 | $\pi \longrightarrow \pi^*$ |
| Carbonyl | H–C=O (aldehyde) | acetaldehyde | vapour | 184 | 10,000 | $\pi \longrightarrow \pi^*$ |
| " | >C=O (ketone) | acetone | *n*-hexane | 180 | 900 | $\pi \longrightarrow \pi^*$ |
| | | | | 279 | 15 | $n \longrightarrow \pi^*$ |
| Carboxyl | —COOH | acetic acid | water | 204 | 60 | $n \longrightarrow \pi^*$ |
| Amide | $—CONH_2$ | acetamide | *n*-hexane | 208 | 9,500 | $n \longrightarrow \pi^*$ |
| Esters | —COOR | ethyl acetate | alcohol | 211 | 57 | $n \longrightarrow \pi^*$ |
| Nitro | $—NO_2$ | nitromethane | alcohol | 271 | 18.6 | $n \longrightarrow \pi^*$ |
| Nitrate | $—ONO_2$ | ethyl nitrate | dioxane | 270 | 12 | $n \longrightarrow \pi^*$ |
| Nitroso | —NO | nitrosobutane | ether | 300 | 100 | $n \longrightarrow \pi^*$ |
| Nitrite | —ONO | amyl nitrite | petroleum | 218.5 | 1,120 | $\pi \longrightarrow \pi^*$ |
| Nitrile | —C≡N | acetonitrile | vapour | 167 | – | $\pi \longrightarrow \pi^*$ |
| Azo | —N=N— | azomethane | alcohol | 338 | 4 | $n \longrightarrow \pi^*$ |
| Diazo | $—\overset{+}{N}\equiv N$ | diazomethane | vapour | 410 | 3 | $n \longrightarrow \pi^*$ |

**Table 4.6** Ultraviolet Absorption Bands for Monosubstituted Benzene

| *Compound* | *Solvent* | *K-band* ($\pi \longrightarrow \pi^*$ *transition)* | | *B-band* | |
|---|---|---|---|---|---|
| | | $\lambda_{max}$ *(nm)* | $\varepsilon_{max}$ | $\lambda_{max}$ *(nm)* | $\varepsilon_{max}$ |
| Benzene | Ethanol | 203.5 | 7,400 | 254 | 204 |
| Chlorobenzene | Ethanol | 209.5 | 7,400 | 263.5 | 190 |
| Phenol | Ethanol | 210.5 | 6,200 | 270 | 1,450 |
| Anisole | Ethanol | 217 | 6,400 | 269 | 1,480 |
| Benzonitrile | Ethanol | 224 | 13,000 | 271 | 1,000 |
| Benzoic acid | Ethanol | 230 | 11,600 | 273 | 970 |
| Aniline | Ethanol | 230 | 8,600 | 280 | 1,430 |
| Acetophoenone | Ethanol | 245.5 | 9,800 | | |
| Styrene | Ethanol | 248 | 14,000 | 282 | 750 |
| Benzaldehyde | Ethanol | 249.5 | 11,400 | | |
| Biphenyl | Ethanol | 251.5 | 18,300 | | |
| Nitrobenzene | Ethanol | 268.5 | 7,800 | | |
| Diphenyl ether | Ethanol | 255 | 11,000 | 272 | 2,000 |
| Stilbene (trans) | Ethanol | 295 | 27,000 | | |
| Stilbene (cis) | Ethanol | 280 | 13,500 | | |

**Table 4.7** Ultraviolet Absorption bands of some Disubstituted Benzenes (Nitrophenols and Nitro anilines)

| *Compound* | $\pi \longrightarrow \pi^*$ transition (K band) $\lambda_{max}$ | $\varepsilon_{max}$ |
|---|---|---|
| Nitro anilines | | |
| o- | 283 | 5,400 |
| m- | 280 | 4,800 |
| p- | 381 | 13,500 |
| Nitro phenols | | |
| o- | 279 | 6,600 |
| m- | 274 | 6,000 |
| p- | 318 | 10,000 |

**Table 4.8** Calculation of the Principal Band ($\pi \longrightarrow \pi^*$ transition) of Substituted Benzene derivatives (Woodward Empirical Rules)

| *Substituted benzene derivatives* *Ar-COG* | | $\lambda_{max}^{EtOH}$ |
|---|---|---|
| Parent chromophore Ar = $C_6H_5$ | | |
| G = Alkyl or ring' residue (e.g. ArCOR) | | 246 |
| G = H (e.g., AR CHO) | | 250 |
| G = OH, O Alk (e.g. Ar $CO_2H$, Ar$CO_2$R) | | 230 |
| Increment for each substituent on Ar | | |
| Alkyl or ring residue | o, m | +3 |
| | p | + 10 |
| —OH, $OCH_3$, —O Alk | o,m | +7 |
| | p | +25 |
| Cl | o, m | +0 |
| | p | +10 |
| Br | o, m | +2 |
| | p | +15 |
| $NH_2$ | o, m | +13 |
| | p | +58 |
| $NHCOCH_3$ | o, m | +20 |
| | p | +45 |
| $NHCH_3$ | p | +73 |
| $N(CH_3)_2$ | o, m | +20 |
| | p | +85 |

**Table 4.9** Calculation of $\lambda_{max}$ (nm) of Dienes (Woodward Empirical Rules)

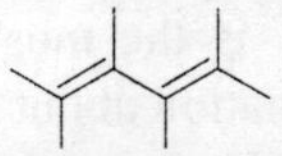

| *Dienes* | $\lambda_{max}$ *(nm)* |
|---|---|
| Parent heteroannular diene or open chain diene | 214 |
| Parent homoannular diene | 253 |
| *Add for* | |
| Double bond extending conjugation | 30 |
| Ring residue or alkyl substituent | 5 |
| Exocyclic double bond | 5 |
| OAc | 0 |
| OAlk | 6 |
| SAlk | 30 |
| Cl, Br | 5 |
| $N(Alk)_2$ | 60 |

**Table 4.10** Calculation of $\lambda_{max}$ (nm) of Enones and Dienones (Woodward Empirical Rules)

(*Enone*) $\beta$—C(β)=C(α)—C=O and (*Dienone*) $\delta$—C(δ)=C(γ)—C(β)=C(α)—C=O

| | | $\lambda_{max}$ (nm) |
|---|---|---|
| Acyclic $\alpha, \beta$-unsaturated ketone | | 215 |
| $\alpha, \beta$-Unsaturated aldehydes | | 210 |
| $\alpha, \beta$-Unsaturated acids and esters | | 195 |
| Six membered cyclic $\alpha, \beta$-unsaturated ketones | | 215 |
| Five membered cyclic $\alpha, \beta$-unsaturated ketones | | 202 |
| *Add for* | | |
| Double bond extending conjugation | | 30 |
| Alkyl groups or ring residue | $\alpha$ | 10 |
| | $\beta$ | 12 |
| | $\gamma, \delta$ | 18 |
| —OAc $\alpha, \beta$ and $\delta$ | | 6 |
| —OH | $\alpha$ | 35 |
| | $\beta$ | 30 |
| | $\gamma$ | 50 |
| $—OCH_3$ | $\alpha$ | 35 |
| | $\beta$ | 30 |
| | $\gamma$ | 17 |
| | $\delta$ | 31 |
| —Cl | $\alpha$ | 15 |
| | $\beta$ | 12 |
| —Br | $\alpha$ | 25 |
| | $\beta$ | 30 |
| $—NR_2$ | $\beta$ | 95 |
| Exocyclic double bond | | 5 |
| Homodiene component | | 39 |

## 4.3 Proton Magnetic Resonance (PMR) Spectrum

The proton magnetic resonance (PMR) spectroscopy is the most important technique used for the characterisation of organic compounds. It gives information about the different kinds of protons in the molecule. In other words it tells one about different kinds of environments of the hydrogen atoms in the molecule. PMR also gives information about the number of protons of each type and the ratio of different types of protons in the molecule.The present discussion is restricted only to PMR spectroscopy.

It is well known that all nuclei carry a positive charge. In some nuclei this charge 'spins' on the nuclear axis, and this circulation of nuclear charge generates a magnetic dipole along the axis. Thus, the nucleus behaves like a tiny bar magnet. The angular momentum of the spinning charge is described in terms of spin number (I) which can have values of 0, $\frac{1}{2}$, 1, $\frac{3}{2}$ etc. (I = 0 denotes no spin). The magnitude of the generated dipole is expressed in terms of nuclear magnetic moment ($\mu$).

The spinning nucleus of a hydrogen atom ($^1$H or proton) is the simplest and is commonly encountered in organic compounds. The hydrogen nucleus has a magnetic moment, $\mu = 2.79268$ and its spin number (I) is $\pm\frac{1}{2}$. Hence, in an applied external magnetic field, its magnetic moment may have two possible orientations. The orientation in which the magnetic moment is aligned with the applied magnetic field is more stable (lower energy) than in which the magnetic moment is aligned against the field (higher energy). The energy required for flipping the proton from its lower energy alignment to the higher energy alignment depends upon the difference in energy ($\Delta E$) between the two states and is equal to $hv$ ($\Delta E = hv$). Just how much energy is needed to flip the proton over depends on the strength of the external field; the stronger the field, the greater the tendency to remain lined up with it, and the higher the frequency of the radiation needed to do the spin flip; $v \alpha H_o$

i.e.,
$$v = \frac{H_o \nu}{2\pi}$$

where, $v$ is the frequency in Hz,
$H_o$ is the strength of the magnetic field in gauss, and
$\nu$ is a nuclear constant, the magnetogyric ratio, 26750 for the proton.

In a field of 14,092 gauss, for example, the energy required to flip a proton corresponds to electromagnetic radiation of frequency 60 MHz (60 megahertz or 60 million cycles per second). For higher frequencies of operation, proportionately higher field has to be used.

In principle, the substance could be placed in a magnetic field of constant strength, and then the spectrum can be obtained in the same way as an infrared or an ultraviolet spectrum by passing radiation of steadily changing frequency through the substance and observing the frequency at which radiation is absorbed. In practice, however, it has been found to be more convenient to keep the radiation frequency constant and to vary the strength of the magnetic field. At some value of the field strength the energy required to flip the proton matches the energy of the radiation, absorption occurs and a signal is obtained. Such a spectrum is called a nuclear magnetic resonance (NMR) spectrum.

### The NMR Spectrometer

It consists of the following components. A powerful magnet with a homogeneous field that can be continuously and precisely varied over a relative narrow range; this is accomplished by a sweep generator. A radiofrequency transmitter and a radio-frequency receiver are the two other components of the spectrometer. Besides these, the instrument has a recorder, a calibrator and an integrator. A sample holder positions the sample between the magnetic poles; the holder also spins the sample by an air-driven turbine to average the magnetic field over the sample. The solution of the sample is normally prepared in a solvent containing no protons. The most common solvents are deuterated chloroform ($CDCl_3$) and carbon tetrachloride ($CC1_4$). A small amount of (1%) tetramethylsilane (TMS) is added as an internal standard.

**Interpretation of the NMR spectra.** The interpretation of the NMR spectra is done by taking into account the number of signals that appear, their position, intensity and splitting pattern.

The number of signals gives the number of different kinds of protons present in a molecule. This is illustrated with the help of the following examples.

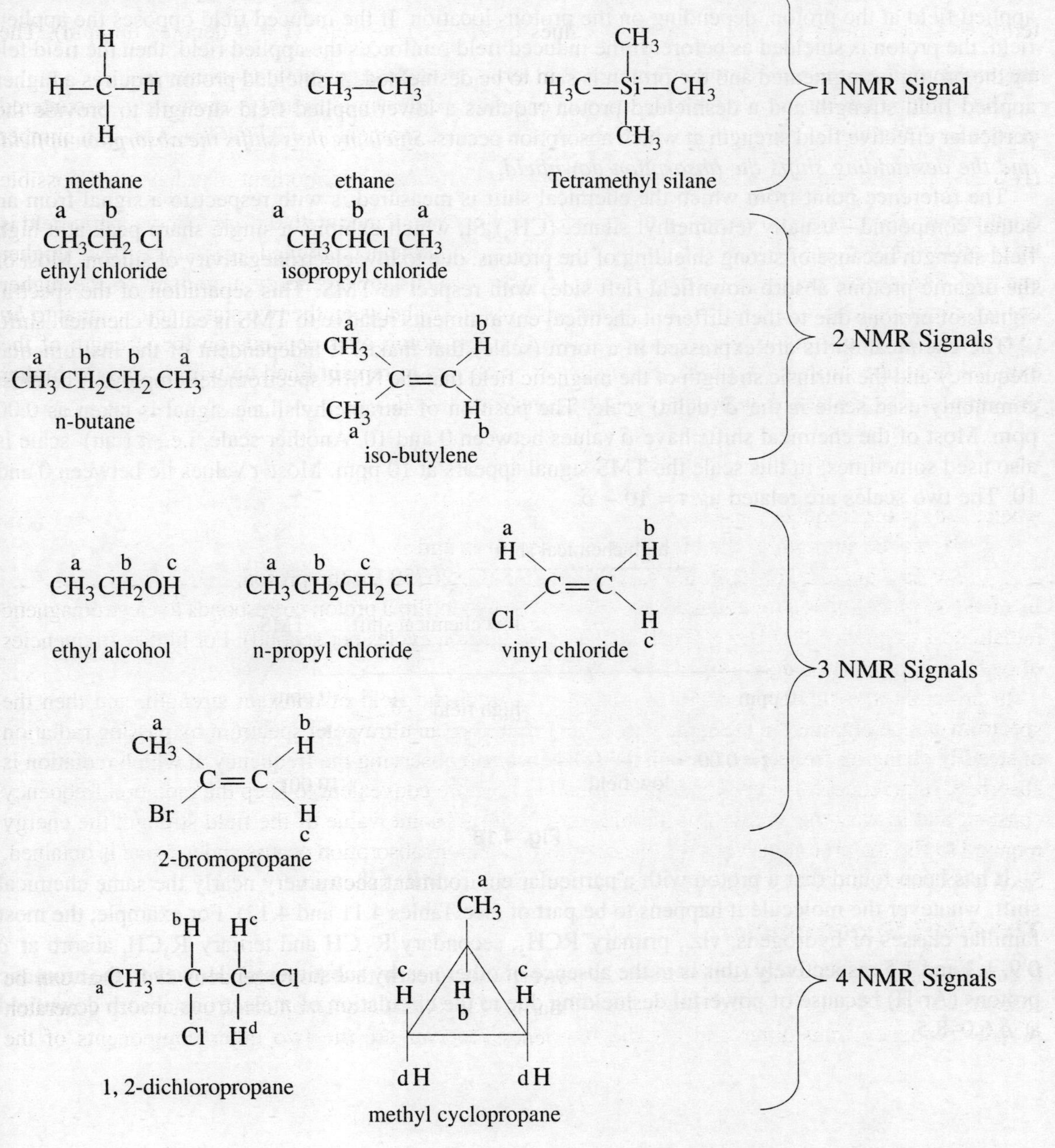

Just as the number of signals in an NMR spectrum indicate how many kinds of protons are present in a molecule, the positions, of the signals give an idea of the nature of the protons, i.e., whether they are aromatic, aliphatic, primary, secondary, tertiary, benzylic, vinylic, acetylenic and adjacent to halogen or other hetero atoms or functional groups. These protons have different electronic environments and hence they show absorption at different positions in the NMR spectrum.

When a compound is placed in a magnetic field, the electrons surrounding hydrogen are caused to circulate and in circulating they generate secondary magnetic fields, i.e., induced magnetic field. The circulation of electrons generates a field aligned in such a way that at the proton it opposes the applied field. Thus, the field felt by the proton is diminished and the proton is said to be shielded. Alternatively, the circulation of electrons about nearby nuclei generates a field that can either oppose or reinforce the applied field at the proton, depending on the protons location. If the induced field opposes the applied field, the proton is shielded as before. If the induced field reinforces the applied field, then the field felt by the proton is augmented and the proton is said to be deshielded. A shielded proton requires a higher applied field strength and a deshielded proton requires a lower applied field strength to provide the particular effective field strength at which absorption occurs. *Shielding thus shifts the absorption upfield, and the deshielding shifts the absorption downfield.*

The reference point from which the chemical shift is measured is with respect to a signal from an actual compound—usually tetramethyl silane, $(CH_3)_4Si$, which exhibits a single sharp peak at a high field strength because of strong shielding of the protons, due to low electronegativity of silicon. Most of the organic protons absorb downfield (left side) with respect to TMS. This separation of the spectral signals of protons due to their different chemical environments relative to TMS is called chemical shift.

The chemical shifts are expressed in a form (scale) that makes it independent of the instrumental frequency and the intrinsic strength of the magnetic field that the NMR spectrometer employs. The most commonly used scale is the $\delta$ (delta) scale. The position of tetramethylsilane signal is taken as 0.00 ppm. Most of the chemical shifts have $\delta$ values between 0 and 10. Another scale, i.e., $\tau$ (tau) scale is also used sometimes; in this scale the TMS signal appears at 10 ppm. Most $\tau$ values lie between 0 and 10. The two scales are related as $\tau = 10 - \delta$.

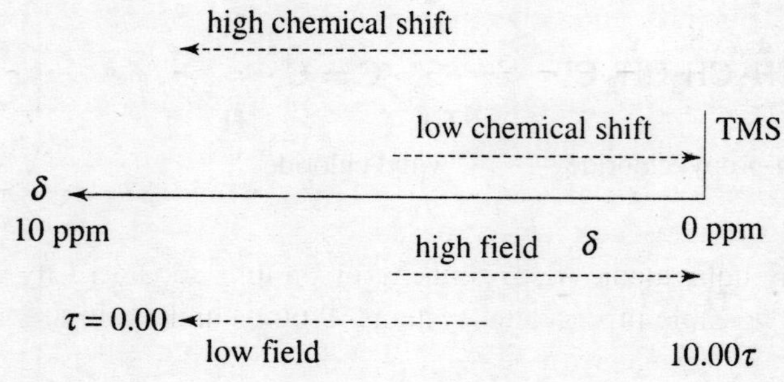


**Fig. 4.18**

It has been found that a proton with a particular environment shows very nearly the same chemical shift, whatever the molecule it happens to be part of (see Tables 4.11 and 4.12). For example, the most familiar classes of hydrogens, viz., primary $RCH_3$, secondary $R_2$ CH and tertiary $R_3CH$, absorb at $\delta$ 0.9, 1.3 and 1.5 respectively (this is in the absence of other nearby substituents). However, the aromatic protons (Ar-H) because of powerful deshielding due to the circulation of $\pi$ electrons absorb downfield at $\delta$ 6.0–8.5.

The intensities of the signals give the ratio of protons of each type in the molecule. Most NMR spectrometers have an integrator which integrates each peak (the area of each peak is found by counting the squares on the chart and is proportional to the number of protons in that signal).

The splitting of a signal into several peaks gives an idea about the environment of a proton with respect to other nearby protons. The understanding of the splitting of a signal is an important step for the structural assignment of an organic compound. Consider the NMR spectra of the three compounds, 1, 1, 2-tribromoethane ($CH_2Br—CHBr_2$), 1, 1-dibromoethane ($CH_3$-$CHBr_2$) and ethyl bromide ($CH_3CH_2Br$). Each of these contains only two kinds of protons. However, instead of two peaks the spectra show five, six and seven peaks respectively. This arises due to splitting of the NMR signals caused by spin-spin coupling. The signals we expect from each set of equivalent protons is appearing as a group of peaks (and not a single peak). Thus, splitting reflects the environment of the absorbing protons with respect to nearby protons. Let us consider the case of adjacent carbon atoms carrying a pair of secondary protons (b) and a tertiary proton (a) respectively $\left(—\overset{|\,a}{C}H—\overset{b}{C}H_2—\right)$. First, consider the absorption by one of the secondary protons. The magnetic field that a secondary proton (b) feels at a particular instant is slightly increased or decreased by the spin of the neighbouring tertiary proton (a) increased if the tertiary proton happens at that instant to be aligned with the applied field, or decreased if the tertiary proton happens to be aligned against the applied field. Thus, for half the molecules, absorption by a secondary proton is shifted slightly downfield and for the other half of the molecules the absorption is shifted slightly up field. The signal is split into two peaks—a doublet with equal peak intensities.

The absorption by a tertiary proton $\left(—\overset{|\,a}{C}H—\overset{b}{C}H_2—\right)$ is affected by the spin of the neighbouring secondary protons; its signal is split into a triplet with relative peak intensities 1 : 2 : 1.

Thus, in the NMR spectrum of a compound containing the grouping —CH—$CH_2$—, we see a 1:1 doublet (for —$CH_2$—) and a 1:2:1 triplet (for —CH—). The total area (both peaks) under the doublet is twice as big as the total area (all three peaks) of triplet, since the doublet is due to twice as many protons as the triplet. Similarly, three equivalent protons will split a signal into four peaks—a quartet—with the intensity pattern 1:3:3:1.

In general, a set of $n$ equivalent neighbouring protons will split an NMR signal into $n + 1$ peaks.

The distance between peaks within a multiplet is a measure of the effectiveness of spin-spin coupling, and is called the coupling constant, J. The value of the coupling constant is expressed in cps or Hertz; it is independent of the field strength and rarely exceeds 20 cps. Spin-spin coupling is not observed for protons that are chemically equivalent, Also, splitting of protons separated by more than two atoms (H—C—C—C—H) is not possible in saturated systems. Protons undergoing rapid chemical exchange e.g. alcohols do not normally couple with adjacent protons. Coupling constant of different types of protons is given in Table. 4.13.

The NMR spectra of some typical compounds are given in Figs. 4.19–4.28.

**Table 4.11** Chemical shifts for methyl, methylene and methine protons

| *Methyl Protons* | $\delta$ | *Methylene protons* | $\delta$ | *Methine protons* | $\delta$ |
|---|---|---|---|---|---|
| $CH_3$—C— | 0.9 | —$CH_2$—C— | 1.4 | —CH—C— | 1.5 |
| $CH_3$—C=C | 1.6 | —$CH_2$—C=C | 2.3 | —CH—C=C | 2.6 |
| $CH_3$—C(=O)— | 2.1 | —$CH_2$—C(=O)— | 2.4 | —CH—C(=O)— | 2.5 |
| $CH_3$—$NR_2$ | 2.2 | —$CH_2$—$NR_2$ | 2.5 | —CH—$NH_2$ | 2.9 |
| $CH_3$—Ar | 2.3 | —$CH_2$—Ar | 2.7 | —CH—Ar | 3.0 |
| $CH_3$—Br | 2.7 | $CH_2$—Br | 3.3 | CH—Br | 4.1 |
| $CH_3$—Cl | 3.1 | $CH_2$—Cl | 3.4 | —CH—Cl | 4.1 |
| $CH_3$—F | 4.3 | –$CH_2$—F | 4.5 | —CH—F | 5.2 |
| $CH_3$—S | 2.1 | —$CH_2$—S | 2.4 | —CH—S | 3.2 |
| $CH_3$—N | 2.3 | $CH_2$—N | 2.5 | —CH—N | 2.8 |
| $C\underline{H}_3OH$ | 3.3 | —$CH_2OH$ | 3.3 | >CHOH | 3.8 |
| $CH_3$—C≡N | 2.0 | —$CH_2$—C≡N | 2.3 | —CH—C≡N | 2.7 |
| $CH_3$—$NO_2$ | 4.3 | —$CH_2$—$NO_2$ | 4.4 | —CH—$NO_2$ | 4.6 |

**Table 4.12** Chemical shifts for the Protons of some common Functional Groups and Unsaturated systems

| *Types of Protons* | $\delta$ | *Types of protons* | $\delta$ |
|---|---|---|---|
| R—C$\underline{H}$O | 9.4-9.7 | $\underline{H}$C=C (isolated) | 5.1-5.8 |
| Ar—C$\underline{H}$O | 9.7-10.0 | $\underline{H}$C=C (conjugated) | 5.8-6.6 |
| R—N$\underline{H}_2$ | 5.0-8.0 | $\underline{H}_2$C=C (isolated) | 4.5-5.0 |
| Ar—N$\underline{H}_2$ | 3.4-6.0 | $\underline{H}_2$C=C (conjugated) | 5.3-5.8 |
| R—O$\underline{H}$ | 0.5-4.0 | $\underline{H}$C≡C (isolated) | 2.4-2.7 |
| Ar—O$\underline{H}$ | 4.4-12 | $\underline{H}$C≡C (conjugated) | 2.7-3.1 |
| R—S$\underline{H}$ | 1.0-2.0 | Ar—$\underline{H}$ | 6.5-8.0 |
| Ar—S$\underline{H}$ | 3.0-4.0 | | |
| R—COO$\underline{H}$ | 10.5-13.0 | | |
| R—C(=O)—OC$\underline{H}_2$R′ | 3.7-4.1 | | |
| R—O—C—$\underline{H}$ | 3.3-4.0 | | |

**Table 4.13** Spin-Spin Coupling constants

| *Structure type* | $J_{ab}$ *Hz* |
|---|---|
| 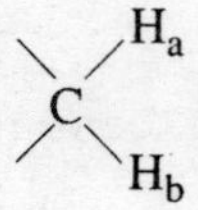 (gem) | 10-18 |
| 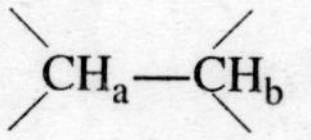 (vic) | 6-8 |
| 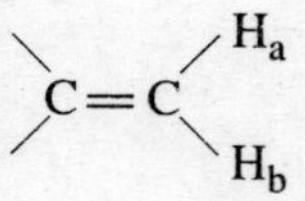 (geminal) | 0-3 |
| 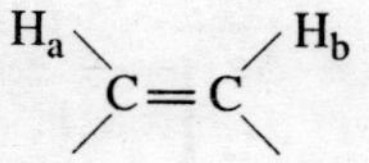 (cis) | 5-12 |
| 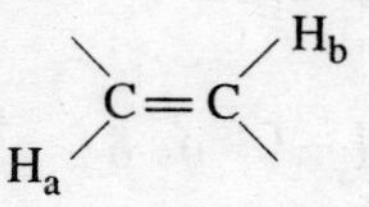 (trans) | 12-19 |
| 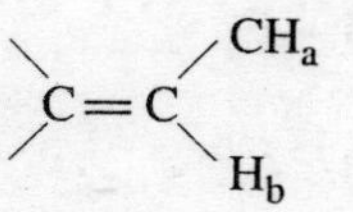 | 4-10 |
| 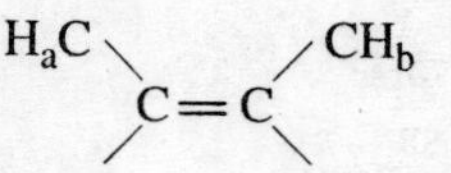 | 1-2 |
| 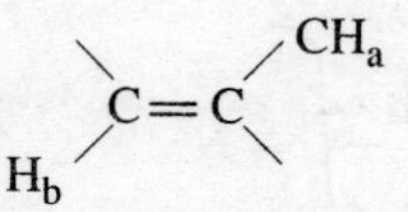 | 0-2 |
| $H_bC—C—C—C—H_a$ | 0 |
| 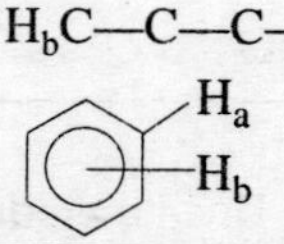 | ortho 7-10<br>meta 2-3<br>para 0-1 |

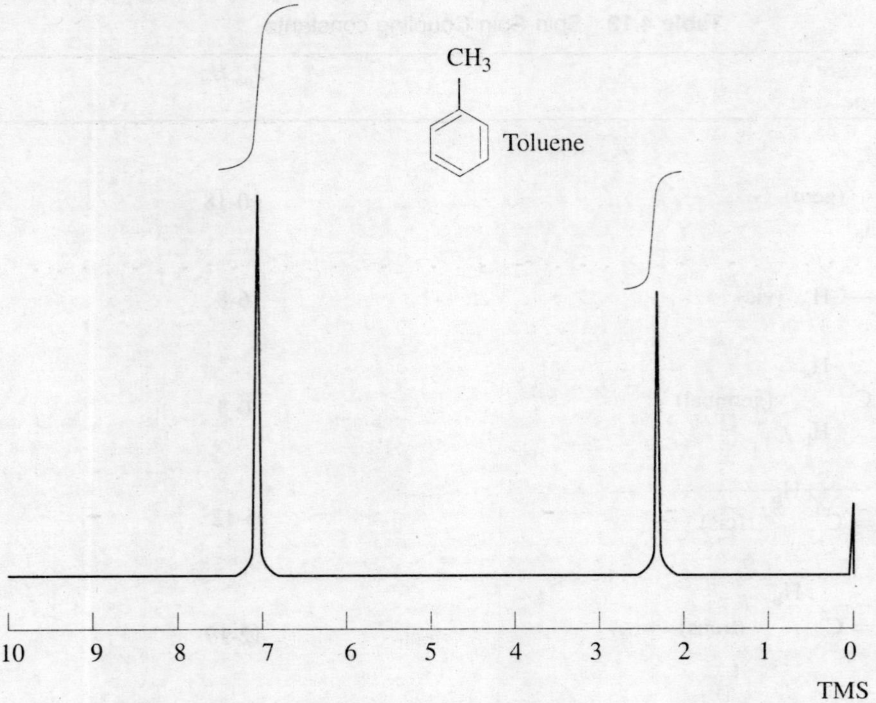


**Fig. 4.19** PMR Spectrum of Toluene

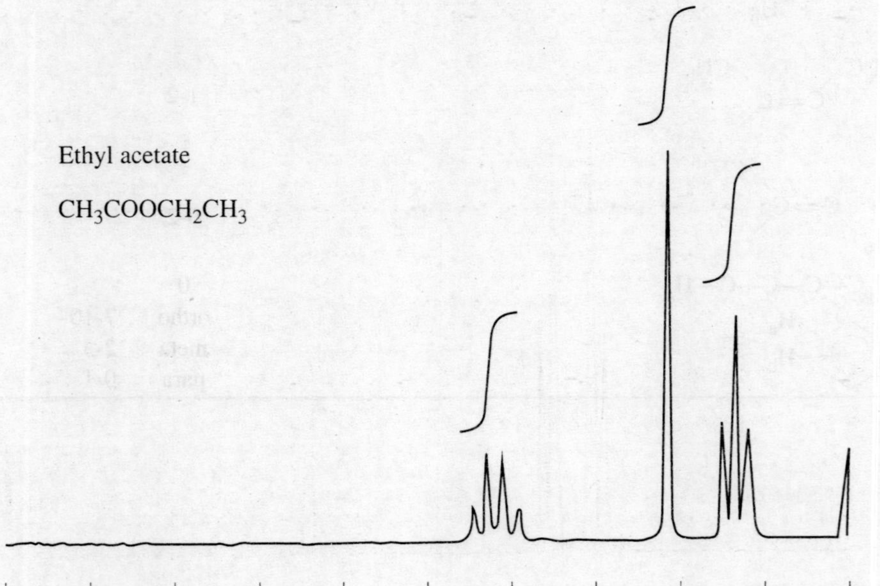


**Fig. 4.20** PMR Spectrum of Ethyl acetate

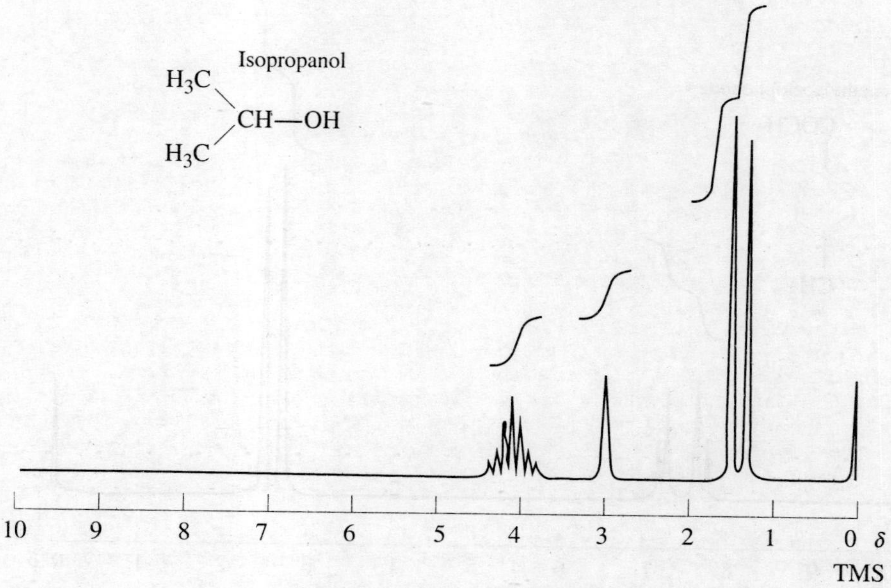


**Fig. 4.21** PMR Spectrum of Isopropanol

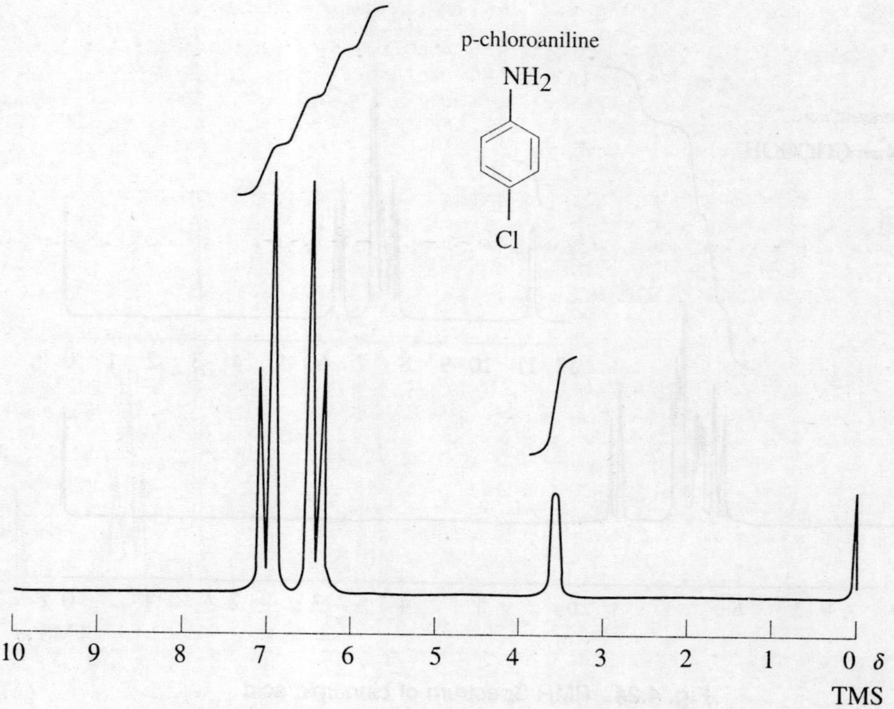


**Fig. 4.22** PMR Spectrum of p-Chloroaniline

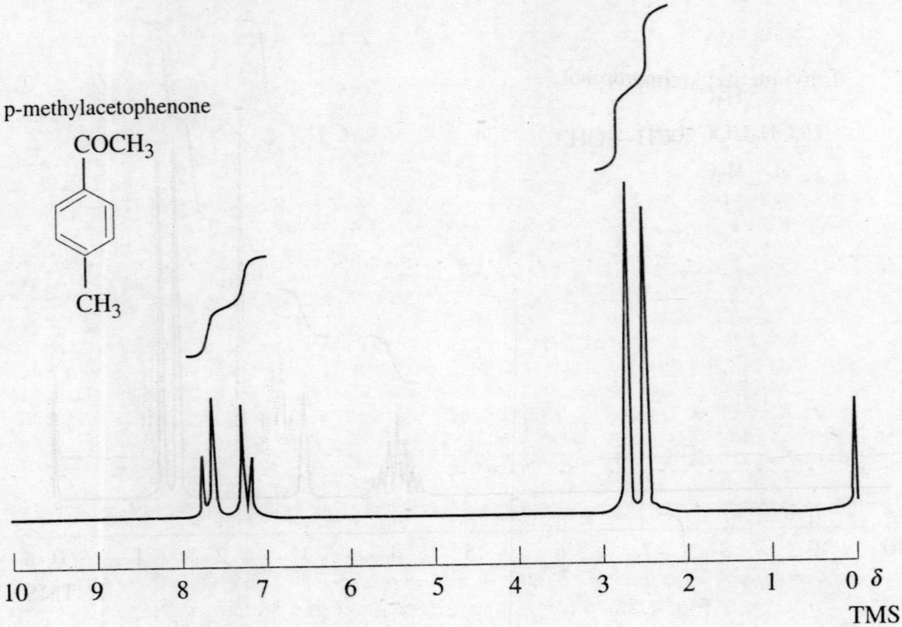


**Fig. 4.23** PMR Spectrum of p-Methylacetophenone

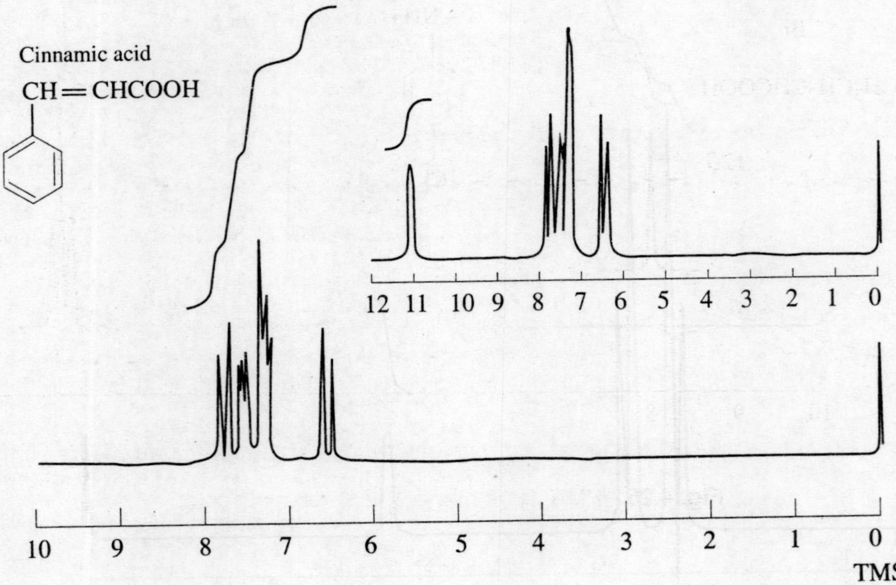


**Fig. 4.24** PMR Spectrum of cinnamic acid

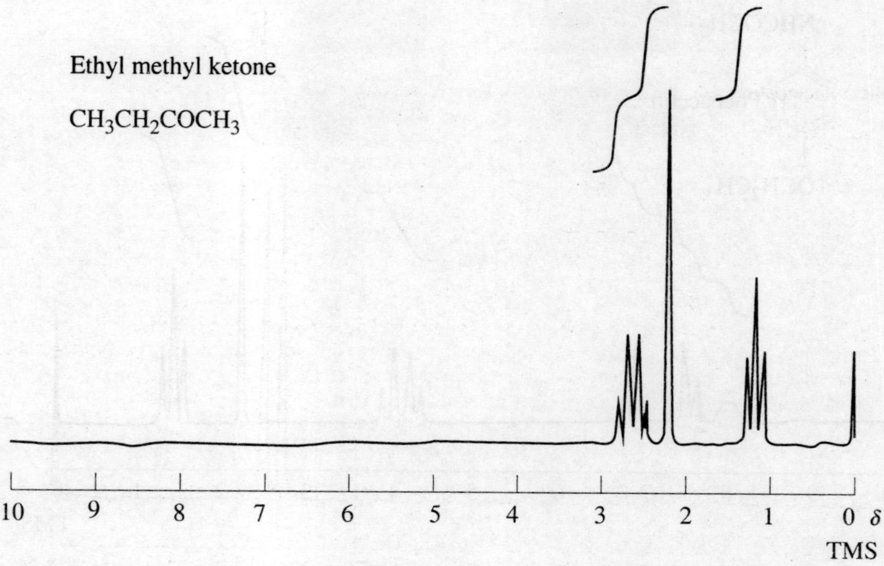


**Fig. 4.25** PMR Spectrum of Ethyl methyl ketone

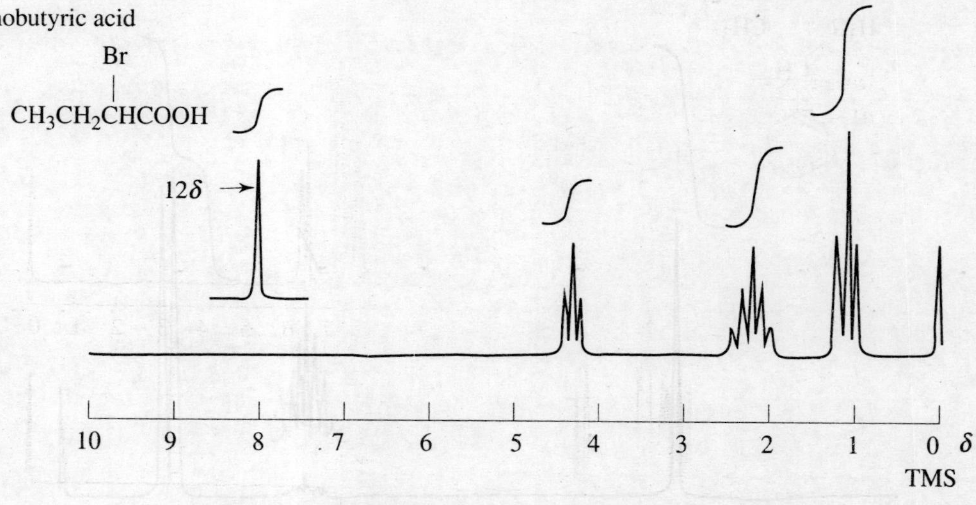


**Fig. 4.26** PMR Spectrum of 2-Bromobutyric acid

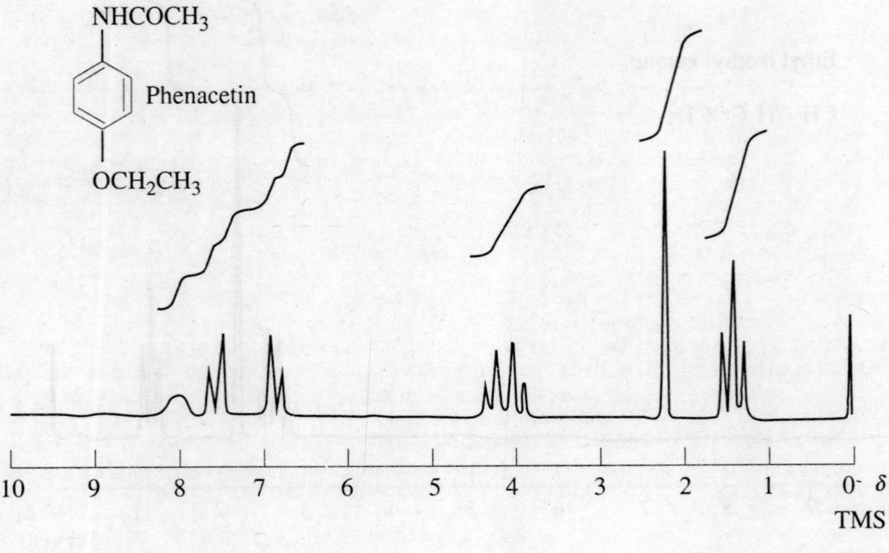


**Fig. 4.27** PMR Spectrum of phenacetin

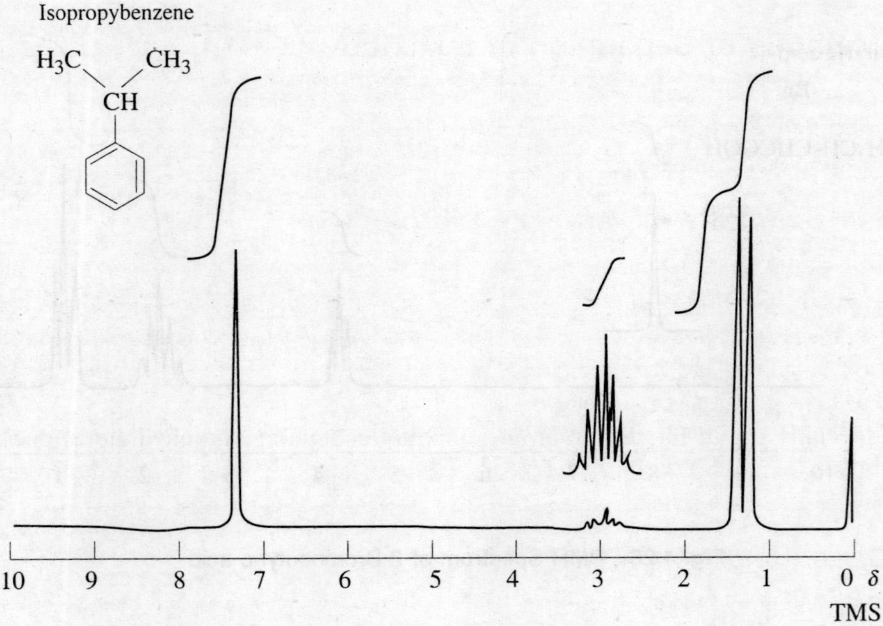


**Fig. 4.28** PMR Spectrum of Isopropylbenzene

# 5

# Separation of Mixtures

First step in the identification of a mixture is the separation of its components. Since the procedure employed for separation depends on the nature of the components, no definite scheme can be compiled to cover all possible ways of separation of mixtures. The following steps will help in the separation of mixtures. The pure components thus separated are identified by following the scheme discussed in chapters 1-4.

## 5.1 Preliminary Examination

Useful information is generally obtained from the preliminary examination of the mixture. It is advisable that the mixture is examined as described in the preliminary tests (Chapter 1) with special reference to the solubility tests. This enables one to get an approximate idea about the nature of the components that may be present in the mixture.

## 5.2 General Methods of Separation of a Mixture

The separation of a mixture may be based on one or a combination of the following:

(a) Difference in solubility in inert solvents.

(b) Methods depending on chemical properties.

### *(a) Difference in Solubility in various Solvents*

Shake a small quantity of the finely powdered mixture with the following solvents: water, ether, alcohol, benzene, acetone and carbon tetrachloride (all solvents except water should be anhydrous). In case one of the components seems to have dissolved in cold or hot, filter the mixture and wash the residue with a little more of the same solvent and dry to get the insoluble component; the second component is obtained by evaporating the solvent (caution) on a water bath.

This method is specially useful for separation of a mixture containing carbohydrate as one of the components. The procedure is illustrated in Scheme-1 (it will be appropriate to test for the presence or absence of a carbohydrate in the mixture itself).

### *(b) Methods Depending on Chemical Properties*

As a first step, it is advisable to test the mixture with moist blue and red litmus solution (or paper). If blue litmus turns red, then at least one of the component is either an acid or a phenol. If red litmus turns blue one of the components is basic in nature and may be an amine. The acid component is most conveniently separated by shaking the mixture with a saturated sodium bicarbonate solution.

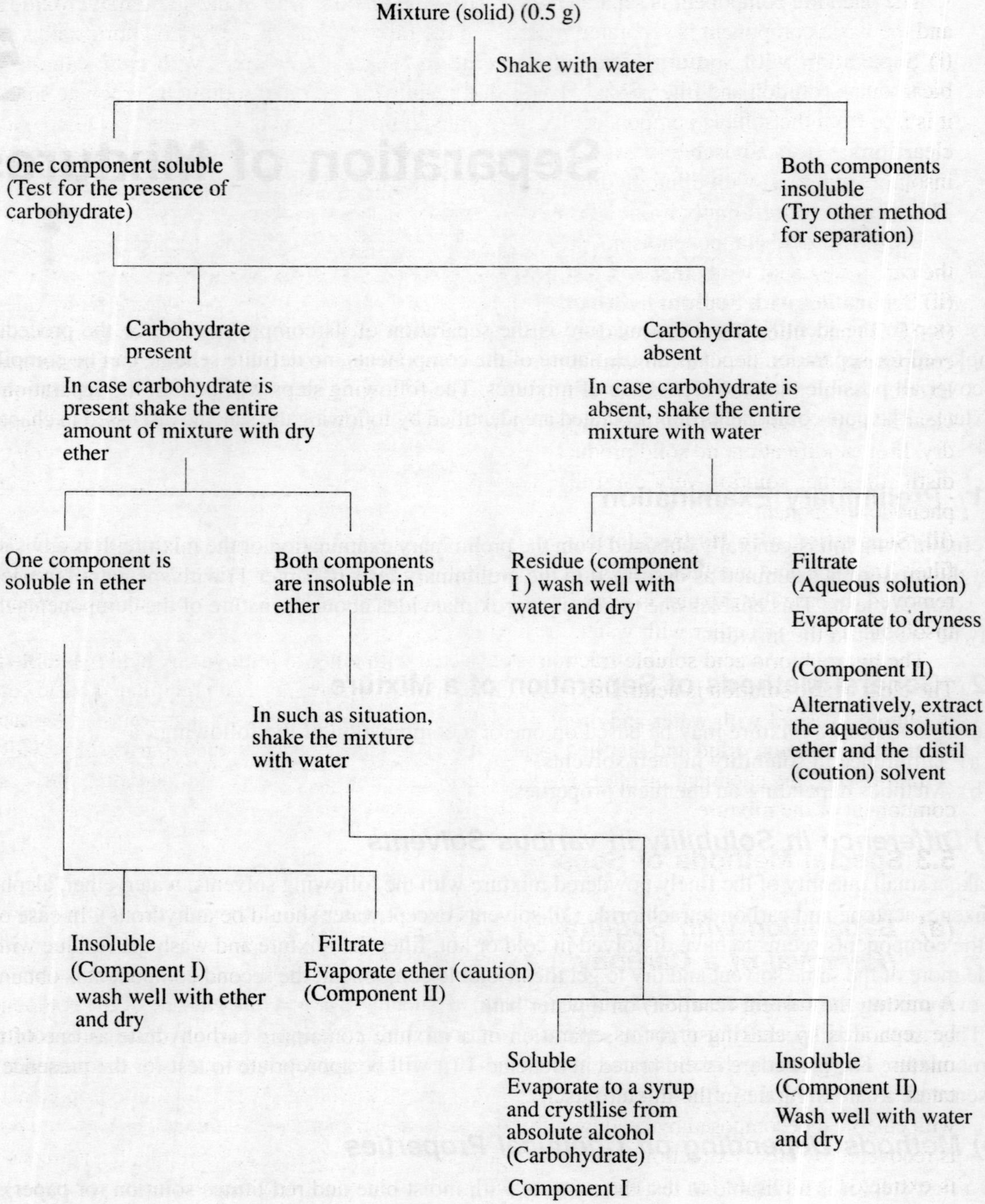
Mixture (solid) (0.5 g)
Shake with water
One component soluble (Test for the presence of carbohydrate)
Both components insoluble (Try other method for separation)
Carbohydrate present
Carbohydrate absent
In case carbohydrate is present shake the entire amount of mixture with dry ether
In case carbohydrate is absent, shake the entire mixture with water
One component is soluble in ether
Both components are insoluble in ether
Residue (component I ) wash well with water and dry
Filtrate (Aqueous solution)
Evaporate to dryness
(Component II)
Alternatively, extract the aqueous solution ether and the distil (coution) solvent
In such as situation, shake the mixture with water
Insoluble (Component I) wash well with ether and dry
Filtrate
Evaporate ether (caution) (Component II)
Soluble
Evaporate to a syrup and crystllise from absolute alcohol (Carbohydrate)
Component I
Insoluble (Component II)
Wash well with water and dry

**Chart 5.1**

The phenolic component is separated by shaking the mixture with dilute sodium hydroxide solution, and the basic component is separated by shaking the mixture with dilute hydrochloric acid.

**(i) Separation with sodium bicarbonate solution.** Shake the mixture with cold saturated sodium bicarbonate solution and filter. Shake the residue again with saturated sodium bicarbonate solution until it is free from the soluble component. The insoluble portion is washed with water and dried. Acidify the clear filtrate (it is advisable to extract the sodium bicarbonate filtrate with 20 ml ether to remove any insoluble material) with dilute hydrochloric acid. Filter the precipitated acid wash with water and dry. This method is useful only if one of the components is a carboxylic acid.

In case the acid component is not precipitated on acidification of sodium bicarbonate solution, extract the carboxylic acid with ether and distil off the solvent.

**(ii) Separation with Sodium hydroxide solution.** Stir the mixture with cold sodium hydroxide solution (8–10 %) and filter. Treat the residue with sodium hydroxide solution till it is free from alkali soluble component. Wash the alkali insoluble portion with water and dry. It is the neutral or basic compound. Extract the alkali soluble fraction with ether to remove any insoluble material. Acidify the remaining clear alkaline solution with dilute hydrochloric acid. Filter the separated product, wash with water and dry. If on acidification, no solid product separates or a turbid solution is obtained, extract with ether and distil the ether solution very carefully. The alkali soluble fraction consists of a carboxylic acid or a phenolic component.

**(iii) Separation with Hydrochloric acid.** Shake the mixture with dilute hydrochloric acid (10%) and filter. Treat the acid insoluble fraction again with dilute hydrochloric until all the acid soluble fraction is removed. In case the mixture is a liquid, dissolve in ether and then extract with dil. HCl. Wash the acid insoluble fraction in ether with water, dry and distil.

The hydrochloric acid soluble fraction is extracted with ether to remove any acid insoluble fraction. The clear acidic solution is neutralised with dilute sodium hydroxide. The precipitated basic component is filtered, washed with water and dried. In case an oil separates on neutralisation, it is extracted with ether, ether solution dried and distilled to give the basic component, which is purified by distillation.

In this way, the chemical methods are used for the separation of an acidic, phenolic, basic or neutral component of the mixture.

## 5.3 Special Methods of Separation

### *(a) Separation with Sodium bisulphite (Removal of a Carbonyl Compound)*

A mixture containing a carbonyl compound (e.g., an aldehyde or a ketone) as one of the components can be separated by shaking it (or its ethereal solution) with a saturated sodium bisulphite solution. The mixture is vigorously shaken. In case the bisulphite addition product of the carbonyl compound is precipitated, it is removed by filteration, (the filterate on ether extraction gives the other component) washed with ether and decomposed by heating with excess of sodium carbonate solution. The carbonyl compound is recovered by ether extraction. In case the bisulphite compound is water soluble, the aqueous solution is extracted with ether for the recovery of the other component and the remaining solution is treated with sodium carbonate solution and the carbonyl compound recovered as already described.

### *(b) Separation of Amines by Hinsberg's Method*

A mixture of primary, secondary and tertiary amines can be separated by treatment with p-toluene sulphonyl chloride. The primary amines gives sulphonamide derivative which is soluble in alkali. The secondary amines give alkali insoluble sulphonamide. The tertiary amines do not react. The separation is carried out as follows.

Shake the mixture of amines (4 g) with p-toluene sulphonyl chloride (8 g) and 120 ml of 2N sodium hydroxide solution for 10–15 minutes. Cool the mixture, add ether (60 ml) and shake. Filter the solid which is insoluble in both alkali and ether. This is the sulphanamide of secondary amine. Separate the two layers of ether and alkali using a separatory funnel. Acidify the alkaline solution with dilute hydrochloric acid. The separated sulfonamide of the primary amine is filtered and crystallised from dilute alcohol. Extract the tertiary amine from the ether layer by shaking with hydrochloric acid. Neutrilise the acid solution to obtain the amine.

The sulphonamides of primary and secondary amines are decomposed separately by boiling with ten-fold excess of 25% hydrochloric acid or 70% sulphuric acid (w/w 8 volumes of acid and 7 volumes of water) for several hours.

### (c) Removal of Nitro Compound

The nitro compound, if present, can be reduced to amine which is removed by dissolving in acid. It is necessary to ascertain that the other component present is not affected by the reducing agent.

## 5.4 Separation by Distillation or Sublimation

It is generally more convenient to use a method of separation based on chemical properties. However, in some cases, fractional distillation is the only suitable procedure. Satisfactory separation by this method can be done only if the boiling points of the components differ by at least 25°; azeotropic mixtures cannot be separated by this procedure.

In some cases, sublimation can be used provided only one of the component sublimes. Sometimes decomposition of the remainder of the mixture may occur. This procedure should be used only as the last resort.

## 5.5 Actual Method to be used for Separation of the Bulk of the Mixture

On the basis of the results of small scale tests (described above) a decision is made as to the most satisfactory method of separation. This method is used to separate the main portion of the mixture. It is advisable to use the following methods of separation:

| *Mixture of two components* | *Suggested method of separation* |
|---|---|
| (i) Both neutral | Solvent extraction, distillation, sometimes a special method by modifying one component, e.g., reduction of a nitro compound etc. Solvent extraction distillation |
| (ii) One of the components is a carboxylic acid | Sodium bicarbonate separation |
| (iii) One of the components is a phenol | Sodium hydroxide separation. |
| (iv) One of the components is basic | Hydrochloric acid separation. |

*N.B.* If one of the component is a carbohydrate, the best method of separation is to extract the other components with dry ether. The insoluble fraction is the carbohydrate.

# 6

# Tables of Organic Compounds

**Table 6.1** Carbohydrates

| *Name and structure* | *M.P.*[1] *(°C)* | *Properties* | *$[\alpha]^{20}_D$ in ($H_2O$)*[5] | *Osazone M.P.*[2] *(°C)* | *Osazone Time of formation (min)* | *Acetate M.P. (°C)* | *Specific tests and derivatives* |
|---|---|---|---|---|---|---|---|
| D-Ribose 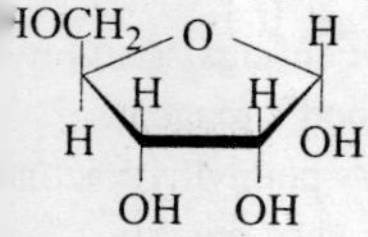 | 87 | Soluble in $H_2O$, sparingly soluble in alcohol | – 23°→ –24° (add trace of $NH_3$) (C = 4 in $H_2O$)[5] | 163 (rapid heat) | | | – Reduces Fehlings Solution and Barfoed's reagent. |
| -Deoxy D-ribose 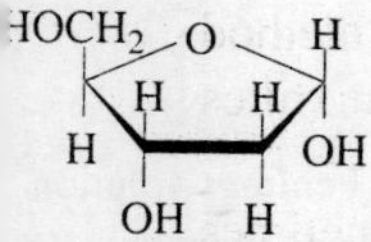 | 90 | Soluble in $H_2O$ | – 56° (final) (C = 1 in $H_2O$)[5] | | | | – Reduces Fehlings solution and Barfoed's reagent.<br>– p-Nitrophenyl hydrazone, m.p. 160°C. |
| D-Fructose (Laevulose Fruit sugar) 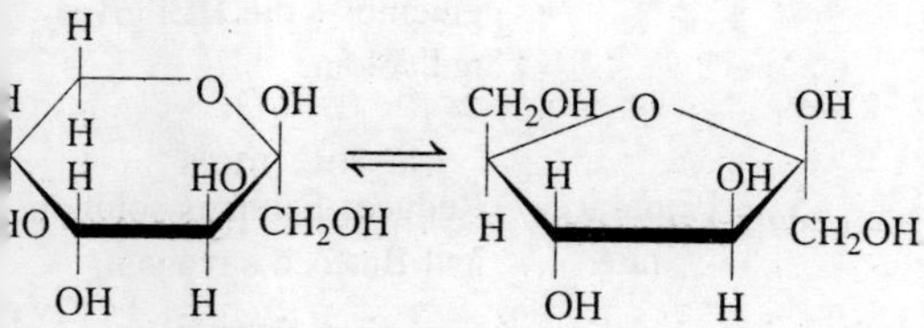 | 102 | Soluble in $H_2O$, fairly soluble in alcohol | – 92° (30 min. after dissolving) (C = 4 in $H_2O$)[5] | 205 | 2 | 70 ($\alpha$)[3] 109 ($\beta$)[4] | – Reduces Fehlings solution in cold and Barfoed's reagent on warming.<br>– On heating with resorcinol + conc. $H_2SO_4$ gives red colour (Seliwanoff test).<br>– Blue colour on warming with ammonium molybdate (specific test).<br>– On boiling with dil. HCl yields furfural (vaporus red-den aniline acetate paper)<br>– p-Nitrophenylhydrazone m.p. 180°C. |
| Maltose $C_{12}H_{22}O_{11}$ 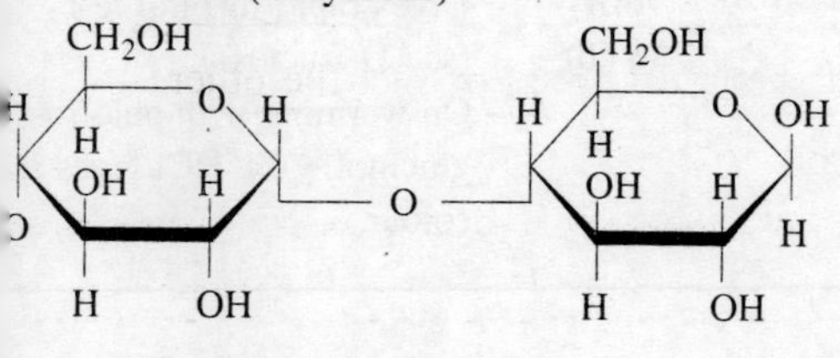 | 102 (mono-hydrate) 160 – 65 (anhydrous) | Soluble in $H_2O$, sparingly soluble in alcohol | +111°→ +130° (C = 4, hydrate in $H_2O$)[5] | 208 | 20–25 (separates on slow cooling) in 5 min. after hydrolysis, see page 51) | 125($\alpha$)[3] 159($\beta$)[4] | - Reduces Fehlings solution but not Barfoed's reagent. |

| *Name and structure* | M.P.[1] (°C) | Properties | $[a]^{20}_D$ in $(H_2O)^5$ | Osazone *M.P.*[2] *(°C)* | Osazone *Time of formation (min)* | *Acetate M.P. (°C)* | *Specific tests and derivatives* |
|---|---|---|---|---|---|---|---|
| Raffinose $C_{18}H_{32}O_{16}$ 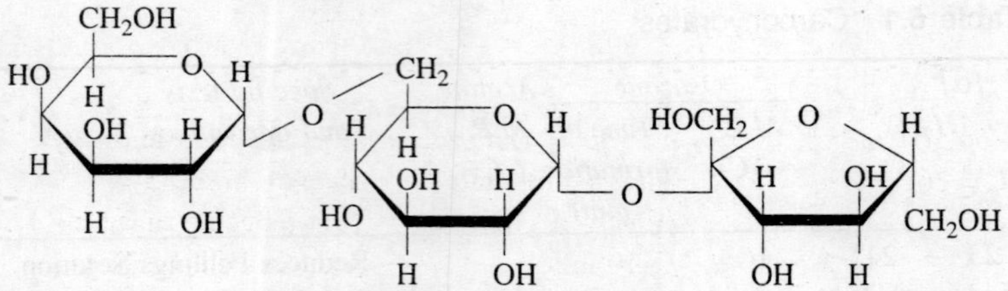 | 118 (anhydrous) 78 (pentahydrate) | Soluble in $H_2O$, insoluble in alcohol | +105° (C = 4)[5] | | | $99(\beta)^4$ (Hendeca-acetate) | – Does not reduce Fehlings Solution and does not reac with phenylhydrazine. – Gives precipitates with lea acetate and with saturated $Ca(OH)_2$ solution. On hydrolysis yields galactose, glucose and fructose. |
| D-Mannose 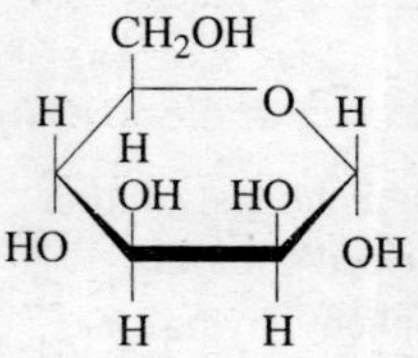 | 132 | Soluble in $H_2O$, sparingly soluble in alcohol | +29.3 →+14.2 | 205 | 0.5 | $74(\alpha)^3$ $115(\beta)^4$ | – Reduces Fehlings solution and Barfoed's reagent. – Colourless phenylhydrazon separates very quickly, m.p. 199°C. |
| α- D-Xylose (wood sugar) 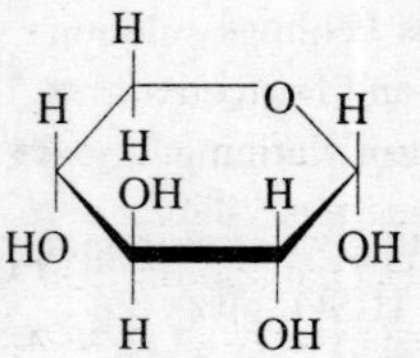 | 145 | Soluble in $H_2O$, alcohol | + 93° → 18.5 (add trace of $NH_3$; (C = 10 in $H_2O)^5$ | 163 | 8–10 | Tetraacetate $59\ (\alpha)^3$ $126\ (\beta)^4$ | – Reduces Fehlings solution and Barfoeds reagent. – Gives rapid furfural test (see D-fructose). – On warming with pholoroglucinol + dil. HCl gives red colour. |
| α-D-Glucose (dextrose, grape sugar) 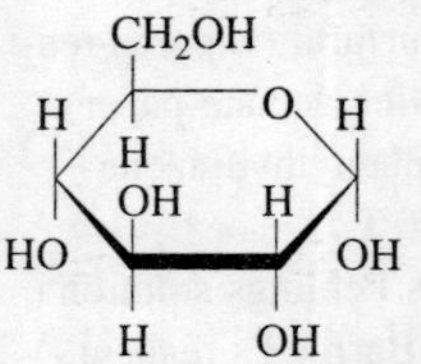 | 146 (anhydrous) 85 (mononydrate) | Soluble in $H_2O$ | + 112°→ + 52.8° (add trace of $NH_3$; C =3.9 anhydrous in $H_2O)^5$ | 205 | 5 | Penta-acetate $112\ (\alpha)^3$ $134\ (\beta)^4$ | – Reduces Fehlings solution and Barfoed's reagent. – On oxidation ($HNO_3$) gives saccharic acid, mp 125°C. |
| β-L-Arabinose 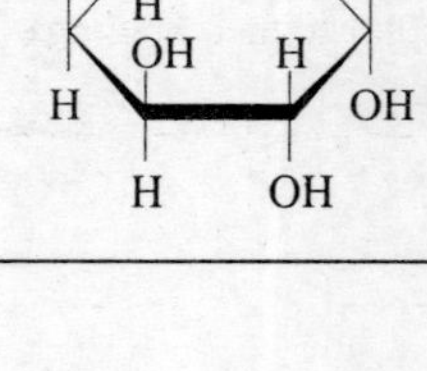 | 160 | Soluble in $H_2O$, alcohol | + 180°→ +105° (add trace of $NH_3$; C = 1 in $H_2O)^5$ | 166 | 10 | Tetra-acetate $96(\alpha)^3$ $86\ (\beta)^4$ | – Reduces Fehilings solution and Barfoeds reagent. – Gives rapid furfural test (see D-fructose). – On warming with phlolroglucinol + dil. HCl gives re colour. |

| *Name and structure* | *M.P.*[1] *(°C)* | *Properties* | $[\alpha]^{20}_{D}$ *in* $(H_2O)^5$ | *Osazone* *M.P.*[2] *(°C)* | *time of formation (min)* | *Acetate M.P. (°C)* | *Specific tests and derivatives* |
|---|---|---|---|---|---|---|---|
| D-Galactose 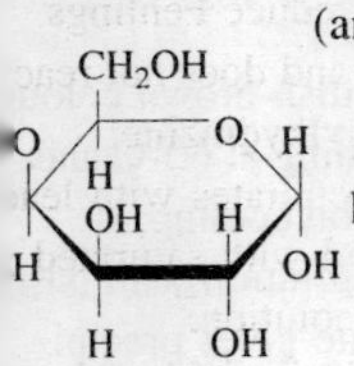 | 168 (anhydrous) 119 (mono-hydrate) | Soluble in $H_2O$, crystallises as hydrate on cooling, insoluble in cold alcohol. | +150°→ +80° (add trace of $NH_3$; C = 4 in $H_2O)^5$ | 196 | 15-20 | Penta-acetate 96 (α)[3] 142 (β)[4] | – Reduces Fehlings solution and Barfoed's reagent. |
| Sucrose (cane sugar) $C_{12}H_{22}O_{11}$ 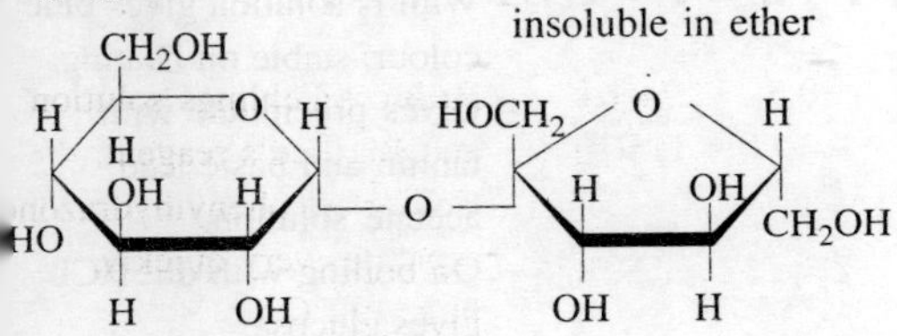 | 169 (185) | Soluble in $H_2O$, sparingly soluble in alcohol, insoluble in ether | + 66.5° (C = 26)[5] | 205 | 30 (after hydrolysis in 5–10 min.) | Octa-acetate 82(β)[4] | – Crystallises from methyl alcohol, m.p.185°C<br>– Non-reducing sugar, no reaction with Fehlings solution or phenylhydrazine.<br>– With $CuSO_4$ + NaOH solution gives blue solution.<br>– On warming with $Co(NO_3)_2$ NaOH gives violet colour.<br>– On oxidation ($HNO_3$) gives saccharic acid m.p. 125°C.<br>– On warming with very dil. HCl gives invert sugar (mixture of glucose + fructose), $[\alpha]^{20}_{D}$ – 37, which reduces Fehlings solution and gives osazone, m.p. 205°C. |
| Lactose (milk sugar) 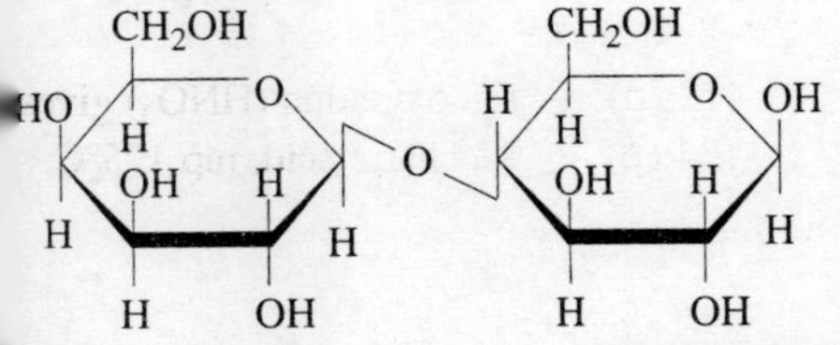 | 203 (hydrate) 223 (anhydrous) | Soluble in hot $H_2O$, insoluble in alcohol | + 85° → +52.5° (add trace of $NH_3$; C = 7.6, hydrate in $H_2O)^5$ | 200 | 20–25 | Octa-acetate 152(α)[3] 100(β)[4] | Loses $H_2O$ at approx. 130°C and turns yellow at approx. 160°.<br>Reduces Fehlings solution but not Barfoeds reagent.<br>With warm dil. HCl gives glucose + galactose. |
| Inulin $(C_6H_{10}O_5)_n$ mol. wt. 5000 (approx.) | d | Soluble in warm $H_2O$ separates very slowly on cooling, more rapidly on addition of alcohol | – 40° (in warm $H_2O$) (C = 2)[5] | | | | – Does not reduce Fehlings solution.<br>– No colour with $I_2$.<br>– On boiling with HCl (or oxalic acid) gives fructose. |
| Glycogen $(C_6H_{10}O_5)_n$ | d | Soluble in $H_2O$ giving opalesant | + 196° | | | | – Does not reduce Fehlings solution. |

| *Name and structure* | *M.P.*[1] *(°C)* | *Properties* | $[\alpha]^{20}_{D}$ *in* ($H_2O$)[5] | *Osazone M.P.*[2] *(°C)* | *Osazone time of formation (min)* | *Acetate M.P. (°C)* | *Specific tests and derivatives* |
|---|---|---|---|---|---|---|---|
| (animal stratch) mol. wt. 1 to 2 × $10^6$ | | solution, separates on adding alcohol | | | | | – Weakly reduces barfoeds reagent. – Gives reddish-brown colour with $I_2$ fading at 60°C and returning on cooling. – Aqueous solution with basic lead acetate gives precipitate. – On boiling with very dil. HCl gives glucose. |
| Starch $(C_6H_{10}O_5)_n$ mol. wt. depends on source | d | Insoluble in cold $H_2O$, colloidal solution on boiling, does not separate on cooling | | | | | – With $I_2$ solution gives blue colour, stable on heating. – Gives precipitate with tannin and basic lead acetate solution. – On boiling with dil. HCl gives glucose. |
| Cellulose $(C_6H_{10}O_5)_n$ mol. wt. 2 to 4 × $10^4$ | d (225) | Insoluble in $H_2O$, alcohol, soluble in ammonical $Cu(OH)_2$ solution | | | | | – Not hydrolysed by dil HCl |

1. The melting points of most carbohydrates vary with the rate of heating and therefore are of little value for identification.
2. The melting points of the osazones given are really decomposition temperatures; it is advisable to determine this decompositio[n] temperature by heating rapidly.
3. The $\alpha$-acetate is prepared by heating with fused zinc chloride and acetic anhydride (see page 50).
4. The $\beta$-acetate is prepared by heating with fused sodium acetate and acetic anhydride. (see page 50).
5. For giving specific rotation $[\alpha]^{10}_{D}$ in water, the concentration is specified. For example, C = 4 in water means 4 g of the carbohydrate is dissolved in water so that the volume of the solution is 10 ml. The concentration is expressed in g/dm$^3$.

## **Table 6.2 (a)** Carboxylic Acids (Liquids)

*Note This table includes carboxylic acids, and substituted carboxylic acids, e.g., simple, saturated, unsaturated, halogen, hydroxy, keto, nitro and mercapto acids*

| *Name and structure* | *B.P. (°C)* | *Properties* | *Derivatives, M.P. (°C)* Anilide | p-Toluidide | Amide | S-Benzylisothiouronium salt | *Specific tests and derivatives* |
|---|---|---|---|---|---|---|---|
| Trifluoroacetic acid $CF_3COOH$ | 72 | Colourless, pungent odour. Extremely reactive | — | — | — | — | — |
| Thioacetic acid $CH_3COSH$ | 93 | Pale yellow, unpleasant odour | 76 | 130 | 108 | — | |
| Formic acid HCOOH | 101 | Colourless, irritating odour, soluble in $H_2O$, alcohol, ether | 47 | 53 | 195 (d) (bp) | 146 | – Reduces Fehlings and Tollens reagent.<br>– On heating with conc. $H_2SO_4$ gives CO gas which burns with a blue flame.<br>– Gives yellow colour with conc. $NaHSO_3$ solution.<br>– Decolourises $KMnO_4$ solution. |
| Acetic acid $CH_3COOH$ | 118 (m.p. 16) | Colourless, pungent vinegar like odour, soluble in $H_2O$, alcohol, ether | 114 | 148 | 82 | 134 | – Gives fruity smell of ethyl acetate on heating with absolute alcohol and a few drops of conc. $H_2SO_4$ and pouring into water. |
| Acrylic acid $CH_2=CHCOOH$ | 140 (m.p. 8) | Colourless, pungent odour, soluble in $H_2O$, alcohol and ether | 104 | 141 | 84 | — | – Decolourises $KMnO_4$ solution and $Br_2$ water.<br>– Gives dibromide, m.p. 67°C with bromine in acetic acid. |
| Propionic acid $CH_3CH_2COOH$ | 140 | Colourless, pungent odour, soluble in $H_2O$, alcohol and ether | 104 | 124 | 79 | 148 | |
| Propiolic acid $HC\equiv CCOOH$ | 144 (d) | Soluble in $H_2O$ | 87 | — | 61 | — | – Bromo derivative, m.p. 87°C. |
| Isobutyric acid $(CH_3)_2CH\cdot COOH$ | 155 | Colourless, unpleasant odour. Sparingly soluble in $H_2O$, soluble in alcohol and ether | 105 | 104 | 129 | 143 | |
| n-Butyric acid $CH_3(CH_2)_2COOH$ | 163 | Colourless, unpleasant odour of rancid butter, soluble in $H_2O$, alcohol, ether | 96 | 72 | 114 | 146 | – Gives fruity odour of ethyl butyrate on heating with absolute alcohol and 2 drops of conc. $H_2SO_4$ and pouring into water. |
| Pyruvic acid $CH_3COCOOH$ | 165 (d) (m.p. 13) | Colourless, soluble in $H_2O$, alcohol, ether | 104 | 109 | 125 | 158 (d) | – Decolourises $Br_2$ water and $KMnO_4$ solution.<br>– Reduces ammoniacal $AgNO_3$ solution. |

| *Name and structure* | *B.P. (°C)* | *Properties* | *Derivatives, M.P. (°C)* Anilide | p-Toluidide | Amide | S-Benzylisothiouronium salt | *Specific tests and derivatives* |
|---|---|---|---|---|---|---|---|
| | | | | | | | – Ammoniacal solution gives violet colour with sodium nitrourusside, which changes to red with KOH and to blue with acetic acid. – Yields iodoform. |
| Crotonic acid (cis) $CH_3CH = CHCOOH$ | 165(d) | Colourless, soluble in $H_2O$ | 102 | 132 | — | — | – Decolourises $Br_2$ water and $KMnO_4$ solution. – On heating (150°C) with a trace of $I_2$ it is converted into $\alpha$-crotonic acid (trans). |
| Isovaleric acid $(CH_3)_2CHCH_2COOH$ | 176 | Colourless, unpleasant odour, slightly soluble in $H_2O$, soluble in alcohol and ether | 110 | 109 | 135 | 153 | |
| $\alpha$-Chloropropionic acid $CH_3CHClCOOH$ | 186 | | 92 | 124 | 80 | — | |
| n-Valeric acid $CH_3(CH_2)_3COOH$ | 186 | Colourless, unpleasant odour, sparingly soluble in $H_2O$, soluble in alcohol, ether | 63 | 70 | 104 (114) | — | |
| Dichloroacetic acid $Cl_2CHCOOH$ | 189 | Colourless, insoluble in $H_2O$, soluble in alcohol, ether | 119 | 153 | 96 | 178 | |
| Methoxyacetic acid $CH_3OCH_2COOH$ | 203 | | 58 | | 96 | | |
| $\alpha$-Bromopropionic acid $CH_3CHBrCOOH$ | 205 (m.p. 24) | | 99 | 125 | 123 | | |
| n-caproic acid (n-Hexanoic acid) $CH_3(CH_2)_4COOH$ | 205 | Colourless insoluble in $H_2O$, soluble in alcohol, ether | 95 | 75 | 100 | 159 | |
| Ethoxyacetic acid $C_2H_5OCH_2COOH$ | 206 | | 92 | 32 | 80 | | |
| n-Heptoic acid $CH_3(CH_2)_5COOH$ | 223 | Colourless, sparingly soluble in cold $H_2O$ | 71 | 80 | 94 | 81 | |
| Octanoic acid n-Caprylic acid $CH_3(CH_2)_6COOH$ | 237 (m.p. 16) | Colourless, sparingly soluble in cold $H_2O$ | 57 | 70 | 105 | 157 | |
| Perlargonic acid $CH_3(CH_2)_7COOH$ | 253 (m.p. 12) | Colourless, insoluble in $H_2O$ | 57 | 84 | 99 | 84 | |

| *Name and structure* | *B.P. (°C)* | *Properties* | *Derivatives, M.P. (°C)* Anilide | *p-Toluidide* | *Amide* | *S-Benzylisothiouronium salt* | *Specific tests and derivatives* |
|---|---|---|---|---|---|---|---|
| n-Capric acid $CH_3(CH_2)_8COOH$ | 269 (m.p. 31) | Colourless, semi-solid | 62 | 78 | 108 | 78 | |
| Oleic acid (cis) (n-Nonoic acid) $CH_3(CH_2)_7CH = CH(CH_2)_7COOH$ | 285 (d) (m.p. 16) | Colourless, insoluble in $H_2O$, soluble in alcohol, ether, tallow (fat) odour. | 41 | 43 | 76 | | |

**Table 6.2 (b)** Carboxylic Acids (Solids)

| *Name and Structure* | *M.P. (°C)* | *Properties* | *Derivatives, M.P. (°C)* Anilide | *p-Toluidide* | *Amide* | *S-Benzylisothiouronium salt* | *Specific tests and derivatives* |
|---|---|---|---|---|---|---|---|
| (±) Lactic acid $CH_3CH(OH)\ COOH$ | 18 b.p. 119 (12 mm) | Colourless, syrupy, soluble in $H_2O$, alcohol, ether | 59 | 107 | 74 | 153 | - Decolourises $KMnO_4$ solution on warming.<br>- Red colour formed on warming (0.2 ml of a 0.1 % solution) with 2 ml conc. $H_2SO_4$ for 2 min. at 100°, cooling and adding 2 drops of 5% alcoholic solution of guaiacol. |
| Thiobenzoic acid 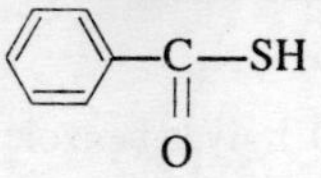 | 24 | — | 102 | 128 | 116 | — | - On oxidation in ether solution by air gives benzoyl disulphide, m.p. 130°C.<br>- On boiling with dil. NaOH solution gives $H_2S$ and benzoic acid, m.p. 121°C.<br>- With aniline gives $H_2S$ and benzanilide, m.p. 164°C. |
| o-Ethoxybenzoic acid 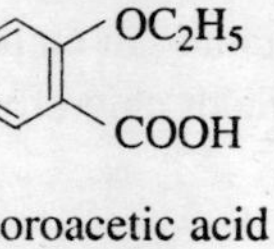 | 25 (b.p. 300 (d)) | — | | | 132 | | |
| Fluoroacetic acid $FCH_2COOH$ | 31 | | | | 108 | | |
| Levulinic acid $CH_3CO(CH_2)_2COOH$ | 33 (b.p. 245) (d) | Colourless, soluble in $H_2O$ | 102 | 109 | 108 | — | - Gives iodoform, m.p. 119°C in cold with iodine solution and NaOH solution. |

| *Name and structure* | *M.P. (°C)* | *Properties* | *Derivatives, M.P. (°C)* Anilide | *p-Toluidide* | *Amide* | *S-Benzyl-isothiouronium salt* | *Specific tests and derivatives* |
|---|---|---|---|---|---|---|---|
| | | | | | | | - Phenylhydrazone, m.p. 108°C.<br>- Oxime, m.p. 95°C.<br>- Semicarbazone, m.p. 184°C (d) |
| Pivalic acid (Trimethyl acetic acid $(CH_3)_3C\ COOH$ | 35 (b.p. 163) | Soluble in $H_2O$ | 128 | 120 | 155 | — | |
| β-Chloropropionic acid $ClCH_2CH_2COOH$ | 42 | — | | — | 101 | — | |
| Lauric acid $CH_3(CH_2)_{10}COOH$ | 44 (b.p. 299) | Insoluble in $H_2O$, steam volatile | 77 | 87 | 99 | 141 | |
| Elaidic acid (trans isomer of oleic acid) $CH_3(CH_2)_7CH = CH(CH_2)_7COOH$ | 44 (51) | Waxy solid, insoluble in $H_2O$, soluble in alcohol | — | — | 90 | — | |
| Hydrocinnamic acid  $-CH_2CH_2COOH$ | 48 | Colourless, insoluble in $H_2O$, soluble in alcohol, ether | 92 | 135 | 105 | — | – oxidation ($CrO_3$) gives benzoic acid, m.p. 121°C. |
| Dibromoacetic acid $Br_2CHCOOH$ | 48 | — | — | — | 156 | | |
| Bromoacetic acid $BrCH_2COOH$ | 50 | Colourless, soluble in $H_2O$, alcohol, ether | 130 | 91 | 91 | | |
| 4-Phenylbutyric acid $C_6H_5CH_2CH_2CH_2COOH$ | 52 | | | | 84 | | – Oxidation ($CrO_3$) gives benzoic acid, m.p. 121°C. |
| Myristic acid $CH_3(CH_2)_{12}COOH$ | 54 | Colourless, insoluble in $H_2O$, soluble in absolute alcohol, ether and benzene | 84 | 93 | 102 | | |
| Trichloroacetic acid $Cl_3CCOOH$ | 58 | Colourless, deliquescent, soluble in $H_2O$, corrosive substance | 95 | 113 | 141 | 149 | - Solution of sodium salt on prolonged boiling yields chloroform. |
| Palmitic acid $CH_3(CH_2)_{14}COOH$ | 62 | Colourless, waxy, insoluble in $H_2O$, soluble in ether, chloroform | 90 | 98 | 106 | 141 | |
| Chloroacetic acid $ClCH_2COOH$ | 63 | Colourless, deliquscent, soluble in $H_2O$, alcohol, chloroform, causes skin blisters | 134 | 162 | 120 | 160 | - With phenol and NaOH yield phenoxyacetic acid, m.p. 98°C.<br>- With β-naphthol and NaOH yields β-naphthoxyacetic acid, m.p. 156°C. |

| *Name and structure* | *M.P. (°C)* | *Properties* | *Derivatives, M.P. (°C)* | | | | *Specific tests and derivatives* |
|---|---|---|---|---|---|---|---|
| | | | *Anilide* | *p-Toluidide* | *Amide* | *S-Benzylisothiouronium salt* | |
| Cyano acetic acid $CNCH_2COOH$ | 66 | Colourless, soluble in $H_2O$, alcohol, ether | 198 | — | 123 | — | - On boiling with conc. HCl yields malonic acid, m.p. 133°C.<br>- On warming with benzaldehyde yields $\alpha$-cyanocinnamic acid, m.p. 180°C. |
| Phenylglyoxalic acid 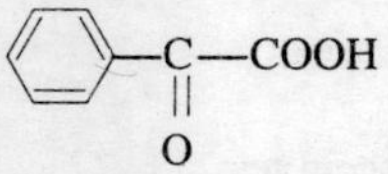 | 66 | | | | 91 | | - 2, 4-Dinitrophenylhydrazone, m.p. 196 (d).<br>- On heating with aniline gives benzylidene aniline m.p. 54°C. |
| Stearic acid $CH_3(CH_2)_{16}COOH$ | 69 | Colourless, waxy, insoluble in $H_2O$, soluble in alcohol | 94 | 102 | 108 | 143 | |
| Crotonic acid (trans) $CH_3CH{=}CHCOOH$ | 72 | Colourless, soluble in $H_2O$ | 116 | 132 | 158 | | - Reduces $KMnO_4$ solution and ammoniacal $AgNO_3$ solution on warming.<br>- With bromine gives $\alpha$, $\beta$-dibromobutyric acid, m.p. 87°C.<br>- With HI yields $\beta$-iodobutyric acid, m.p. 110°C. |
| Phenylacetic acid 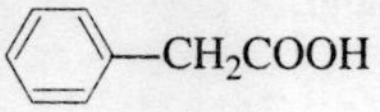 | 76 | Colourless, strong odour like that of perfumes, sparingly soluble in $H_2O$, soluble in hot $H_2O$, alcohol | 118 | 136 | 154 | 160 | - With acidic $KMnO_4$ gives benzaldehyde.<br>- With alkaline $KMnO_4$ gives benzoic acid, m.p. 121°C. |
| Glycolic acid $CH_2OHCOOH$ | 80 | Colourless, tends to deliquesce, soluble in $H_2O$ | 96 | 143 | 120 | 142 | - Gives violet colour in guaiacol test (see lactic acid) |
| Cyclohexanone-2-carboxylic acid 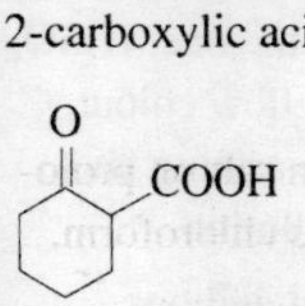 | 81 (d) | | | | | | |
| $\beta$-Iodopropionic acid $ICH_2CH_2COOH$ | 82 | Colourless, soluble in hot $H_2O$ | 101 | | | | - With $Br_2$-water gives $\beta$-Bromopropionic acid, m.p. 62°C. |
| Iodoacetic acid $ICH_2COOH$ | 83 | | 143 | — | 95 | — | |
| Citraconic acid (cis) $HOOCCH{=}(CCH_3)COOH$ | 92 | | 175 | | 187 (d) | | - Decolourises $Br_2$ water.<br>- Dibromide, m.p. 150°C. |

| *Name and structure* | *M.P. (°C)* | *Properties* | *Derivatives, M.P. (°C)* *Anilide* | *p-Toluidide* | *Amide* | *S-Benzyl-isothiouronium salt* | *Specific tests and derivatives* |
|---|---|---|---|---|---|---|---|
| p-Sulphobenzoic acid  | 94 (hydrated) 259 (anhydrous) | Colourless soluble in $H_2O$, alcohol, ether | 253 (di) | — | 236 (di) | | - Nitro derivative m.p. 130°C.<br>- On fusion with KOH yields p-hydroxybenzoic acid, m.p. 213°C. |
| Phenoxyacetic acid $C_6H_5OCH_2COOH$ | 98 | | 99 | — | 101 | — | |
| Glutaric acid $HOOC(CH_2)_3COOH$ | 98 (b.p. 303) | Soluble in $H_2O$ | 224 | 217 | 174 | | - Gives fluorescein test.<br>- Amide on heating gives imide, m.p. 154°C. |
| Citric acid<br>$CH_2COOH$<br>\|<br>$C(OH)COOH$<br>\|<br>$CH_2COOH$ | 100 (hydrated) | Colourless, soluble in $H_2O$, alcohol, ether loses water of crystallisation at 130°C forming anhydous acid, m.p. 153°C. | 199 | 189 | 210 | | - Neutral solution on heating with aqueous $CaCl_2$ solution gives a white pricipitate.<br>- With warm conc. $H_2SO_4$ gives a yellow solution; the solution on neutralising with alkali and addition of sod. nitroprusside gives wine red colour.<br>- With $\beta$-naphthol in conc. $H_2SO_4$ gives blue colour. |
| o-Methoxybenzoic acid 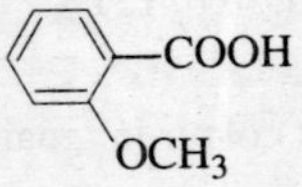 | 100 | | 131 | | 128 | — | |
| (–) Malic acid HOOC $CH_2$ CH (OH)COOH | 100 | Colourless, hygroscopic, soluble in $H_2O$, alcohol. | 197 | 207 | 157 (di) | 124 | - With cobalt nitrate solution and excess NaOH gives blue colour.<br>- With $\beta$-naphthol in conc. $H_2SO_4$ gives greenish-yellow colour, clear yellow on warming and pale yellow on dilution.<br>- With $PCl_5$ in cold and on subsequent treatment with $H_2O$ yields chlorosuccinic acid, m.p. 176°C (decom.) |
| Oxalic acid $(COOH)_2 \cdot 2H_2O$ | 101 | Colourless, soluble in $H_2O$ | 246 (di) | 267 (di) | 419 (di) | 196 | - With urea gives urea oxalate, m.p. 171°C. |
| o-Toluic acid 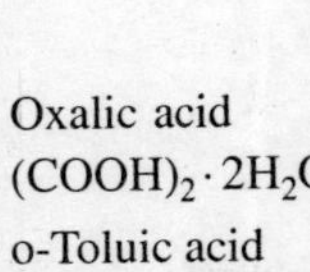 | 104 | Colourless, slightly soluble in $H_2O$, soluble in alcohol | 125 | 144 | 142 | | - On oxidation ($KMnO_4$) gives phthalic acid, m.p. 195°C. |

| *Name and structure* | *M.P. (°C)* | *Properties* | *Derivatives, M.P. (°C)* Anilide | *p-Toluidide* | *Amide* | *S-Benzylisothiouronium salt* | *Specific tests and derivatives* |
|---|---|---|---|---|---|---|---|
| Pimelic acid<br>$HOOC(CH_2)_5COOH$ | 105 | Colourless, slightly soluble in $H_2O$ | 155 | 206 | | | |
| Azelaic acid<br>$HOOC(CH_2)_7COOH$ | 106 | Colourless, sparingly soluble in cold $H_2O$ | 186 | 201 | 175 | 130 | |
| m-Toluic acid<br>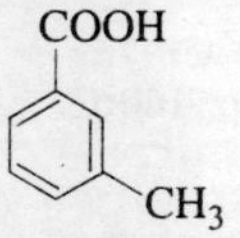 | 110 | Colourless, insoluble in $H_2O$, sparingly soluble in alcohol, ether | 126 | 118 | 95 | 164 (140) | - On oxidation ($KMnO_4$) gives isophthalic acid, m.p. >300°C (sublimes). |
| Ethyl malonic acid<br>$C_2H_5CH(COOH)_2$ | 111 | Colourless, soluble in $H_2O$ | 150 | | 214 | | |
| Methyl succinic acid<br>$CH_2COOH$<br>\|<br>$CH(CH_3)COOH$ | 115 | Colourless, soluble in $H_2O$ | 200 | 164 (mono) | 165<br>225 (di) | | |
| Benzyl malonic acid<br>$C_6H_5CH_2CH(COOH)_2$ | 117 | Soluble in $H_2O$ | 217 | — | 225 | — | - On heating above its m.p. gives hydrocinnamic acid m.p. 48°C. |
| (±) Mandelic acid<br>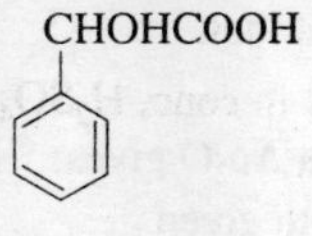 | 118 | Colourless, sparingly soluble in $H_2O$, soluble in alcohol, ether | 151 | 172 | 133 | 166 | - Gives smell of bitter almonds on heating with $KMnO_4$ and dil. $H_2SO_4$.<br>- On heating with $\beta$-naphthol and conc. $H_2SO_4$ (1 min.) and cooling gives wine red colour. |
| 5-Sulphosalicylic acid<br>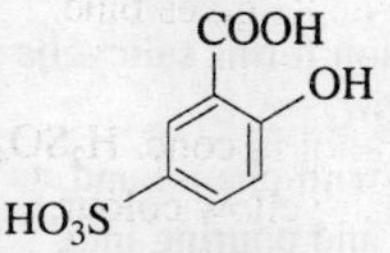 | 120 (anhydrous) | Colourless, soluble in $H_2O$ | | | | 204 | - Gives red colour with $FeCl_3$. |
| Benzoic acid<br>$C_6H_5COOH$ | 122 | Colourless, soluble in hot $H_2O$, alcohol | 164 | 158 | 128 | 167 | - Neutral solution of the compound with $FeCl_3$ gives buff coloured precipitate. |
| 3-Nitrosalicylic acid<br>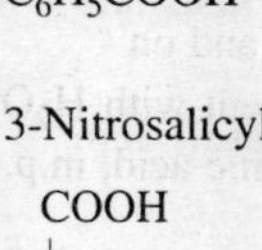 | 125 (hydrated) | Yellow, sparingly soluble in $H_2O$, soluble in alcohol, ether | — | — | 145 | — | - Gives red colour with aqueous $FeCl_3$. |
| dl-Saccharic acid<br>$HOOC(CHOH)_4COOH$ | 125 | Soluble in $H_2O$, alcohol | — | — | 172 | — | |

| *Name and structure* | *M.P. (°C)* | *Properties* | *Derivatives, M.P. (°C)* | | | | *Specific tests and derivatives* |
|---|---|---|---|---|---|---|---|
| | | | *Anilide* | *p-Toluidide* | *Amide* | *S-Benzylisothiouronium salt* | |
| o-Benzoylbenzoic acid 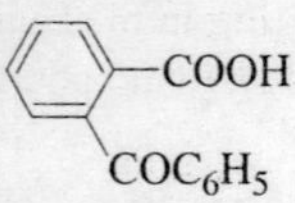 | 126 | Soluble in $H_2O$ | 195 | — | 165 | | |
| Maleic acid (cis) HOOCH=CHCOOH | 130 to 140 (varies with rate of heating) | Colourless, soluble in $H_2O$, alcohol, ether | 187 (di) 198 (mono) | 142 | 153 | 175 163 | - Decolourises bromine water only on warming.<br>- Gives dibromide, m.p. 160°C. |
| Cinnamic acid 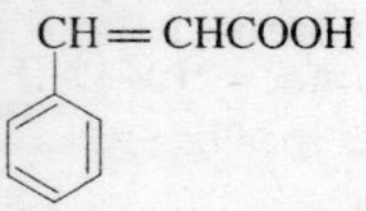 | 133 | Cream coloured, insoluble in $H_2O$ soluble in alcohol | 151 | 168 | 147 | 175 | - On oxidation (acidic $KMnO_4$) yields benzaldehyde.<br>- On oxidation (alkaline $KMnO_4$) yields benzoic acid, m.p. 122°C.<br>- With fuming $HNO_3$ (cold) yields p-nitro derivative, m.p. 285°C. |
| Furoic acid (Pyromucic acid)  | 133 | Colourless, soluble in hot $H_2O$, soluble in alcohol | 123 | 108 | 142 | 211 | |
| Malonic acid $CH_2(COOH)_2$ | 133 (d) | Colourless, soluble in $H_2O$, alcohol, ether | 225 | 252 | 170 | 147 | - On warming with $Ac_2O$ gives orange colour with green fluorescence. |
| Sebacic acid $HOOC(CH_2)_8COOH$ | 133 | Colourless, sparingly soluble in hot water | 201 | 201 | 210 | 155 | |
| Sorbic acid $CH_3(CH{=}CH)_2COOH$ | 134 | White crystalline, soluble in $H_2O$. | 153 | — | — | — | |
| o-Sulphobenzoic acid 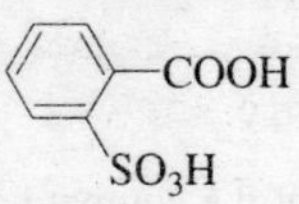 | 134 (anhydrous) 94 (hydrated) | Colourless, soluble in $H_2O$, alcohol, insoluble in ether | 190 | — | — | 206 | - On KOH fusion forms salicyclic acid m.p. 159°C.<br>- On warming with phenol and conc. $H_2SO_4$ and pouring into NaOH solution gives red colour which turns yellow in acidic medium.<br>- Gives imide (saccharin) m.p. 229°C. |
| (+) Mandelic acid | 135 | See (±) mandelic acid m.p. 118°C | | | | | |

| *Name and structure* | *M.P. (°C)* | *Properties* | *Derivatives, M.P. (°C)* | | | | *Specific tests and derivatives* |
|---|---|---|---|---|---|---|---|
| | | | *Anilide* | *p-Toluidide* | *Amide* | *S-Benzylisothiouronium salt* | |
| Acetyl salicylic acid (aspirin)<br>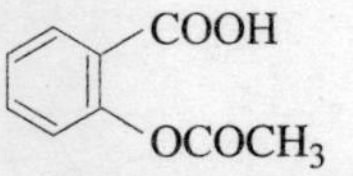 | 135 | Colourless, insoluble in $H_2O$ soluble in ether, alcohol (hot) | 136 | — | 138 | 144 | - Gives smell of oil of winter green on heating in methanol with conc. $H_2SO_4$ and pouring in $H_2O$. |
| Acetone dicarboxylic acid<br>$HOOCCH_2COCH_2COOH$ | 135 | Soluble in $H_2O$ | 135 | — | — | — | - Oxime, m.p. 54°C. |
| Phenyl propiolic acid<br>$C_6H_5C{\equiv}CCOOH$ | 136 | Insoluble in $H_2O$ | 126 | — | 100 | — | |
| Picolinic acid<br>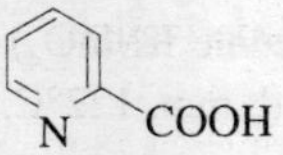 | 136 | Colourless, soluble in $H_2O$ | — | — | — | — | N-oxide, m.p. 161°C.<br>- Hydrochloride, m.p. = 210–12°C.<br>- On heating with sodalime gives smell of pyridine. |
| Methyl malonic acid<br>$CH_3CH(COOH)_2$ | 137 | | 182 | — | 217 | — | |
| m-Nitrobenzoic acid<br>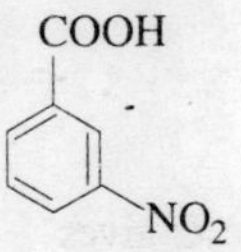 | 141 | Colourless, soluble in hot $H_2O$ | 153 | 162 | 142 | 163 | - With Sn + HCl gives m-aminobenzoic acid, m.p. 174°C. |
| m-Sulphobenzoic acid<br>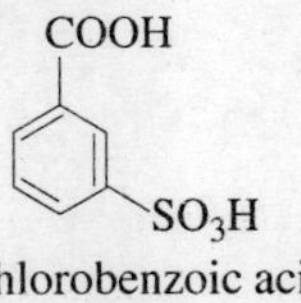 | 141 | Colourless, soluble in $H_2O$, alcohol, ether | — | — | 170 | 163 | - On KOH fusion, gives m-hydroxybenzoic acid, m.p. 201°C. |
| o-Chlorobenzoic acid<br>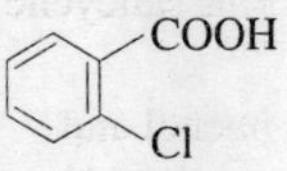 | 142 | | 118 | 131 | 142 | — | - With sodium amalgam yields benzoic acid, m.p. 122°C.<br>- On KOH fusion gives o-hydroxybenzoic acid, m.p. 159°C. |
| Suberic acid<br>$HOOC(CH_2)_6COOH$ | 144 | White, soluble in $H_2O$ | 186 | 218 | 216 | — | |
| o-Nitrobenzoic acid<br>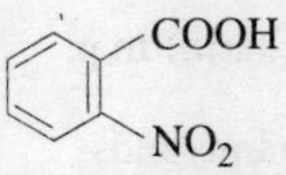 | 146 | Pale yellow, insoluble in $H_2O$, slightly soluble in alcohol | 155 | 203 | 174 | 159 | - Gives smell of bitter almonds.<br>- On reduction (Sn + HCl) gives anthranilic acid, m.p. 144°C. |
| Diphenylacetic acid<br>$(C_6H_5)_2CH$ COOH | 148 | Pale yellow, insoluble in $H_2O$ | 180 | 173 | 168 | 145 | |

| *Name and structure* | *M.P. (°C)* | *Properties* | *Derivatives, M.P. (°C)* | | | | *Specific tests and derivatives* |
|---|---|---|---|---|---|---|---|
| | | | *Anilide* | *p-Toluidide* | *Amide* | *S-Benzylisothiouronium salt* | |
| 3-Nitrosalicylic acid 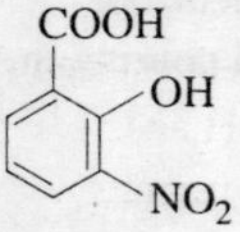 | 148 (anhyd) | See 3-Nitrosalicylic acid (hydrated), m.p. 125°C | | | | | |
| Benzilic acid $(C_6H_5)_2C(OH)COOH$ | 150 | Colourless, soluble in $H_2O$, alcohol, ether | 175 | 190 | 154 | — | - Dissolves in conc. $H_2SO_4$ giving red colour.<br>- On oxidation ($CrO_3$) gives benzophenone, m.p. 48°C.<br>- Acetyl derivative, m.p. 98°C. |
| o-Bromobenzoic acid 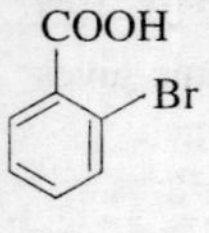 | 150 | Colourless, insoluble in $H_2O$, soluble in alcohol | 141 | — | 155 | 171 | - With conc. $HNO_3$ and $H_2SO_4$ at 130° gives dinitroderivative, m.p. 213°C. |
| Adipic acid $HOOC(CH_2)_4COOH$ | 151 | Colourless, insoluble in $H_2O$, soluble in alcohol | 239 | 241 | 220 | 159 | |
| p-Nitrophenylacetic acid 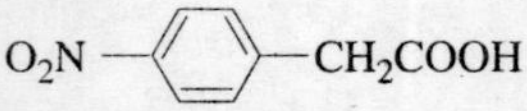 | 152 | | 212 | 208 | 198 | — | |
| Phenylmalonic acid $C_6H_5CH(COOH)_2$ | 153 | Soluble in $H_2O$ | — | — | 233 | — | |
| 2, 5-Dichlorobenzoic acid 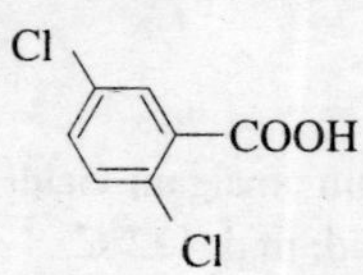 | 153 | | — | — | 155 | — | |
| Citric acid | 153 (anhydrous) | See citric acid (hydrated), m.p. 100°C. | | | | | |
| m-Bromobenzoic acid 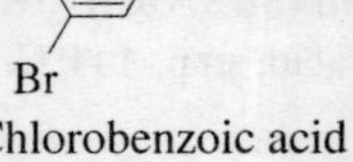 | 155 | Colourless, insoluble in $H_2O$, soluble in alcohol | 146 | — | 155 | 168 | - On KOH fusion gives m-hydroxybenzoic acid, m.p. 201°C. |
| m-Chlorobenzoic acid 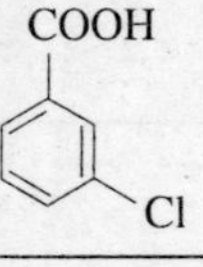 | 158 | Colourless, insoluble in $H_2O$, soluble in alcohol | 124 | — | 134 | 155 | - With sodium amalgam yields benzoic acid, m.p. 122°C.<br>- On KOH fusion gives m-hydroxybenzoic acid, m.p. 201°C. |

| *Name and structure* | *M.P. (°C)* | *Properties* | *Derivatives, M.P. (°C)* Anilide | p-Toluidide | Amide | S-Benzylisothiouronium salt | *Specific tests and derivatives* |
|---|---|---|---|---|---|---|---|
| Salicylic acid 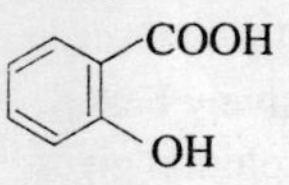 | 159 | Colourless, insoluble in $H_2O$, soluble in alcohol | 134 | 156 | 139 | 148 | - Gives deep violet colour with aqueous $FeCl_3$.<br>- Gives smell of oil of wintergreen on heating with $CH_3OH$ and conc. $H_2SO_4$ and pouring into water.<br>- With 15% $HNO_3$ yields 5-nitroderivative m.p. 230°C.<br>- Acetyl derivative m.p. 135°C. |
| o-Iodobenzoic acid 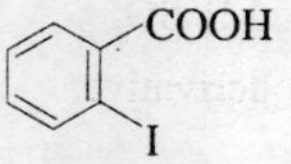 | 162 | Colourless, sparingly soluble in hot $H_2O$, soluble in alcohol | 142 | — | 184 | — | |
| α-Naphthoic acid 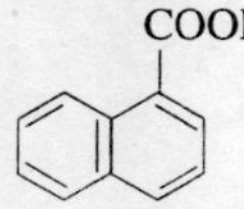 | 162 | Colourless, sparingly soluble in hot $H_2O$, soluble in alcohol | 160 | — | 205 | — | – On heating with sodalime. gives naphthalene, m.p. 80°C. |
| Thiosalicylic acid 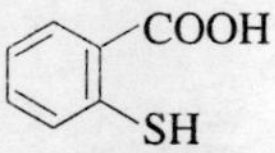 | 165 | Yellow, sparingly soluble in hot $H_2O$, Soluble in alcohol | — | — | — | — | - Acetyl derivative m.p. 125°C. |
| 4-Nitrophthalic acid 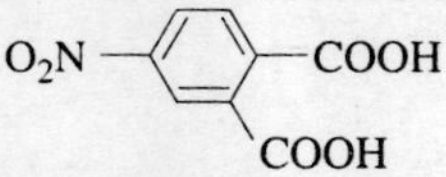 | 165 | Yellow soluble in $H_2O$, alcohol, ether | 192 | 172 | 200 | — | |
| Itaconic acid<br>HOOC C($=CH_2$) $CH_2$ COOH | 165 | soluble in $H_2O$ | 190 | 90 | 191 | — | |
| α, β-Dibromosuccinic acid (±)<br>HOOCCHBrCHBrCOOH | 167 | | | | | | - p-Nitrobenzyl ester, m.p. 168°C. |
| Tartaric acid (+) or (–)<br>COOH–CHOH–CHOH–COOH | 170<br>140 (meso)<br>206 (±) | Colourless, soluble in $H_2O$, alcohol | 180 - | 264 | 195 | — | - On heating with conc. $H_2SO_4$ it chars, effervesces and gives smell of burnt sugar.<br>- With cobalt nitrate and excess NaOH solution gives blue solution in cold which becomes colourless on warming. |

| Name and structure | M.P. (°C) | Properties | Derivatives, M.P. (°C) | | | | Specific tests and derivatives |
|---|---|---|---|---|---|---|---|
| | | | Anilide | p-Toluidide | Amide | S-Benzylisothiouronium salt | |
| Tartaric acid (contd.) | | | | | | | - With $\beta$-naphthol and conc. $H_2SO_4$ gives green colour.<br>- On adding successively $FeSO_4$, $H_2O_2$ and NaOH solution gives violet colour. |
| 3, 5-Dinitrosalicylic acid 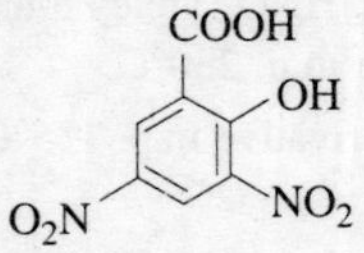 | 173 | Colourless, soluble in $H_2O$, alcohol, ether | 181 | — | 197 | — | - Gives red colour with $FeCl_3$ solution.<br>- With $PCl_5$ gives chloride, m.p. 69°C. |
| p-Toluic acid 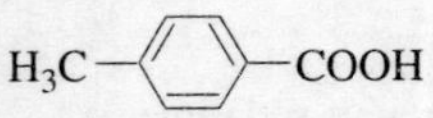 | 178 | Colourless insoluble in $H_2O$, soluble in alcohol, ether | 140 | 160 | 159 | 190 | - On oxidation (alkaline $KMnO_4$) gives terephthalic acid m.p. > 300°C. |
| Acetylene dicarboxylic acid HOOCC≡CCOOH | 179 | | | | 294 (di) | | |
| Veratric acid 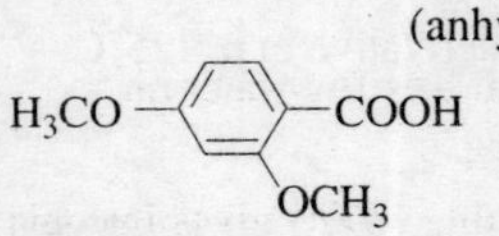 | 180 (anhyd) | Colourless, insoluble in $H_2O$, soluble in alcohol | 166 | | 164 | | |
| 2, 4-Dinitrobenzoic acid 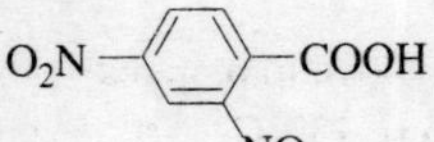 | 183 | Colourless, insoluble in $H_2O$, soluble in alcohol, ether | | | 203 | | - On reduction (Sn + HCl) gives m-phenylenediamine, m.p. 63°C. |
| p-Fluorobenzoic acid 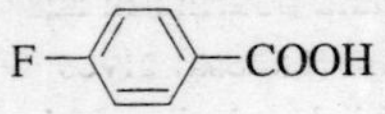 | 183 | | | | 154 | | |
| Anisic acid  | 184 | Colourless, insoluble in $H_2O$, soluble in alcohol | 168 | 186 | 162 | 185 | - On boiling with 48% HBr or on KOH fusion yields p-hydroxybenzoic acid, m.p. 213°C. |
| $\beta$-Naphthoic acid 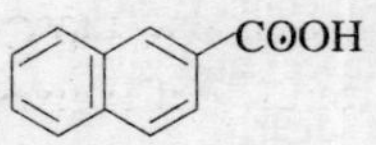 | 184 | Colourless, insoluble in $H_2O$, soluble in alcohol | 170 | 191 | 192 | – | - On heating with sodalime yields naphthalene, m.p. 80°C. |
| Succinic acid $HOOCCH_2CH_2COOH$ | 185 | Colourless, soluble in hot $H_2O$, sparingly soluble in alcohol | 228 (di) 145 (mono) | 255 (di) | 260 (di) | 149 | - Gives brown ppt. with neutral aq. $FeCl_3$ solution.<br>- Gives fluorescein test on heating with resorcinol and conc. $H_2SO_4$ and pouring into excess NaOH solution. |

| *Name and structure* | *M.P. (°C)* | *Properties* | *Derivatives, M.P. (°C)* Anilide | *p-Toluidide* | *Amide* | *S-Benzylisothiouronium salt* | *Specific tests and derivatives* |
|---|---|---|---|---|---|---|---|
| (+) Camphoric acid 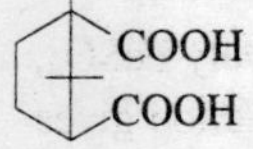 | 187 | Colourless, insoluble in $H_2O$, soluble in alcohol | 226 | — | 192 | — | - On warming with $Ac_2O$ yields anhydride, m.p. 221°C. |
| m-Iodobenzoic acid 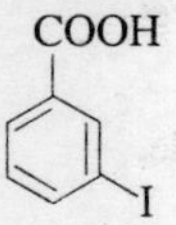 | 187 | Colourless, sparingly soluble in $H_2O$, soluble in alcohol | — | — | 186 | — | |
| Oxalic acid Anhydrous | 189 | See oxalic acid hydrated m.p. 101 | | | | | |
| L-Ascorbic acid (vitamin-c) 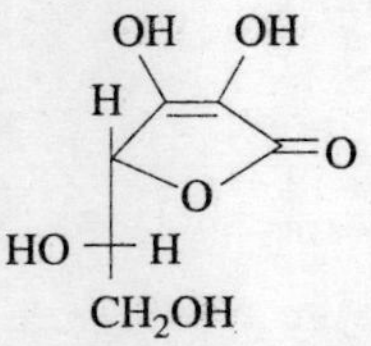 | 190 (d) | See lactones Table 6.8(b) | | | | | |
| Aconitic acid (trans) 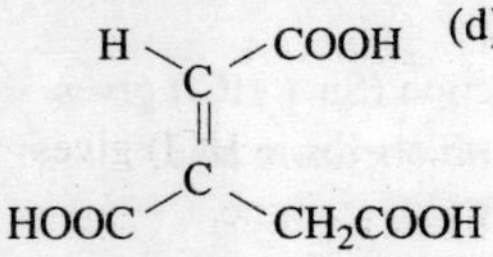 | 191 (d) | Crystalline solid, sparingly soluble in $H_2O$ | — | 76 | 250 | — | - Decolourises $Br_2$ water on warming.<br>- On heating (200°) gives Itaconic anhydride, m.p. 68°C and $CO_2$ |
| Phthalic acid 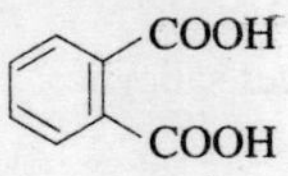 | 195 | Colourless, insoluble in $H_2O$, sparingly soluble in alcohol. m.p. varies from 195-213 (d) depending on rate of heating | 250 (di) 169 (mono) | 201 (di) | 220 (di) 149 (mono) | — | - On heating above m.p. gives anhdride, m.p. 132°C.<br>- On heating with resorcinol and conc. $H_2SO_4$ and pouring in aq. NaOH solution (excess) gives orange or reddish solution with intense green flourescence. |
| m-Nitrocinnamic acid (trans) 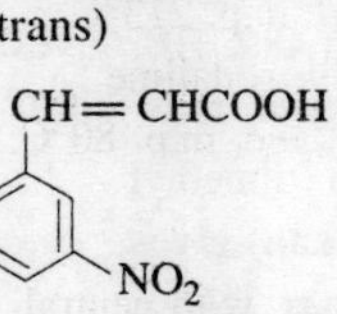 | 200 | Colourless solid | — | — | 196 | — | - Decolourises $KMnO_4$ solution and $Br_2$ water.<br>- On oxidation ($KMnO_4$) it gives m-nitrobenzoic acid, m.p. 140°C.<br>- On reduction ($SnCl_2$ +HCl) gives m-aminocinnamic acid, m.p. 191°C. |
| Protocatechuic acid (3, 4-dihydroxy-benzoic acid) | 200 (d) | Soluble in $H_2O$ | 166 | — | 212 | — | - Gives blue green colour with neutral $FeCl_3$ solution.<br>- On heating yields catechol. |

| *Name and structure* | *M.P. (°C)* | *Properties* | *Derivatives, M.P. (°C)* *Ani-lide* | *p-Tolui-dide* | *Amide* | *S-Benzyl-isothiouro-nium salt* | *Specific tests and derivatives* |
|---|---|---|---|---|---|---|---|
| 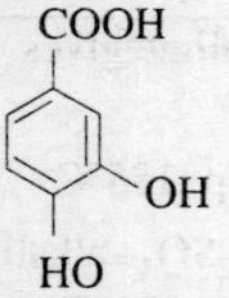  | | | | | | | - Reduces Tollen's reagent.<br>- With Methyl sulphate +NaOH gives veratric acid, m.p. 180°C.<br>- Diacetyl derivative, m.p. 162°C. |
| Fumaric acid (trans) 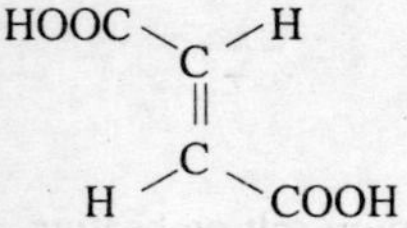  | 200 sub-limes | M.P. 300 (sealed tube) sparingly soluble in $H_2O$ | 314 | — | 266 (di) | 195 | - Decolourises $Br_2$ water. |
| m-Hydroxybenzoic acid 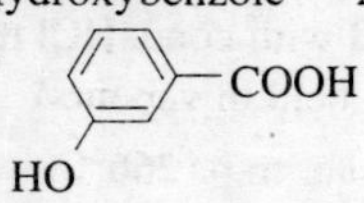  | 201 | Colourless or light colour, sparingly soluble in alcohol | 155 | 163 | 170 | — | - No colour with $FeCl_3$.<br>- On heating with sodalime gives phenol.<br>- Acetate, m.p. 127°C. |
| Mesaconic acid Methyl fumaric acid 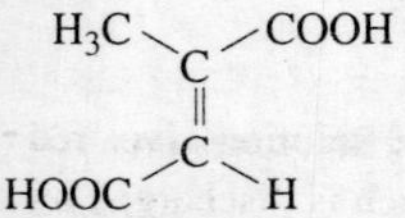  | 204 (sub-limes) | Sparingly soluble in $H_2O$ | 185 | — | 176 | | |
| 3, 5 Dinitrobenzoic acid 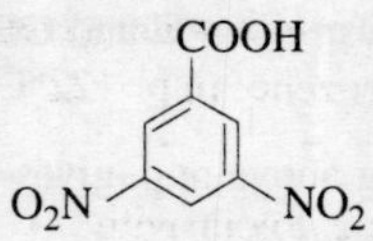  | 205 | Pale yellow, insoluble in $H_2O$, soluble in alcohol | 234 | — | 183 | | - On reduction (Sn + HCl) gives 3, 5-diaminobenzoic acid m.p. 240°C. |
| o-Coumaric acid 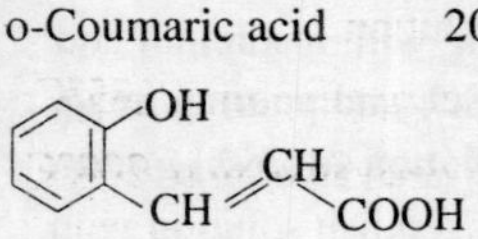  | 207 | Sparingly soluble in water | | | 209 (d) | | - On KOH fusion gives salicylic acid.<br>- With HBr gives coumarin, m.p. 67°C<br>- Acetyl derivative m.p. 146°C. |
| Vanillic acid 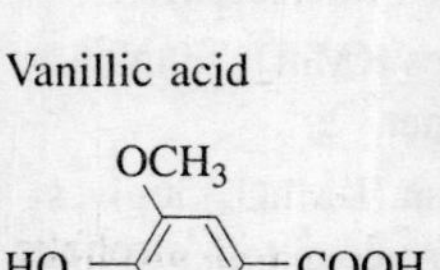  | 207 | Colourless, sparingly soluble in water, soluble in alcohol | | | | | - Gives no colour with $FeCl_3$.<br>- On boiling with 48% HBr gives protocatechuic acid, m.p. 200°C.<br>- On methylation (dimethyl sulphate and alkali) gives veratric acid, m.p. 180°C.<br>- Acetyl derivative m.p. 146°C |
| p-Hydroxybenzoic acid | 213 | Sparingly soluble in $H_2O$, soluble in alcohol | 202 | — | 162 | — | - Gives red colour with $FeCl_3$ solution. |

| Name and structure | M.P. (°C) | Properties | Derivatives, M.P. (°C) Anilide | p-Toluidide | Amide | S-Benzyl-isothiouronium salt | Specific tests and derivatives |
|---|---|---|---|---|---|---|---|
| 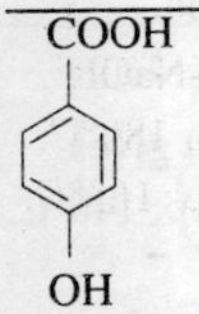 | | | | | | | - On heating with sodalime gives smell of phenol.<br>- Acetyl derivative, m.p. 185°C.<br>- On methylation ($Me_2SO_4$+alkali) gives anisic acid, m.p. 184°C. |
| β-Resorcylic acid 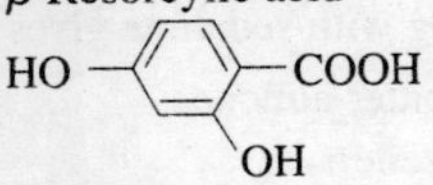 | 213 | Sparingly soluble in $H_2O$, soluble in alcohol | 127 | — | 222 | — | |
| Mucic acid $HOOC(CHOH)_4COOH$(d) | 213 | Sparingly soluble in $H_2O$, soluble in alcohol | — | — | — | 178 | - The ammonium salt on heating gives pyrrole (a wooden match moistened with conc. HCl turns red when held in vapours)<br>Tetraacetate, m.p. 266°C. |
| 3-Nitrophthalic acid 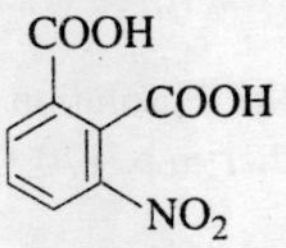 | 218 | Colourless, insoluble in $H_2O$, soluble in alcohol | 234 (di) 181 (mono) | 224 | 201 | — | - On heating above m.p. yields the anhydride, m.p. 162°C. |
| 2, 4, 6-Trinitrobenzoic acid 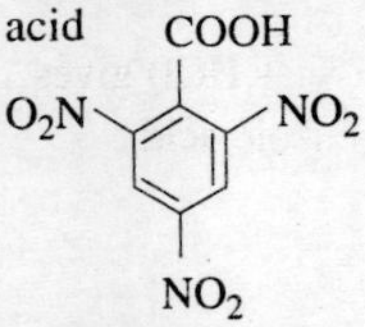 | 220 (d) | Light pale yellow, insoluble in $H_2O$ | — | — | 264 (d) | — | - With NaOH solution gives red colour, which is discharged on acidification.<br>- On boiling with water gives trinitrobenzene, m.p. 122°C. |
| 3-Hydroxy-β-napthoic acid 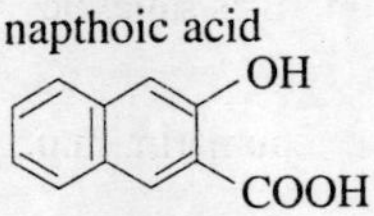 | 222 | Sparingly soluble $H_2O$ | 243 | 222 | 217 | — | - Gives blue colour with $FeCl_3$ solution.<br>- Acetyle derivative m.p. 185°C.<br>- Benzoyl derivative m.p. 208°C. |
| Diphenic acid 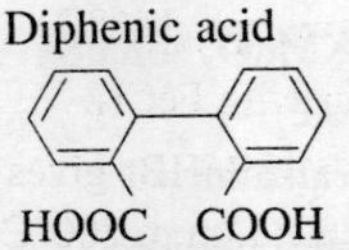 | 228 | Light cream colour, insoluble in $H_2O$ | 229 | — | 212 | — | |
| 5-Nitrosalicylic acid 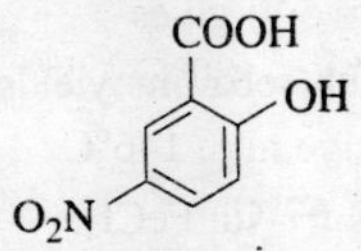 | 230 | Colourless, insoluble in $H_2O$, soluble in alcohol | 224 | — | 225 | — | - Gives red colour with alcoholic $FeCl_3$ solution |
| Nicotinic acid 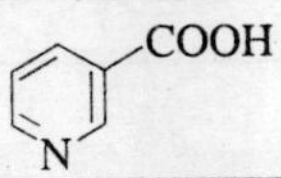 | 236 | Sparingly soluble in $H_2O$ | 85 | — | 122 | — | - N-oxide, m.p. 254°(d). |

| *Name and structure* | *M.P. (°C)* | *Properties* | *Derivatives, M.P. (°C)* | | | | *Specific tests and derivatives* |
|---|---|---|---|---|---|---|---|
| | | | *Anilide* | *p-Toluidide* | *Amide* | *S-Benzylisothiouronium salt* | |
| o-Nitrocinnamic acid 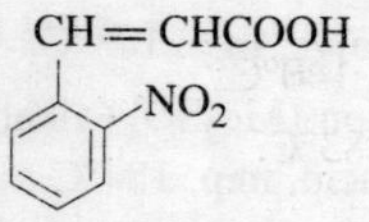 | 240 | Colourless, insoluble in $H_2O$, alcohol | | | 185 | | - Decolourises $KMnO_4$ solution.<br>- On oxidation ($KMnO_4$) gives o-nitrobenzoic acid, m.p. 146°C. |
| p-Nitrobenzoic acid 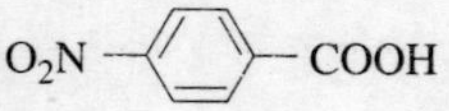 | 240 | Pale yellow, insoluble in $H_2O$, alcohol | 217 (204) | 201 | 201 | 182 | - On heating with sodalime gives smell of bitter almonds (nitrobenzene).<br>- On reduction (Sn + HCl) gives p-aminobenzoic acid, m.p. 186°C. |
| Gallic acid 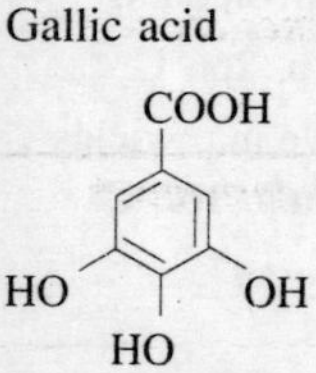 | 240 (d) | Sparingly soluble in cold $H_2O$, more soluble in hot $H_2O$ | 207 | | 244 | | - Gives blue-black ppt. with neutral $FeCl_3$ solution.<br>- On heating gives pyrogallol m.p.133°C.<br>- With $Me_2SO_4$ + NaOH solution gives trimethyl ether m.p. 170°C. |
| p-Chlorobenzoic acid 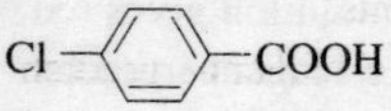 | 243 | Colourless, insoluble in $H_2O$, soluble in alcohol | 194 | | 179 | 186 | |
| p-Bromobenzoic acid 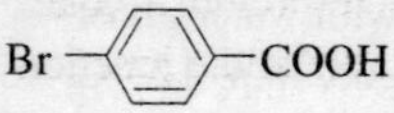 | 251 | Colourless, sparingly soluble in hot $H_2O$, soluble in alcohol | 197 | — | 189 | — | |
| p-Iodobenzoic acid 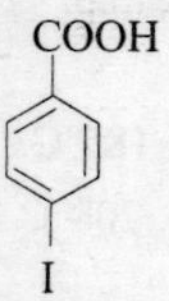 | 270 | Colourless, sparingly soluble in hot $H_2O$, soluble in alcohol | 210 | — | 217 | | |
| p-Nitrocinnamic acid 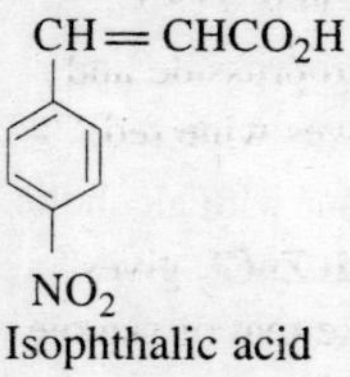 | 285 | Colourless, sparingly soluble in hot $H_2O$, insoluble in alcohol | — | — | 204 | — | - Decolourises $KMnO_4$ solution and $Br_2$ solution.<br>- On oxidation (alkaline $KMnO_4$)— p-nitrobenzoic acid, m.p. 240°C. |
| Isophthalic acid 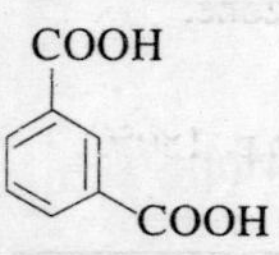 | >300 | Insoluble in $H_2O$ | 250 | — | 280 | 215 | - On heating with sodalime yields benzene.<br>- Me ester, m.p. 67°C. |

| *Name and structure* | *M.P. (°C)* | *Properties* | *Derivatives, M.P. (°C)* | | | | *Specific tests and derivatives* |
|---|---|---|---|---|---|---|---|
| | | | *Anilide* | *p-Toluidide* | *Amide* | *S-Benzylisothiouronium salt* | |
| Terephthalic acid  | >300 | Insoluble in $H_2O$ and $CHCl_3$ | 337 | — | — | — | - On heating with sodalime, yields benzene.<br>- Me ester, m.p. 140°C. |
| Isonicotinic acid 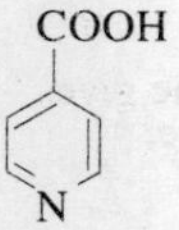 | 319 | Insoluble in $H_2O$ | 156 | — | — | — | - N-oxide, m.p. 83°C. |

### **Table 6.3(a)** Carbonyl Compounds (Liquid Aldehydes and Ketones)

***Note*** *Sometimes it is difficult to differentiate between aldehydes and ketones, therefore both aldehydes and ketones are grouped together under carbonyl compounds*

| *Name and structure* | *B.P (°C)* | *Properties (colour, odour, solubility)* | *2, 4-Dinitrophenylhydrazone* | *Semicarbazone* | *Dimethone* | *Oxime* | *Specific tests and derivatives* |
|---|---|---|---|---|---|---|---|
| Formaldehyde<br>HCHO | Gas | Soluble in $H_2O$, alcohol, ether, aq. solution known as formalin (b.p. 97–98°C) pungent, irritating odour. | 168 | 169 (d) | 189 | — | - Reduces Fehling's and Tollens reagents.<br>- With Nessler's reagent—reddish brown ppt.<br>- Aq. solution + resorcinol + conc. $H_2SO_4$—red layer at the junction of two liquids<br>p-Nitrophenyl hydrazone, m.p. 182°C. |
| Acetaldehyde<br>$CH_3CHO$ | 21 | Colourless, disagreeable pungent odour, soluble $H_2O$, alcohol, ether | 168 | 163 | 140 | 47 | - Reduces Tollen's reagent (silver mirror on heating in waterbath).<br>- Reduces Fehling's solution<br>- Gives iodoform, m.p. 119°C.<br>- With sodium nitroprusside and excess NaOH gives wine red colour. |
| Propionaldehyde<br>$CH_3CH_2CHO$ | 49 | Colourless, disagreeable odour like acetaldehyde, soluble in $H_2O$ | 155 | 154 (89) | 155 | 40 | - On warming with $ZnCl_2$ gives intense odour like that of skatole<br>p-Nitrophenylhydrazone, m.p. 123°C. |
| Glyoxal<br>OHC—CHO | 50 | Usually available as an aq. solution | 328 | 270 | — | 178 | - Phenylhydrazone, m.p. 180°C. |

| *Name and structure* | *B.P (°C)* | *Properties (colour, odour, solubility)* | *2, 4-Dinitrophenylhydrazone* | *Semicarbazone* | *Dimethone* | *Oxime* | *Specific tests and derivaties* |
|---|---|---|---|---|---|---|---|
| Acrolein $CH_2=CHCHO$ | 52 | Colourless, irritating odour, soluble in $H_2O$, alcohol and ether | 165 | 171 | 192 | — | - Gives tests for unsaturation ($KMnO_4$ and bromine water). - Reduces Tollen's reagent. |
| Acetone $CH_3COCH_3$ | 56 | Colourless, characteristic odour, soluble in $H_2O$, alcohol and ether | 126 | 188 | — | 59 | - Gives red colour with sodium nitroprusside (fresh) and excess NaOH solution. - Gives iodoform test. - p-Nitrophenylhydrazone, m.p. 148°C. |
| iso-Butyraldehyde $(CH_3)_2CHCHO$ | 63 | Characteristics odour, soluble in cold $H_2O$ | 187 (182) | 125 | 154 | — | - p-Nitrophenylhydrazone, m.p. 131°C. |
| n-Butyraldehyde $CH_3CH_2CH_2CHO$ | 74 | Characteristic odour, soluble in $H_2O$, alcohol, ether | 123 | 104 | 135 | — | - Gives Fehling's and Tollen's test. - p-Nitrophenylhydrazone, m.p. 91°C. |
| Trimethylacetaldehyde (Pivaldehyde) $(CH_3)_3CCHO$ | 75 | | 210 | 190 | — | — | - P-Nitrophenylhydrazone m.p. 119°C. |
| Ethyl methyl ketone $CH_3COCH_2CH_3$ | 80 | Colourless, pleasant odour, soluble in $H_2O$, alcohol, ether | 115 | 147 | — | — | - Gives iodoform test. - Gives sodium nitroprusside test. (see acetone) |
| Methyl vinyl ketone $CH_2 = CHCOCH_3$ | 80 | | | 141 | | | |
| Diacetyl $CH_3COCOCH_3$ | 88 | | 315 (di) | 278 (di) | — | 234 (di) | - Gives quinoxaline derivative, m.p. 106°C with o-phenylene diamine. |
| Isovaleraldehyde $(CH_3)_2CHCH_2CHO$ | 92 | | 123 | 132 (107) | 155 | 48 | - p-Nitrophenylhydrazone, m.p. 109°C. |
| Isopropylmethyl ketone $CH_3COCH(CH_3)_2$ | 94 | | 117 | 112 | — | 109 | - Gives iodoform test. - With sodium nitroprusside gives red colour. |
| Chloral (Trichloroacetaldehyde) $CCl_3CHO$ | 98 | Colourless, pungent odour, soluble in $H_2O$, alcohol, ether | 131 | — | — | 56 | - On shaking with conc. NaOH solution it gives smell of chloroform. - Reduces Fehlings and Tollen's regent. |
| Diethyl ketone $(C_2H_5)_2CO$ | 102 | Colourless, pleasant odour, Sparingly soluble in $H_2O$, soluble in alcohol, ether | 156 | 139 | — | 69 | - Does not give iodoform test. - p-Nitrophenylhydrazone, m.p. 144°C. |

| *Name and structure* | *B.P (°C)* | *Properties (colour, odour, solubility)* | *2,4-Dinitrophenylhydrazone* | *Semicarbazone* | *Dimethone* | *Oxime* | *Specific tests and derivaties* |
|---|---|---|---|---|---|---|---|
| Methyl n-propyl ketone (Pentanone-2) $CH_3COC_3H_7(n)$ | 102 | Colourless, sparingly soluble in $H_2O$, soluble in alcohol, ether | 142 | 110 | — | 58 | - Gives iodoform test. |
| Crotonaldehyde $CH_3CH{=}CHCHO$ | 103 | Colourless, pungent odour, sparingly soluble in $H_2O$, soluble in alcohol, ether | 189 | 201 | 193 | 119 | - Reduces Tollens reagent.<br>- On heating with dil. KOH solution gives yellow ppt. changing to orange with disagreeable odour.<br>- With sodium nitroprusside and NaOH solution gives wine red colour. |
| n-Valeraldehyde $CH_3(CH_2)_3CHO$ | 103 | | 107 | — | 108 | 52 | |
| Pinacolone (Methyl t-butyl ketone) $CH_3COC(CH_3)_3$ | 106 | Soluble in $H_2O$, camphor like odour | 125 | 157 | — | 75 | - Gives iodoform reaction.<br>- Gives red colour with sodium nitroprusside. |
| Isobutyl methyl ketone $CH_3COCH_2CH(CH_3)_2$ | 117 | Colourless, insoluble in $H_2O$, soluble in alcohol, ether, benzene | 95 | 135 | — | 58 | - Gives iodoform test. |
| Di-isopropyl ketone $[(CH_3)_2CH]_2CO$ | 124 | Colourless, sparingly soluble in $H_2O$, soluble in alcohol, ether | 96 | 160 | — | 34 | |
| Paraldehyde 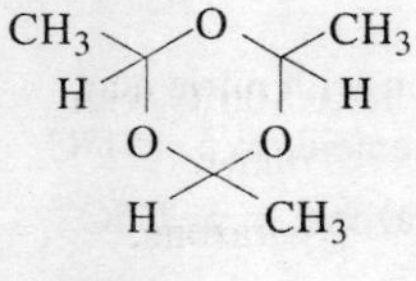 | 124 | See acetaldehyde | | | | | - Gives acetaldehyde on warming with dil. $H_2SO_4$. |
| n-Butyl methyl ketone (Hexanone-2) $CH_3COC_4H_9(n)$ | 129 | Colourless, sparingly soluble in $H_2O$, soluble in alcohol, ether | 106 | 122 | — | 49 | - Gives iodoform test. |
| Cyclopentanone 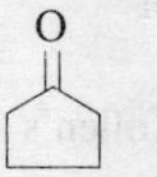 | 130 | Colourless, sparingly soluble in $H_2O$ | 142 | 206 | — | 56 | - Phenylhydrazone, m.p. 50°C. |
| Mesityl oxide $CH_3COCH{=}C(CH_3)_2$ | 130 | Sparingly soluble in $H_2O$, peppermint-like odour | 203 | 164 | — | 49 | - Decolourises bromine water.<br>- On boiling with dil $H_2SO_4$ yields acetone.<br>- p-Nitrophenylhydrazone, m.p. 133°C. |

| Name and structure | B.P (°C) | Properties (colour, odour, solubility) | 2,4-Dinitrophenyl-hydrazone | Semi-carbazone | Dimethone | Oxime | Specific tests and derivatives |
|---|---|---|---|---|---|---|---|
| Caproaldehyde (n-Hexaldehyde) $CH_3(CH_2)_4CHO$ | 131 | | 104 | 106 | 108 | 51 | |
| Methyl pyruvate $CH_3COCOOCH_3$ | 135 | | 187 | 208 | — | 69 | - Gives iodoform test. |
| Acetylacetone $CH_3COCH_2COCH_3$ | 139 | Colourless, sparingly soluble in $H_2O$, soluble in alcohol, ether, chloroform | 209 (di) | | | 149 (di) | - It is a keto-enolic compound. - Gives red to violet colour with neutral aq. $FeCl_3$. - Gives light blue ppt. of Cu salt with Cu acetate solution. |
| Tetrahydrofurfural 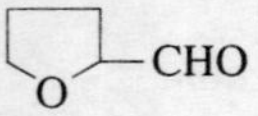 | 143 | | 134 | 166 | 123 | — | - On warming with conc. HCl gives a bright red colour. |
| 4-Heptanone (Di-n-propyl ketone) $C_3H_7COC_3H_7$ | 145 | Colourless, insoluble in $H_2O$, soluble in alcohol, ether | 75 | 133 | — | — | |
| Acetoin $CH_3CHOHCOCH_3$ | 148 (m.p. 15°) | | 318 | 185 | — | — | |
| n-Amyl methyl ketone (2-Heptanone) $CH_3CO(CH_2)_4CH_3$ | 151 | | 89 | 127 | — | — | |
| Cyclohexanone 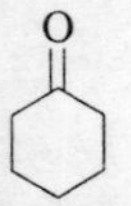 | 155 | Sparingly soluble in $H_2O$ | 162 | 166 | — | 90 | - On oxidation with nitric acid yields adipic acid, m.p. 151°C. - Phenylhydrazone, m.p. 77°C. |
| Ethyl pyruvate $CH_3COCOOC_2H_5$ | 155 | | 155 | 206(d) | — | 94 | - Gives iodoform test. |
| n-Heptaldehyde (Oenanthol) $CH_3(CH_2)_5CHO$ | 156 | | 108 | 109 | 103 | 57 | |
| Furfural 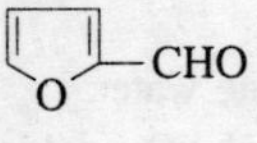 | 161 | Colourless, turns brown in air | 229 | 203 | 162 (d) | 89 (74) | - With alcoholic KOH gives Cannizzaro reaction. - On oxidation ($KMnO_4$) gives Furoic acid, m.p. 133°C. |
| Diacetone alcohol $(CH_3)_2C(OH)CH_2COCH_3$ | 164 | Colourless, soluble in $H_2O$, alcohol, ether | 202 | — | — | 56 | - Yields acetone on distillation with dil. NaOH solution. |

| *Name and structure* | *B.P (°C)* | *Properties (colour, odour, solubility)* | *2,4-Dinitrophenylhydrazone* | *Semicarbazone* | *Dimethone* | *Oxime* | *Specific tests and derivatives* |
|---|---|---|---|---|---|---|---|
| 2-Methylcyclohexanone 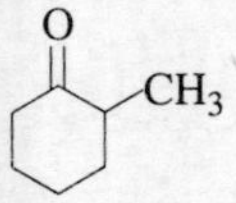 | 165 | | 136 | 197 | | 43 | |
| Di-isobutyl ketone $[(CH_3)_2CHCH_2]_2CO$ | 168 | | 92 | 124 | | | |
| 3-Methylcyclohexanone 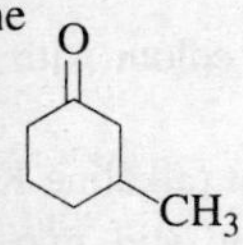 | 169 | | 155 | 190 | — | — | |
| Succinaldehyde $OHCCH_2CH_2CHO$ | 169 | | 280 | — | — | 172 (Di) | |
| 4-Methylcyclohexanone 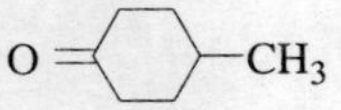 | 170 | | 134 | 199 | — | 38 | |
| Methyl acetoacetate $CH_3COCH_2COOCH_3$ | 170 | | — | 152 | — | — | - It is a keto-enolic compound.<br>- Gives red colour with aq. $FeCl_3$ solution. |
| n-Hexyl methyl ketone $CH_3CO(CH_2)_5CH_3$ | 172 | | 58 | 123 | — | — | |
| Benzaldehyde 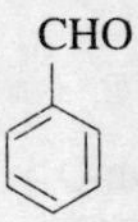 | 179 | Colourless or light yellow, odour of bitter almonds, insoluble in $H_2O$, soluble in alcohol, ether | 237 | 222 | 195 | 35 | - On oxidation ($KMnO_4$) gives benzoic acid, m.p. 121°C.<br>- On warming with aniline gives benzalaniline, m.p. 54°C<br>- Phenylhydrazone, m.p. 158°C.<br>- p-Nitrophenylhydrazone m.p. 192°C. |
| Ethyl acetoacetate $CH_3COCH_2COOC_2H_5$ | 181 | Sparingly soluble in $H_2O$, soluble in dil. NaOH solution | 93 | 130 | — | — | - It is a keto enolic compound.<br>- Gives red colour with aq. $FeCl_3$ solution.<br>- On warming with phenylhydrazine it yields methyl phenyl pyrazolone, m.p. 127°C. |
| Cycloheptanone 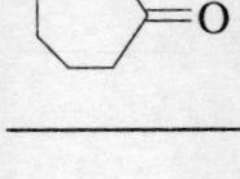 | 181 | | 148 | 163 | — | — | |

| *Name and sttucture* | *B.P (°C)* | *Properties (colour, odour, solubility)* | *2,4-Dinitrophenylhydrazone* | *Semicarbazone* | *Dimethone* | *Oxime* | *Specific tests and derivaties* |
|---|---|---|---|---|---|---|---|
| Di-n-butyl ketone $(C_4H_9)_2CO$ | 187 | Colourless, insoluble in $H_2O$, soluble in chloroform, either | — | 90 | — | — | |
| Phenylacetaldehyde 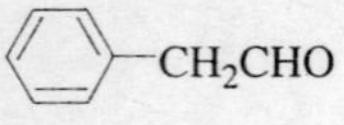 | 193 (m.p. 34°) | | 121 | 156 | 165 | 103 | |
| Acetonyl acetone (hexane-2, 5-dione) $CH_3COCH_2CH_2COCH_3$ | 194 | Colourless, soluble in $H_2O$, alcohol, ether | 255 | 220 (di) | — | 138 (di) | |
| Salicylaldehyde 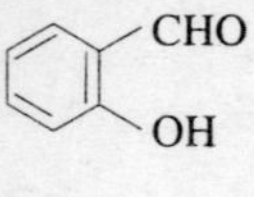 | 196 | Colourless, characteristic odour, sparingly soluble in $H_2O$, soluble in alcohol, ether | 252 | 232 | 211 | 57 | - Gives violet colour with $FeCl_3$ solution.<br>- On oxidation (alkaline $KMnO_4$) gives salicylic acid, m.p. 159°C<br>- On shaking with $Ac_2O$ and a drop of conc. $H_2SO_4$ yields triacetyl derivative m.p. 100°C. |
| Phorone $[(CH_3)_2C{=}CH]_2CO$ | 198 | | 112 | 221 | | 48 | |
| m-Tolualdehyde 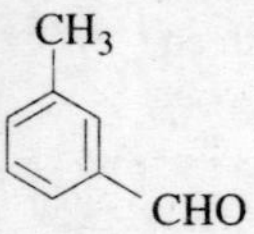 | 199 | | 211 | 213 | 172 | 60 | - Phenylhydrazone, m.p. 84°C. |
| o-Tolualdehyde 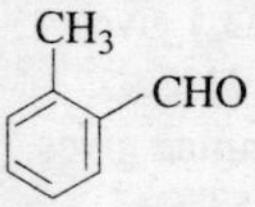 | 200 | | 194 | 212 | 167 | 49 | - Phenylhydrazone, m.p. 105°C. |
| Acetophenone 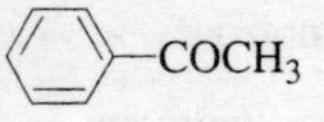 | 202 (m.p. 20) | Colourless, sweet smell, insoluble in $H_2O$, soluble in alcohol, ether | 250 | 198 | — | 59 | - On oxidation (alkaline $KMnO_4$) gives benzoic acid, m.p. 121°C.<br>- Phenylhydrazone, m.p. 105°C.<br>- Gives iodoform test. |
| p-Tolualdehyde 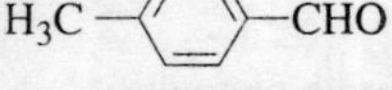 | 205 | | 239 | 234 | — | 80 | |
| Ethyl levulinate $CH_3COCH_2CH_2COOC_2H_5$ | 206 | Soluble in $H_2O$ | 101 | 148 | — | — | - On boiling with NaOH solution yields sodium levulinate (free acid, m.p. 33°, b.p. 245°C)<br>- Phenylhydrazone, m.p. 104°C. |
| (–) Menthone 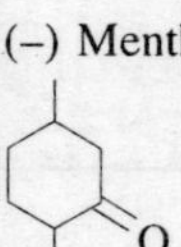 | 207 | Pippermint odour | 146 | 189 | — | 59 | |

| *Name and structure* | *B.P (°C)* | *Properties (colour, odour, solubility)* | *2,4-Dinitrophenylhydrazone* | *Semicarbazone* | *Dimethone* | *Oxime* | *Specific tests and derivaties* |
|---|---|---|---|---|---|---|---|
| Citronellal (Rhodinal) 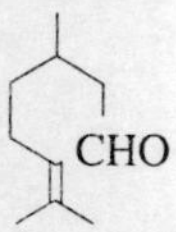 | 207 | Optically active, $[\alpha]_D^{25} \pm 11.5$ | 78 | 84 | 78 | — | - Decolourises $Br_2$ water. |
| o-Chlorobenzaldehyde 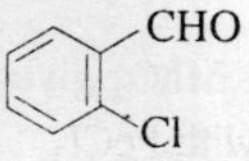 | 208 (m.p. 11) | Colourless, insoluble in $H_2O$, soluble in alcohol, ether | 209 | 226 | 205 | 75 | - On oxidation ($KMnO_4$) gives o-chlorobenzoic acid, m.p. 142°C.<br>- Phenylhydrazone, m.p. 86°C. |
| m-Chlorobenzaldehyde 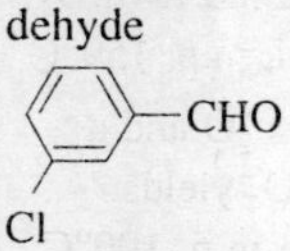 | 213 | | 255 | 228 | — | 70 | - On oxidation ($KMnO_4$) gives m-chlorobenzoic acid, m.p. 158°C. |
| Isophorone 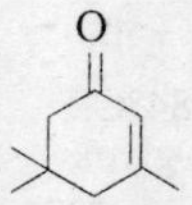 | 215 | Characteristic odour | 130 | 200 | — | 76 | |
| o-Hydroxyacetophenone 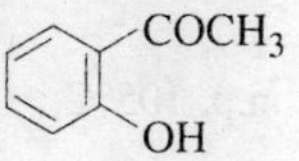 | 215 | Colourless, sparingly soluble in $H_2O$, soluble in alcohol, ether | — | 209 | — | 117 | - Gives red to violet colour with neutral $FeCl_3$ solution.<br>- Phenylhydrazone, m.p. 110°C.<br>- Acetate, m.p. 89°C. |
| Benzyl methyl ketone $CH_3COCH_2C_6H_5$ | 216 (m.p. 27°). | Colourless, insoluble in $H_2O$, soluble in alcohol, ether | 156 | 198 | — | 69 | - Phenylhydrazone, m.p. 87°C. |
| Propiophenone $C_6H_5COCH_2CH_3$ | 218 | Pleasant odour | 191 | 174 | — | 53 | |
| Hydrocinnamaldehyde $C_6H_5CH_2CH_2CHO$ | 224 | | 149 | 127 | — | 94 | - On oxidation gives hydrocinnamic acid, m.p. 48°C. |
| n-Butyrophenone $C_6H_5CO(CH_2)_2CH_3$ | 228 | | 190 | 186 | — | 50 | |
| Citral 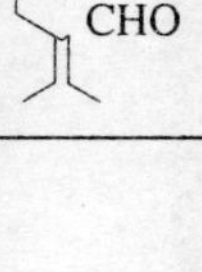 | 228 | Odour of lemon | 110 | 164 | — | — | - Decolourises $Br_2$ water and $KMnO_4$ solution. |

| *Name and structure* | *B.P (°C)* | *Properties (colour, odour, solubility)* | *2,4-Dinitrophenylhydrazone* | *Semicarbazone* | *Dimethone* | *Oxime* | *Specific tests and derivatives* |
|---|---|---|---|---|---|---|---|
| d-Carvone 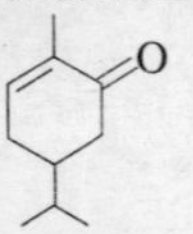 | 230 | Odour of caraway | 191 | 162 (142) | — | 72 | - p-Nitrophenylhydrazone, m.p. 175°C.<br>- Decolourises $Br_2$ water. |
| p-Chloroaceto-phenone 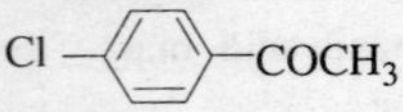 | 236 (m.p. 20°) | | 231 | 203 | — | 95 | |
| Anisaldehyde 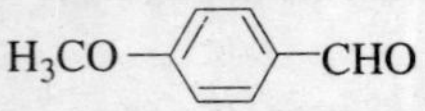 | 247 | | 253 | 210 | 145 | | - On oxidation with $KMnO_4$ gives anisic acid, m.p. 184°C. |
| Cinnamaldehyde 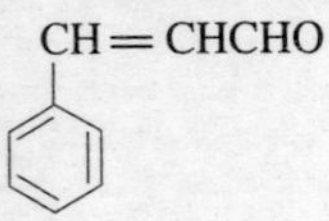 | 252 (d) | | 255 (d) | 216 (208) | 219 (213) | 138 (69) | |
| Ethylbenzoyl acetate $C_6H_5COCH_2COOC_2H_5$ | 269 (d) | | | 125 (d) | — | 152 | - Keto-enolic compound. Gives red violet colour with neutral aq. $FeCl_3$.<br>- With phenylhydrazine, gives diphenyl pyrazolone, m.p. 137°C. |
| α-Ionone 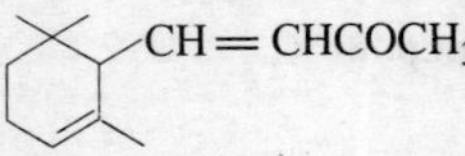 | 130/ 13 mm | | 151 | 143 108 | — | 90 | - p-Nitrophenylhydrazone, m.p. 113°C. |
| β-Ionone 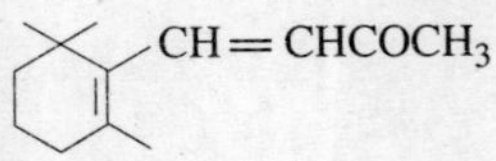 | 140/ 18 mm | odour of cedar wood | 128 | 148 | — | — | - p-Nitrophenylhydrazone, m.p. 173°C. |

**Table 6.3 (b)** Carbonyl Compounds (Solids)

| *Name and structure* | *M.P (°C)* | *Properties (colour, odour, solubility)* | *2, 4-Dinitrophenylhydrazone* | *Semicarbazone* | *Dimethone* | *Oxime* | *Specific tests and derivatives* |
|---|---|---|---|---|---|---|---|
| p-Methylacetophenone 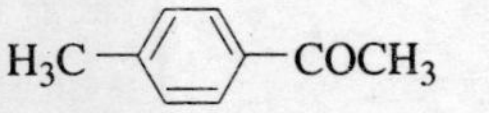  | 28 (b.p. 226) | | 258 | 205 | — | 88 | |
| Dibenzyl ketone $(C_6H_5CH_2)_2C{=}0$ | 34 (b.p. 331) | | 100 | 146 | — | 124 | - On warming with benzil in alcoholic KOH solution gives deep violet colour. |
| Methyl $\alpha$-naphthyl ketone 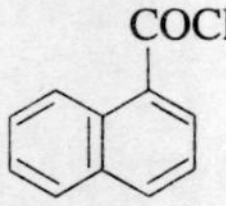  | 34 | | | 229 | — | 137 | - Picrate, m.p. 116°C.<br>- Phenyl hydrazone, m.p. 146°C. |
| 1-Naphthyl aldehyde 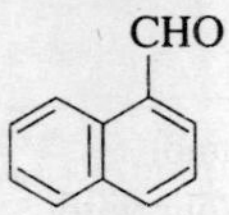  | 34 (b.p. 292) | | — | 221 | — | 90 | - On oxidation ($KMnO_4$) gives 1-naphthoic acid, m.p. 162°C. |
| p-Methoxyacetophenone 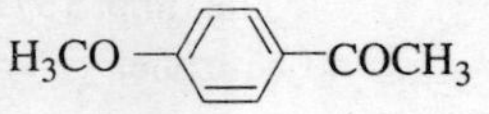  | 38 | | 220 | 198 | — | 87 | - p-Nitrophenylhydrazone, m.p. 195°C. |
| Piperonal 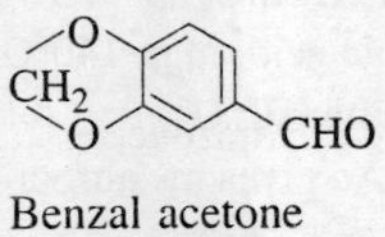  | 38 (b.p. 263) | Odour of heliotrope | 265 | 230 | 178 | 110 | - On oxidation ($KMnO_4$) gives piperonylic acid, m.p. 228°C. |
| Benzal acetone (Benzylidene acetone) $C_6H_5CH{=}CHCOCH_3$ | 42 (b.p. 262) | Light yellow, insoluble in $H_2O$ | 223 | 186 | — | 115 | - Decolourises bromine colour.<br>- With $Br_2$ in $CCl_4$—di-bromide, m.p. 125°C.<br>- Phenylhydrazone, m.p. 156°C. |
| 1-Indanone 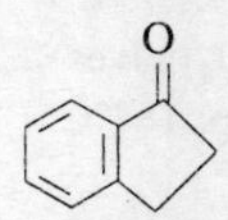  | 42 | | 258 | 233 | — | 145 | |
| o-Nitrobenzaldehyde 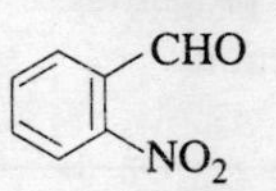  | 44 | Yellow, insoluble in $H_2O$, soluble in hot $H_2O$, alcohol, ether | 250 | 256 | — | 103 | - On heating with aq. NaOH gives reddish brown solution.<br>- On oxidation ($KMnO_4$) o-nitrobenzoic acid, m.p. 146°C. |
| p-Chlorobenzaldehyde   | 47 (b.p. 214) | Colourless, soluble in hot $H_2O$, alcohol and ether | 270 | 231 | — | 110 | - On oxidation ($KMnO_4$)—p-chlorobenzoic acid, m.p. 243°C. |

| *Name and structure* | *M.P (°C)* | *Properties (colour, odour, solubility)* | *2,4-Dinitrophenyl-hydrazone* | *Semi-carbazone* | *Dime-thone* | *Oxime* | *Specific tests and derivatives* |
|---|---|---|---|---|---|---|---|
| Benzophenone $C_6H_5COC_6H_5$ | 48 | Fragrant odour | 238 | 164 | — | 142 | - On heating with freshly cut small piece of Na gives deep blue colouration (THF is used as a solvent). |
| ω-Bromoacetophenone (Phenacyl bromide) 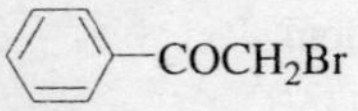  | 50 | Colourless, irritating odour, insoluble in $H_2O$, soluble in alcohol, ether. | — | 146 | — | 97 (89) | |
| p-Bromoacetophenone 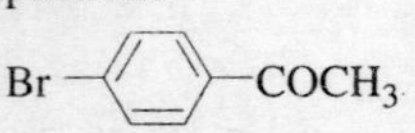  | 51 | Colourless, insoluble in $H_2O$, soluble in alcohol, ether | 230 | 208 | — | 128 | |
| Methyl β-naphthyl-ketone (2-Acetylnaphthalene) 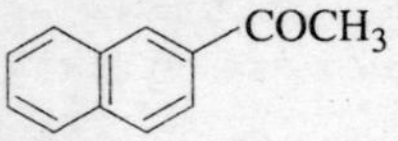  | 53 | | 262(d) | 236 | — | 145 | - Phenylhydrazone m.p. 171°C. |
| Chloral hydrate $Cl_3C\,CH(OH)_2$ | 57 | See chloral | | | | | |
| Benzal acetophenone (chalcone) $C_6H_5CH{=}CHCOC_6H_5$ | 58 | Yellow, insoluble in $H_2O$, sparingly soluble in alcohol, soluble in ether | 244 (d) | 168 | — | 73 | |
| m-Nitrobenzaldehyde 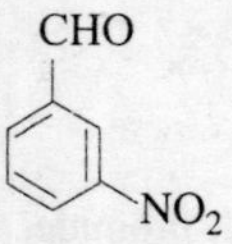  | 58 | Yellow, insoluble in $H_2O$, soluble in hot alcohol, ether, chloroform. | 293 (d) | 246 | 198 | 118 | - On oxidation ($KMnO_4$)— m-nitrobenzoic acid, m.p. 140°C - On heating with $AC_2O$ + fused NaOAc gives m- nitrocinnamic acid, m.p. 200°C. - Phenylhydrazone, m.p. 120°C. |
| Desoxybenzoin (Phenyl benzyl ketone) $C_6H_5CH_2COC_6H_5$ | 60 | | 204 | 148 | — | 98 | |
| β-Naphthaldehyde 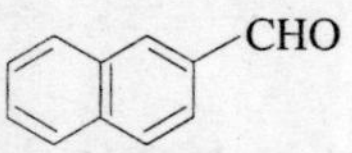  | 60 | | 270 | 245 | — | 156 | - On oxidation ($KMnO_4$) gives β-naphthoic acid, m.p. 184°C. |
| Benzoyl acetone $C_6H_5COCH_2COCH_3$ | 61 (b.p. 261) | | 151 | | | | - Exhibits keto-enol properties. - positive ferric reaction. |

| *Name and structure* | *M.P (°C)* | *Properties (colour, odour, solubility)* | *2,4-Dinitrophenylhydrazone* | *Semicarbazone* | *Dimethone* | *Oxime* | *Specific tests and derivatives* |
|---|---|---|---|---|---|---|---|
| p-Chlorobenzophenone 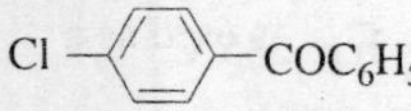  | 78 | Colourless, insoluble in $H_2O$, soluble in benzene | 184 | — | — | 163 | - Phenylhydrazone, m.p. 106°C. |
| p-Nitroacetophenone 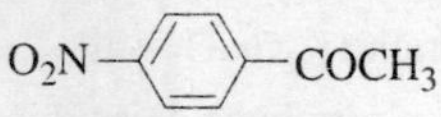  | 80 | Colourless, insoluble in $H_2O$ | — | — | — | 174 | - Phenylhydrazone, m.p. 132°C. |
| m-Nitroacetophenone 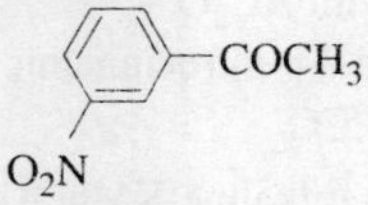  | 81 | Colourless, insoluble in $H_2O$, soluble in alcohol | 233 | 261 | — | 132 | - Phenlhydrazone, m.p. 128°C. |
| Vanillin (3-Methoxy-4-hydroxybenzaldehyde) 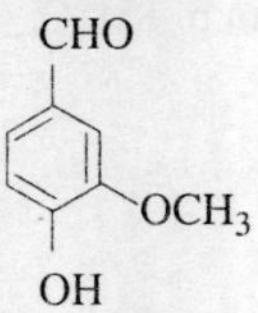  | 81 | Colourless, sparingly soluble in $H_2O$, soluble in ether and benzene, odour of vanilla | 271 | 240 (d) | 197 | 117 | - Gives blue colour with $FeCl_3$. - On oxidation (alkaline $KMnO_4$) gives vanillic acid, m.p. 207°C. - With bromine water gives bromo derivative m.p. 160°C. |
| Fluorenone 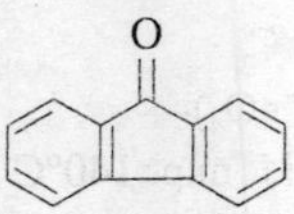  | 83 | Yellow, insoluble in $H_2O$ | 283 | 234 | — | 196 | - Phenylhydrazone, m.p. 152°C. |
| Benzil 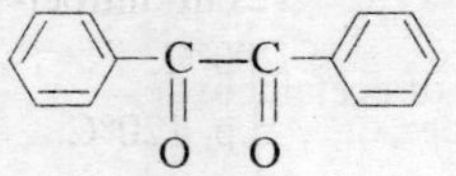  | 95 | Yellow, insoluble in $H_2O$, soluble ether, alcohol | 189 | 224 (di) 182 (mono) | — | 237 (di) 137 (mono) | - On boiling with alcoholic KOH solution yields benzilic acid, m.p. 150°C. - Quinoxaline derivative m.p. 126°C. |
| Glycolaldehyde (Hydroxyacetaldehyde) $HOCH_2CHO$ | 96 | Soluble in $H_2O$ | — | — | — | — | - Phenylhydrazone, m.p. 162°C. - With $Br_2/H_2O$ gives glycollic acid, m.p. 80°C. |
| m-Hydroxy acetophenone 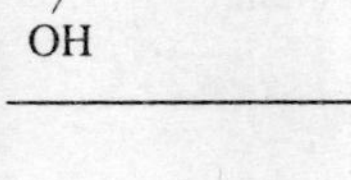  | 96 | | — | 195 | — | — | |

| *Name and structure* | *M.P (°C)* | *Properties (colour, odour, solubility)* | *2,4-Dinitrophenylhydrazone* | *Semicarbazone* | *Dimethone* | *Oxime* | *Specific tests and derivatives* |
|---|---|---|---|---|---|---|---|
| m-Hydroxybenzaldehyde 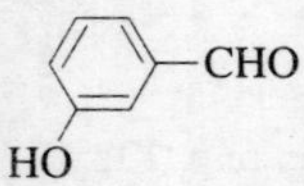 | 104 | Sparingly soluble in $H_2O$ | 260 (d) | 199 | — | 88 | - Negative ferric reaction.<br>- Methyl ether, b.p. 230°C.<br>- Ethyl ether, b.p. 245°C. |
| p-Nitrobenzaldehyde 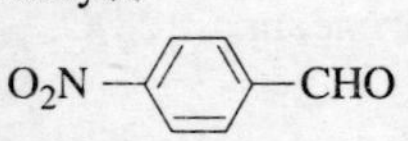 | 106 | Yellow, sparingly soluble in $H_2O$, soluble in alcohol | 320 | 220 | 190 | 129 | - On oxidation ($K_2Cr_2O_7$ +dil. $H_2SO_4$)—p-nitrobenzoic acid, m.p. 240°C.<br>- On heating with $AC_2O$ + NaOAc fused—p-nitrocinnamic acid, m.p. 285°C. |
| p-Bromophenacyl bromide 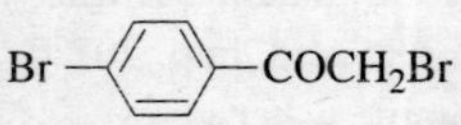 | 109 | Irritating odour, insoluble in $H_2O$, soluble in hot $H_2O$, alcohol | — | — | — | 115 | - On oxidation (alkaline $KMnO_4$) gives p-bromobenzoic acid acid, m.p. 251°C. |
| p-Hydroxyacetophenone 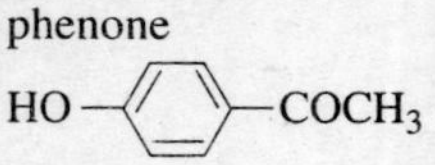 | 109 | Colourless, insoluble in $H_2O$, soluble in alcohol, ether | 261 | 199 | — | 145 | - Phenylhydrazone, m.p. 151°C. |
| Dibenzalacetone $(C_6H_5CH{=}CH)_2CO$ | 113 | Yellow, insoluble in $H_2O$, soluble in hot alcohol, acetone. | 180 | 190 | — | 143 | |
| Anisoin (pp′-dimethoxy benzoin) 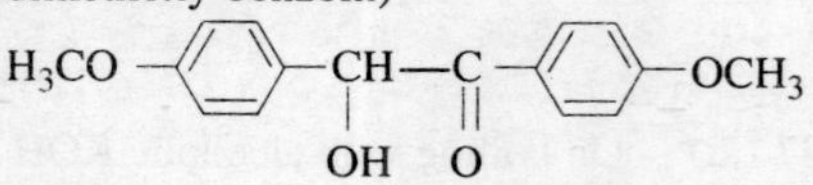 | 113 | | — | 185 | — | — | - Acetate, m.p. 95°C. |
| Metaldehyde | 115 | Insoluble in $H_2O$, sublimes | | | | | - Polymer of acetaldehyde, |
| 1, 2-Naphthoquinone 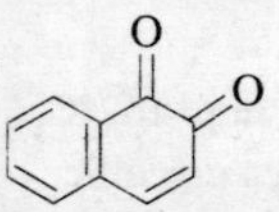 | 115 (d) | See quinones (Table 6.5) | | | | | |

| *Name and structure* | *M.P (°C)* | *Properties (colour, odour, solubility)* | *2,4-Dinitrophenylhydrazone* | *Semicarbazone* | *Dimethone* | *Oxime* | *Specific tests and derivatvies* |
|---|---|---|---|---|---|---|---|
| p-Hydroxybenzaldehyde 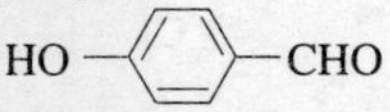 | 117 | Pink, sparingly soluble in $H_2O$, soluble in hot water, alcohol | 280 (d) | 223 | 188 | 73 | - Gives faint violet colour with aq. $FeCl_3$ solution. |
| 1, 4-Naphthoquinone 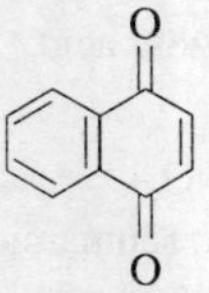 | 125 | See quinones (Table 6.5) | | | | | |
| Benzoin (dl) 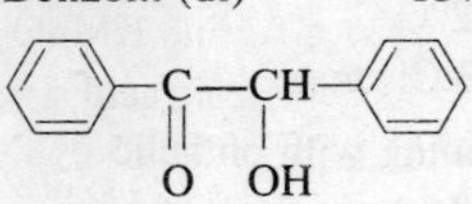 | 137 | Pale yellow, insoluble in $H_2O$, sparingly soluble in hot water, alcohol | 234 (245) | 206 (d) | — | 151 | - On heating with Fehling's solution gives red ppt.<br>- benzoate, m.p. 125°C.<br>- On oxidation (Conc. $HNO_3$) gives benzil, m.p. 95°C. |
| Furoin 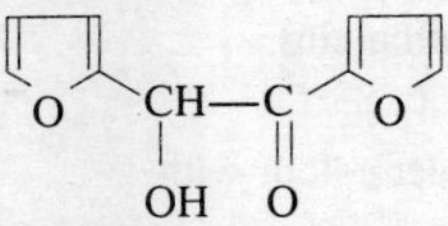 | 139 | | 216 | — | — | 161 | - On mild oxidation with cupric acetate and $NH_4NO_3$ in aq. acetic acid. (reflux) gives furil (see below) m.p. 162°C. |
| *β*-Resacetophenone (2, 4-Dihydroxyacetophenone) 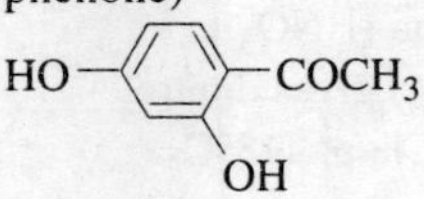 | 147 | | — | 218 | | 199 | - With $FeCl_3$ gives red to violet colour.<br>- Dibenzoate, m.p. 81°C. |
| Furil 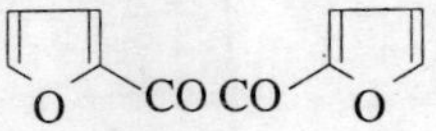 | 162 | | 215 | — | — | | - p-Nitrophenylhydrazone, m.p. 199°C.<br>- Quinoxaline derivative, m.p. 134°C. |
| Polyoxymethylene (paraformaldehyde) | 170-180 | Linear high polymer of formaldehyde which is generated by heating or by the action of dil. alkali. | | | | | |
| (±) Camphor 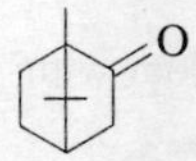 | 179 | Characteristic odour, colourless, soluble in alcohol, sublimes | 177 | 237 | — | 118 | |

## Oxidation of furoin to furil

Take copper acetate (0.05 g), ammonium nitrate (2 g), furoin (3.8 g) in acetic acid (70 ml, aq. $CH_3COOH$) in a round bottomed flask. Heat occasionally to make a clear solution (nitrogen gas is vigorously evolved during dissolution). Now reflux for about 1.5 hours, cool, filter the furil and recrytalline from methanol.

**Table 6.4 (a)** Phenols (Liquids)

*Note* *This table includes phenols, and substituted phenols e.g. simple, halogen and nitrosubstituted phenols*

| *Name and Structure* | *B.P (°C)* | *Properties* | *Derivatives* | | | | *Specific tests and derivatives* |
|---|---|---|---|---|---|---|---|
| | | | *Benzoate* | *3, 5-Dinitro benzoate* | *P-Toluene sulphonate* | *Bromo derivative* | |
| o-Chlorophenol 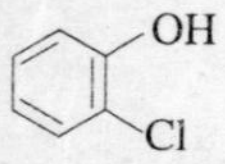 | 175 | Colourless, sparingly soluble in $H_2O$, soluble in alcohol | — | 143 | 74 | 95 | - p-Nitrobenzoate, m.p. 115°C.<br>- Nitration ($HNO_3$ in acetic acid)—dinitro derivative, m.p. 115°C.<br>- Picrate, m.p. 81°C. |
| Phenol 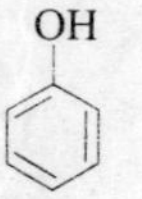 | 181 m.p. 42° | Colourless or pinkish, deliquescent, carbolic odour, sparingly soluble in $H_2O$, soluble in alcohol | 69 | 145 | 96 | 95 (tri) | - p-Nitrobenzoate, m.p. 126°C.<br>- Nitration gives Picric acid, m.p. 122°C.<br>- Violet $FeCl_3$ test.<br>- On warming with phthalic anhydride + one drop of conc. $H_2SO_4$ and pouring in alkaline solution gives red colour (phenolphthalein).<br>- Picrate m.p. 83°C |
| o-Cresol 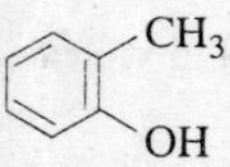 | 190 m.p. 31° | Colourless or light pink, carbolic odour, sparingly soluble in $H_2O$, soluble in alcohol. | — | 138 | 55 | 56 (di) | - Gives violet colour with $FeCl_3$ (aq.).<br>- Picrate, m.p. 88°C. |
| o-Bromophenol 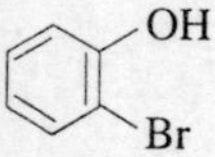 | 195 | Colourless, sparingly soluble in $H_2O$, soluble in alcohol | | | 78 | 95 | - On nitration ($HNO_3$ in $CH_3COOH$) gives dinitro derivative, m.p. 118°C. |
| Salicylaldehyde 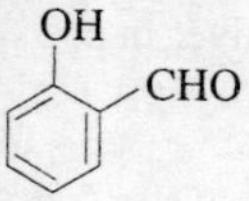 | 196 | See carbonyl compounds (Table 6.3a) | | | | | |
| p-Cresol 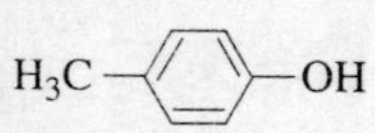 | 201 m.p. 36° | Colourless or pink colour; carbolic odour, sparingly soluble in $H_2O$, soluble in alcohol | 70 | 188 | 70 | 199 (tetra) 49 (di) | - Gives blue colour with aq. $FeCl_3$ solution. |
| m-Cresol 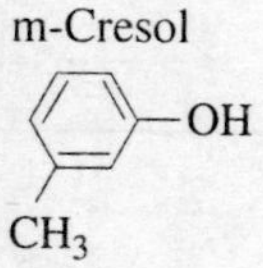 | 203 m.p. 12° | Colourless or pinkish, carbolic odour, sparingly soluble in $H_2O$, soluble in alcohol | 55 | 165 | 51 | 84 (tri) | - Gives bluish violet colour with aq. $FeCl_3$.<br>- Phthalein test-bluish purple colour.<br>- Nitration (conc. $HNO_3$ + conc. $H_2SO_4$)—trinitro derivative m.p. 109°C. |

| *Name and structure* | *B.P (°C)* | *Properties* | *Derivative* | | | *Bromo derivative* | *Specific tests and derivatives* |
|---|---|---|---|---|---|---|---|
| | | | *Benzoate* | *3,5-Dinitro benzoate* | *P-Toluene sulphonate* | | |
| o-Methoxyphenol (Guaicol) 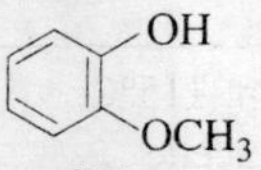 | 205 m.p. 28° | Colourless, tarry odour, sparingly soluble in $H_2O$, soluble in alcohol | 57 | 141 | 85 | 116 (tri) | - Gives greenish blue colour with alcoholic $FeCl_3$ solution.<br>- p-Nitrobenzoate, m.p. 93°C. |
| m-Chlorophenol 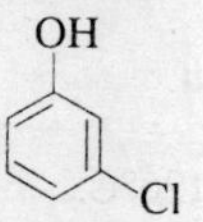 | 214 m.p. 32° | Colourless, sparingly soluble in $H_2O$, soluble in alcohol | 71 | 156 | — | — | - Gives violet colour with aq. $FeCl_3$ solution. |
| p-Chlorophenol 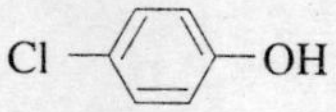 | 217 m.p. 38 | | 88 | 186 | 71 | — | - Nitration ( conc. $HNO_3$)— dinitro derivative m.p. 81°C. |
| o-Hydroxyacetophenone 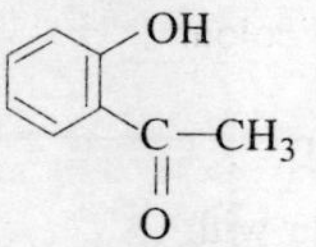 | 215 | See carbonyl compounds (Table 6.3a) | | | | | |
| Methylsalicylate 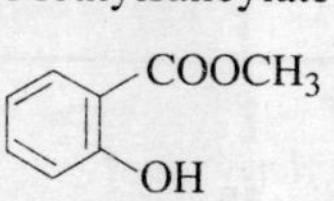 | 224 | See esters (Table 6.7a) | 92 | | | | |
| Ethyl salicylate 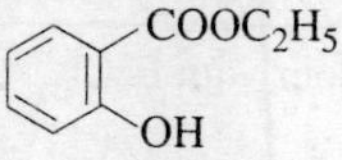 | 234 | See esters (Table 6.7a) | 79 | | | | |
| m-Bromophenol 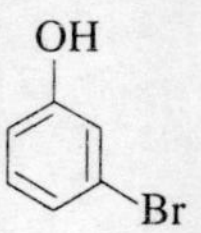 | 236 m.p. 32° | Colourless, sparingly soluble in $H_2O$, soluble in alcohol | 86 | — | 53 | — | - Gives violet colour with aq. $FeCl_3$ solution. |
| Carvacrol 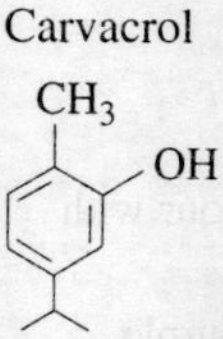 | 237 | Colourless, insoluble in $H_2O$, soluble in alcohol | — | 78 | — | 46 | - p-Nitrobenzoate, m.p. 51°C. |
| m-Methoxyphenol 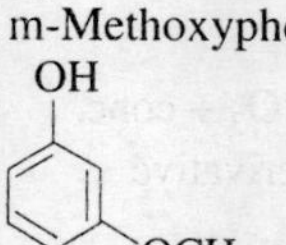 | 243 | | | | | 104 (tribromo) | - Gives faint violet colour with $FeCl_3$ solution. |

| *Name and structure* | *B.P (°C)* | *Properties* | *Derivative* | | | | *Specific tests and derivatives* |
|---|---|---|---|---|---|---|---|
| | | | *Benzoate* | *3,5-Dinitro benzoate* | *P-Toluene sulphonate* | *Bromo derivative* | |
| Eugenol (4-allyl-2-methoxy phenol) 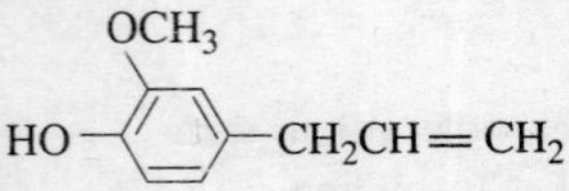 | 255 | Colourless, insoluble in $H_2O$, soluble in alcohol, odour of cloves | 70 | 131 | 85 | 118 | - Gives blue ferric reaction. - Picrate, m.p. 81°C. |
| Isoeugenol | 266 | Colourless, insoluble in $H_2O$, soluble in alcohol, odour of cloves | 103 | 158 | | 94 | - Gives green colour with neutral alcoholic $FeCl_3$. - Picrate, m.p. 109°C. |
| m-Acetoxyphenol (Resorcinol monoacetate) 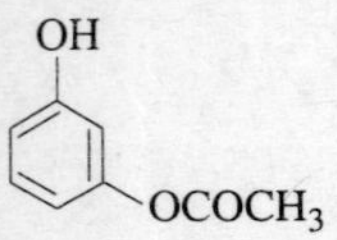 | 283 | | | | | | |

**Table 6.1 (b)** Phenols (Solids)

| *Name and Structure* | *M.P (°C)* | *Properties* | *Derivatives* | | | | *Specific tests and derivatives* |
|---|---|---|---|---|---|---|---|
| | | | *Benzoate* | *3, 5-Dinitro benzoate* | *P-Toluene sulphonate* | *Bromo derivative* | |
| 2, 4-Dibromophenol 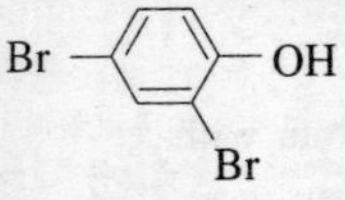 | 40 | Colourless, insoluble in $H_2O$, soluble in alcohol | 97 | | 120 | 95 | - Gives violet colour with $FeCl_3$ solution. |
| m-Iodophenol 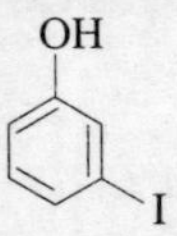 | 40 | | 73 | 183 | | | |
| Phenylsalicylate (Salol) 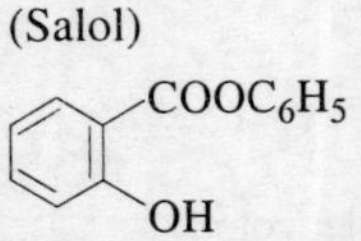 | 42 | See esters (Table 6.7b) | 80 | | | | |
| 2, 4-Dichlorophenol 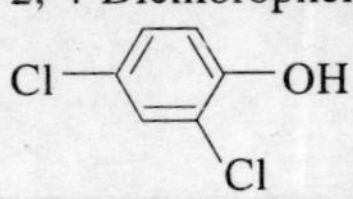 | 45 b.p. 209° | Colourless, insoluble in $H_2O$, soluble in alcohol | 97 | | 125 | 68 | |

| *Name and structure* | *M.P (°C)* | *Properties* | *Derivative* Benzoate | *Derivative* 3,5-Dinitro benzoate | *Derivative* P-Toluene sulphonate | *Derivative* Bromo derivative | *Specific tests and derivatives* |
|---|---|---|---|---|---|---|---|
| o-Nitrophenol 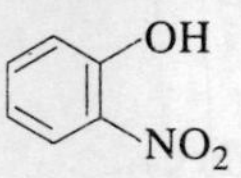  | 45 b.p. 215 | Yellow, tarry odour, insoluble in $H_2O$, soluble in hot $H_2O$, alcohol | 59 | 155 | 83 | 117(di) | - Dissolves in dil NaOH to give orange solution.<br>- With Zn dust + dil. $CaCl_2$ gives o-aminophenol, m.p. 174°C.<br>- On warming with 60% $HNO_3$ gives 2, 4-dinitrophenol, m.p. 114°C. |
| 4-Ethyl phenol 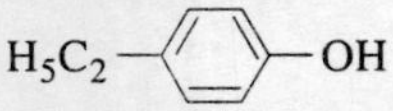  | 46 | | 60 | 133 | | | |
| 2, 6-Dimethylphenol 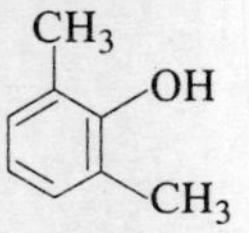  | 49 | | — | 159 | — | 79 (di) | |
| Thymol 2-Isopropyl-5-methyl phenol 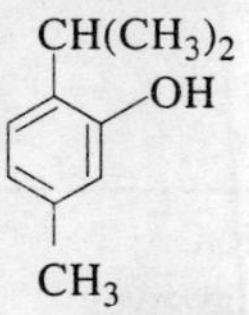  | 50 b.p. 230 | Colourless, characteristic odour, sparingly soluble in hot $H_2O$ | 32 | 103 | 71 | 55 | - Alcoholic $FeCl_3$ gives transient green colour.<br>- On boiling with $MnO_2$ + dil $H_2SO_4$ gives thymoquinone, m.p. 45°C. |
| p-Methoxyphenol 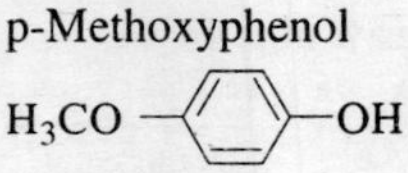  | 56 b.p. 243 | Colourless, odourless, soluble in $H_2O$, alcohol | 87 | — | — | — | - With ammonical $AgNO_3$ on heating gives grey or black ppt. |
| Orcinol 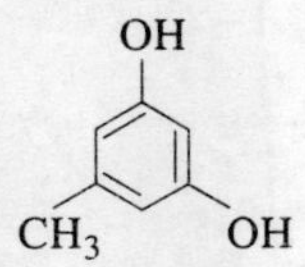  | 58 (hydrated) | Colourless, odourless, soluble in $H_2O$, alcohol | 88 (di) | 190 | — | 104 (tri) | - With aq. $FeCl_3$ gives blue violet colour.<br>- Reduces ammonical $AgNO_3$ on heating.<br>- Solution in $NH_4OH$ turns red on standing in air. |
| 2-Hydroxybiphenyl (O-Phenyl phenol) 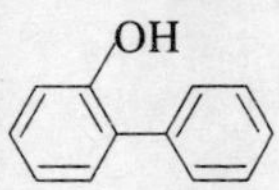  | 58 | | 76 | — | 65 | — | - Acetate, m.p. 62°C. |
| 3, 4-Dimethylphenol 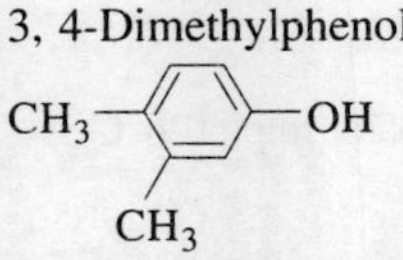  | 62 | | 58 | 181 | — | 171 (tri) | |

| *Name and structure* | *M.P (°C)* | *Properties* | *Derivative* | | | *Bromo derivative* | *Specific tests and derivatives* |
|---|---|---|---|---|---|---|---|
| | | | *Benzoate* | *3,5-Dinitro benzoate* | *P-Toluene sulphonate* | | |
| p-Bromophenol 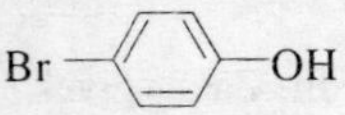 | 64 | Colourless, sparingly soluble in $H_2O$, soluble in alcohol | 102 | 191 | 94 | 95 | - Gives violet colour with aq. $FeCl_3$. |
| 2, 4, 6-Trichlorophenol 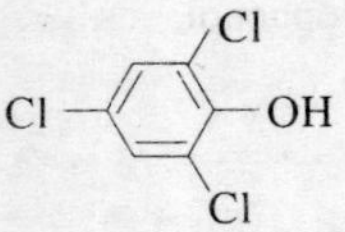 | 68 | Colourless, insoluble in $H_2O$, soluble in alcohol, ether | 75 | 136 | — | — | - Gives no colour with $FeCl_3$.<br>- p-Nitrobenzoate, m.p. 106°C. |
| 1, 3, 5-Xylenol (3, 5-dimethyl phenol) 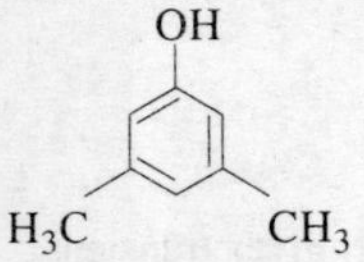 | 68 | | — | 195 | 83 | 166 (tri) | |
| Methyl m-hydroxy benzoate 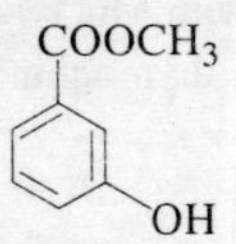 | 70 | See esters (Table 6.7b) | | | | | |
| Pseudocumenol (2, 4, 5-trimethyl-phenol) 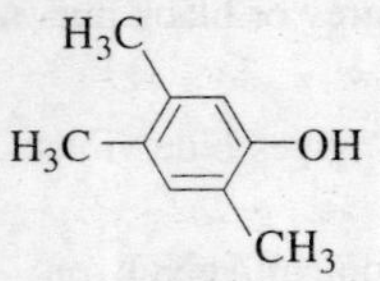 | 71 (b.p. 230) | | 63 | — | — | 35 (mono) | - Negative ferric reaction.<br>- With fuming nitric acid gives mono nitroderiv, m.p. 48°C. |
| Ethyl m-hydroxy benzoate 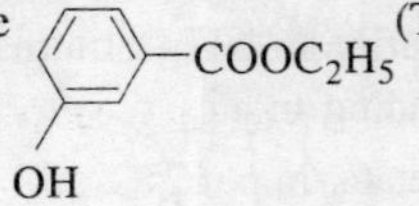 | 72 | See esters (Table 6.7b) | | | | | |
| 2, 5-Dimethylphenol 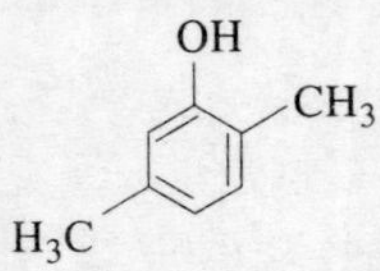 | 75 | | 61 | 137 | — | 178 (Tri) | |
| 2, 3-Dimethylphenol 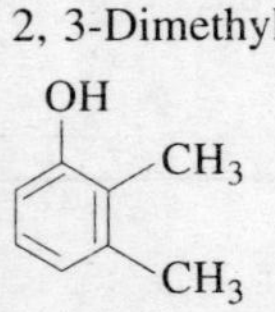 | 75 | | — | — | — | — | - p-Nitrobenzoate, m.p. 126°C |

| *Name and structure* | *M.P (°C)* | *Properties* | *Derivative* Benzoate | *Derivative* 3,5-Dinitro benzoate | *Derivative* P-Toluene sulphonate | *Bromo derivative* | *Specific tests and derivatives* |
|---|---|---|---|---|---|---|---|
| p-Dimethylamino-phenol 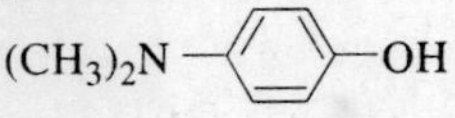 | 75 | Colourless, soluble in alcohol, ether | 158 | — | 130 | — | - With aq. $FeCl_3$ no colour but smells like quinone on warming.<br>- With NaOH gives dark brown solution with unplesant odour.<br>- On treatment with NaOAc + $Ac_2O$ in HCl at room temp gives acetate, m.p. 78°C. |
| 8-Hydroxyqui-noline 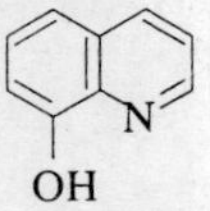 | 76 | Soluble in dil. HCl and dil. NaOH solution | 118 | — | 115 | — | - Picrate, m.p. 204°C. |
| 2, 4-Diaminophenol 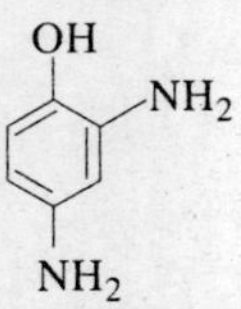 | 79 (d) | Colourless, soluble in $H_2O$ | 231 (tribenzoate) | — | — | — | - Gives intense red colour with aq. $FeCl_3$ solution.<br>- On treatment with NaOAc + $Ac_2O$ in HCl at room temp gives triacetate, m.p. 180°C. |
| Vanillin | 81 | See carbonyl compounds (Table 6.3b) | | | | | |
| p-Methylamino-phenol 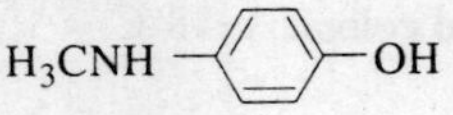 | 85 | Colourless, sparingly soluble in $H_2O$, soluble in ether | 175 | — | — | — | - With aq. $FeCl_3$ gives purple colour which on warming gives yellow colour of quinone.<br>- On warming with NaOH solution gives dark brown colour with unpleasant odour.<br>- On treatment with NaOAc+ $Ac_2O$ in presence of HCl at room temp. gives monoacetate, m.p. 240°C. |
| $\alpha$-Naphthol 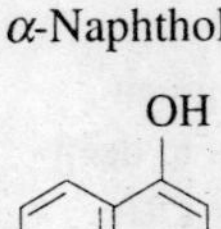 | 94 | Colourless or violet, faint odour, sparingly soluble in alcohol | 56 | 217 | 88 | 105 (di) | - Gives white ppt with aq $FeCl_3$ solution.<br>- On warming with dil. NaOH and pure $CCl_4$ + Cu powder gives blue colour.<br>- On adding dil. $HNO_3$ to a solution of conc. $H_2SO_4$ and warming Yields 2, 4-dinitro-derivative, m.p. 138°C.<br>- Picrate, m.p. 189°C. |

| *Name and structure* | *M.P (°C)* | *Properties* | *Derivative* *Benzoate* | *3,5-Dinitro benzoate* | *P-Toluene sulphonate* | *Bromo derivative* | *Specific tests and derivatives* |
|---|---|---|---|---|---|---|---|
| p-Iodophenol 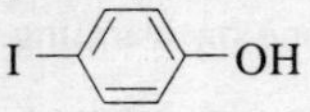 | 94 | | 119 | — | 99 | — | |
| 2-Naphthylsalicylate | 95 | See ester (Table 6.7b) | | | | | |
| 2, 4, 6-Tribromophenol 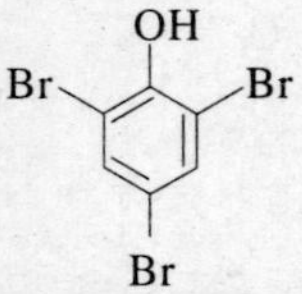 | 95 | Colourless, insoluble in $H_2O$, soluble in alcohol, ether | 81 | 174 | 113 | 120 (tetra) | - Acetate, m.p. 82°C. |
| m-Hydroxyacetophenone | 96 | See Carbonyl compounds (Table 6.3b) | | | | | |
| o-Methylaminophenol 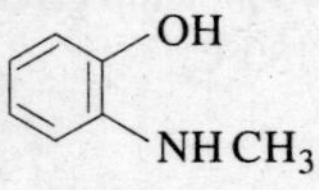 | 96 | Sparingly soluble in $H_2O$, soluble in alcohol | 150 (di) | | | | - Purple colour will $FeCl_3$.<br>- Acetyl derivative, m.p. 150°C (mono). |
| m-Nitrophenol 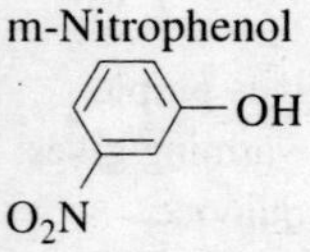 | 97 | Pale yellow, sparingly soluble in hot $H_2O$, soluble in alcohol, ether | 95 | 159 | 113 | 91 (di) | - On heating with aq. $FeCl_3$ solution gives violet red colour.<br>- With NaOH solution gives orange red colour. |
| p-tert-Butylphenol 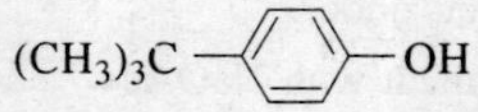 | 98 | | 83 | 156 | — | 50 | |
| Catechol 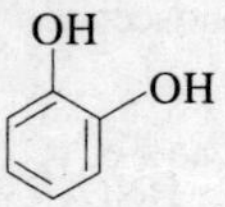 | 104 | Colourless, faint carbolic odour, soluble in $H_2O$, benzene, ether | 84 (di) | 152 | — | 192 (tetra) | - Gives red colour on heating with Fehling solution.<br>- Reduces Tollens reagent.<br>- Diacetate, m.p. 63°C.<br>- With aq. $FeCl_3$ gives green colour, which changes to deep red on addition of $NH_4OH$ solution. |
| Chlorohydroquinone 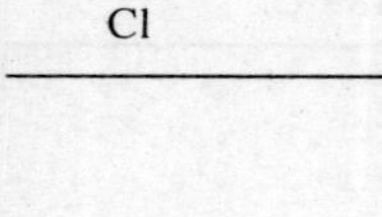 | 106 | Soluble in cold $H_2O$, sparingly soluble in benzene and $CCl_4$ | 130 (di) | — | — | — | - On oxidn ($K_2Cr_2O_7 + H_2SO_4$) in cold yields chloroquinone, m.p. 57°C.<br>- Diacetate, m.p. 72°C. |

| *Name and structure* | *M.P (°C)* | *Properties* | *Derivative* | | | | *Specific tests and derivatives* |
|---|---|---|---|---|---|---|---|
| | | | *Benzoate* | *3,5-Dinitro benzoate* | *P-Toluene sulphonate* | *Bromo derivative* | |
| Orcinol (anhydrous) | 107 | See orcinol hydrated m.p. 58°C | | | | | |
| p-Hydroxyacetophenone 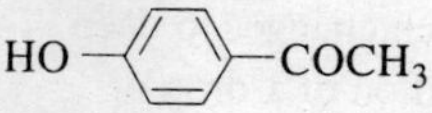 | 109 | See carbonyl compounds (Table 6.3b) | | | | | |
| Resorcinol 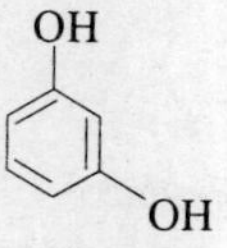 | 110 | Colourless or pinkish, soluble in $H_2O$, alcohol, ether | 117 (di) | 201 | 80 | 112 (di) (4, 6) | - With aq. $FeCl_3$ gives blue violet colour.<br>- Phthalein test- yellowish green fluorescence<br>- On heating with aq. NaOH and chloroform gives red or violet-red colour. |
| Bromohydroquinone 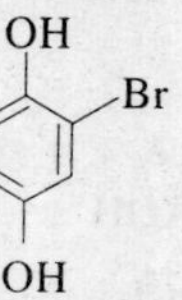 | 110 | Soluble in $H_2O$, sparingly soluble in $CHCl_3$ | | | | 186 (di) | - With aq. $FeCl_3$ gives bromoquinone, m.p. 55°C.<br>- Diacetate, m.p. 72°C. |
| p-Nitrophenol 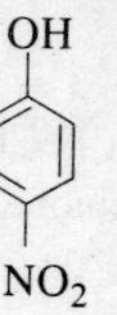 | 114 | Colourless or pale yellow, soluble in hot $H_2O$, alcohol, ether | 142 | 188 | 97 | 142 (di) (2, 6) | - With aq. NaOH gives deep yellow colour.<br>- With aq. $FeCl_3$ solution on warming gives violet-red colour.<br>- On nitration (conc. $HNO_3$ + conc. $H_2SO_4$) yields picric acid, m.p. 122°C.<br>- Acetate, m.p. 81°C. |
| 2, 4-Dinitrophenol 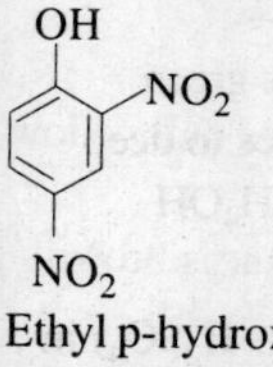 | 114 | | | | | 118 | - Picric acid, m.p. 122°C. (On nitration with conc. $HNO_3$+ conc. $H_2SO_4$) |
| Ethyl p-hydroxybenzoate 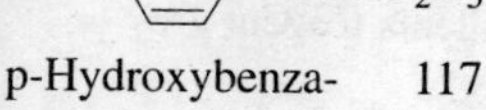 | 116 | See esters (Table 6.7b) | | | | | |
| p-Hydroxybenzaldehyde  | 117 | See carbonyl compounds (Table 6.3b) | | | | | |

| *Name and structure* | *M.P (°C)* | *Properties* | *Derivative* Benzoate | *Derivative* 3,5-Dinitro benzoate | *Derivative* P-Toluene sulphonate | *Bromo derivative* | *Specific tests and derivatives* |
|---|---|---|---|---|---|---|---|
| 5- Sulphosalicylic acid | 120 | See sulphonic acids (Table 6.26) | | | | | |
| S-Trinitrophenol (Picric acid) 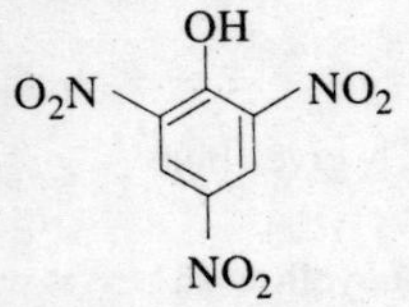 | 122 | Yellow, sparingly soluble in hot $H_2O$, soluble in alcohol, benzene | — | — | — | — | - With aq. NaOH (heated just to boil) gives deep yellow colour which changes to deep red on addition of 1 drop of ammonium sulphide solution.<br>- Acetate, m.p. 76°C. |
| m-aminophenol 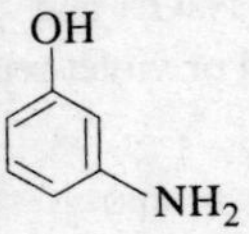 | 122 | Colourless, sparingly soluble in $H_2O$, benzene, soluble in hot $H_2O$, alcohol | 153 (di) | — | 157 | — | - On treatment with NaOAc + HOAc in presence of HCl at room temp. gives mono acetate, m.p. 148°C.<br>- Diacetate, m.p. 101°C.<br>- On heating with aq. $FeCl_3$ gives dark solution. |
| β-Naphthol 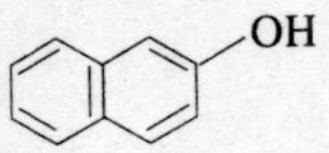 | 122 | Colourless or pinkish brown, faint odour, insoluble in $H_2O$, sparingly soluble in hot $H_2O$, soluble in alcohol, ether, benzene | 107 | 210 | 125 | 84 (1-bromo) | - With $FeCl_3$ solution gives white opalescence.<br>- On warming in conc. KOH solution with $CHCl_3$ blue colour is obtained.<br>- Acetate, m.p. 70°C.<br>- Picrate, m.p. 156°C. |
| Toluhydroquinone 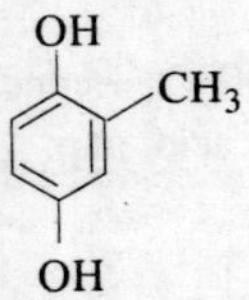 | 124 | Soluble in $H_2O$ | 119 (Di) | | | | - Gives blue colour with aq. $FeCl_3$<br>- On oxidation gives toluquinone, m.p. 68°C.<br>- Diacetate, m.p. 52°C |
| Methyl p-hydroxy benzoate | 131 | See esters (Table 6.7b) | | | | | |
| Pyrogallol 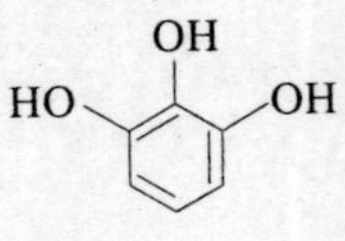 | 133 | Colourless, darkens on exposure to air, soluble in $H_2O$, alcohol, ether | 90 (tri) | 205 | — | 158 | - With aq. $FeCl_3$ gives dull yellow colour.<br>- Soluble in NaOH, turns brown in air.<br>- with $FeSO_4$ solution gives blue ppt.<br>- With Tollen's reagent grey or black ppt.<br>- Triacetate, m.p. 165°C. |

| *Name and structure* | *M.P (°C)* | *Properties* | *Benzoate* | *3,5-Dinitro benzoate* | *P-Toluene sulphonate* | *Bromo derivative* | *Specific tests and derivatives* |
|---|---|---|---|---|---|---|---|
| 2, 4-Dinitro-1-naphthol 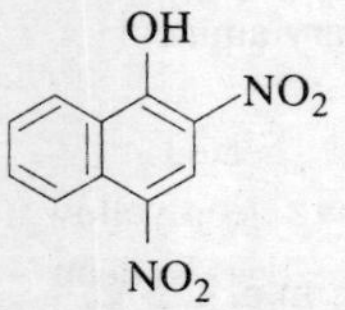 | 138 | Yellow, insoluble in $H_2O$, sparingly soluble in alcohol, ether and benzene | 174 | | | | - On reduction (Sn + HCl)—2, 4-diamino-$\alpha$-naphthol (triacetyl deriv, of reduction product m.p. 280°C (d). |
| 1, 2, 4-Trihydroxybenzene (Hydroxy hydroquinone) 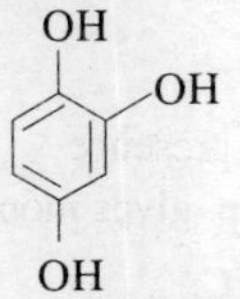 | 140 | | 120 (tri) | | | | - Triacetate, m.p. 96°C. |
| 3- Nitrosalicylic acid (anhydr.) | 148 | See carboxylic acids (Table 6.2b) | | | | | |
| 2, 4, 6-Triiodophenol 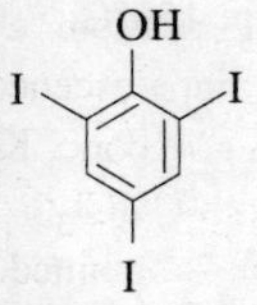 | 158 | | 137 | | | | - Acetate, m.p. 156°C. |
| Salicylic acid | 159 | See carboxylic acids (Table 6.2b) | | | | | |
| 4-Hydroxybiphenyl (p-Phenyl phenol) 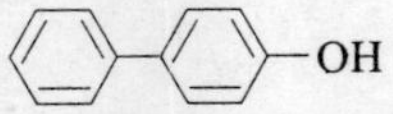 | 165 | Colourless, insoluble in $H_2O$, soluble in alcohol | 149 | — | 179 | — | |
| Hydroquinone (quinol) 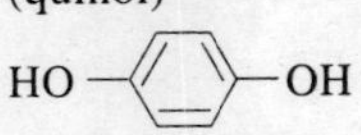 | 169 | Colourless, soluble in $H_2O$, alcohol, ether, insoluble in benzene | 205 (di) | 317 | 159 | 186 (di) | - Gives transient blue colour with aq. $FeCl_3$.<br>- Yields benzoquinone, m.p. 116°C, on warming with excess conc. $FeCl_3$ solution.<br>- Diacetate, m.p. 123°C. |
| $\alpha$-Naphthol-4-sulphonic acid | 170 (d) | See sulphonic acids (Table 6.26) | | | | | |
| 3, 5-Dinitrosalicylic acid | 173 | See carboxylic acids (Table 6.2b) | | | | | |
| o-Aminophenol 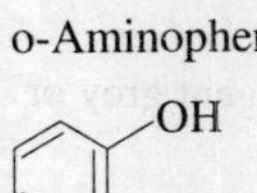 | 174 (d) | Colourless, sparingly soluble in $H_2O$, alcohol and ether | 183 (di) | — | 146 | — | - Gives dark brown ppt. with aq. $FeCl_3$.<br>- On treatment with NaOAc + acetic acid in HCl at room temp. gives monoacetate, m.p. 201°C. |

| *Name and structure* | *M.P (°C)* | *Properties* | *Derivative* *Benzoate* | *3,5-Dinitro benzoate* | *P-Toluene sulphonate* | *Bromo derivative* | *Specific tests and derivatives* |
|---|---|---|---|---|---|---|---|
| 4-Amino-2-methylphenol 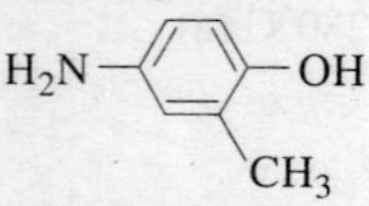 | 175 | Soluble in water and ether | 194 (di) | — | 109 | — | - Acetate, m.p. 103° (di). - (Also see primary amines) (Table 6.15b) |
| 2, 4, 6-Trinitroresorcinol (Styphinic acid) 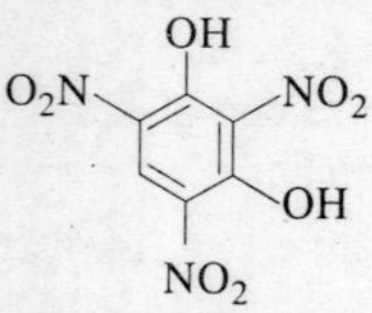 | 176 | | | | | | - Dimethyl ether, m.p. 125°C. |
| p-Aminophenol 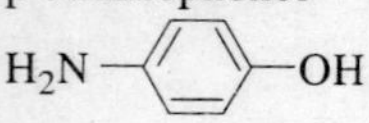 | 184 (d) | Colourless, insoluble in $H_2O$, sparingly soluble in alcohol | 234 (di) | — | — | — | - HCl salt, m.p. 306°C. - Diacetate, m.p. 150°C. - On treatment with NaOAc + HOAc in HCl at room temp. gives monoacetate m.p. 168°C. |
| Pentachlorophenol 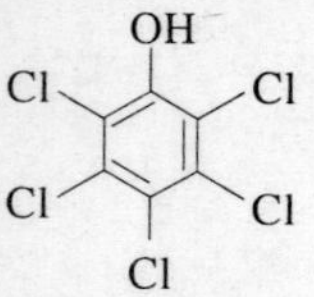 | 190 | | 159 | — | 145 | — | - Acetate, m.p. 150°C. |
| Carbostyril (2-Hydroxyquinoline) 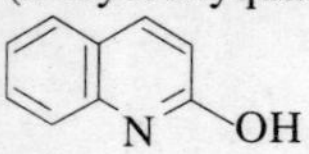 | 199 | Insoluble in cold $H_2O$ | — | — | 82 | — | - On Nitration (conc. $HNO_3$ + conc. $H_2SO_4$) -gives 6-nitro derivative, m.p. 280°C. |
| m-Hydroxybenzoic acid | 201 | See carboxylic acid (Table 6.2b) | | | | | |
| p-Hydroxybenzoic acid | 213 | See carboxylic acids (Table 6.2b) | | | | | |
| 4-Amino-1-naphthol 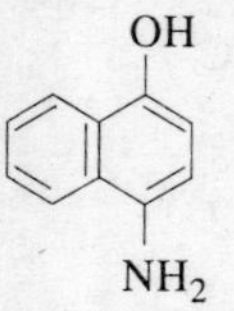 | d | | 215 (di) | | | | - Diacetate, m.p. 158°C. - Hydrochloride salt m.p.-280°C. |
| 1-Amino-2-naphthol 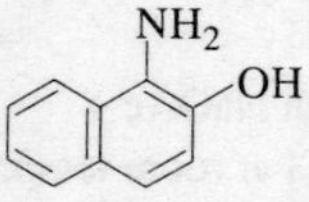 | d | | 235 (di) | | | | - Diacetate, m.p. 206°C. - Picrate, m.p. 110°C. |

| *Name and structure* | *M.P (°C)* | *Properties* | *Derivative* | | | *Bromo derivative* | *Specific tests and derivatives* |
|---|---|---|---|---|---|---|---|
| | | | *Benzoate* | *3,5-Dinitro benzoate* | *P-Toluene sulphonate* | | |
| Phloroglucinol 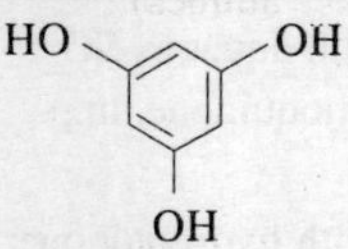 | 218 | Colourless, soluble in $H_2O$, alcohol, ether | 174 (tri) | 162 | — | 151 (tri) | - With $FeCl_3$ solution gives transient violet colour.<br>- On nitration—trinitro derivative m.p. 165°C.<br>- Triacetate, m.p. 105°C |
| Phenolphthalein 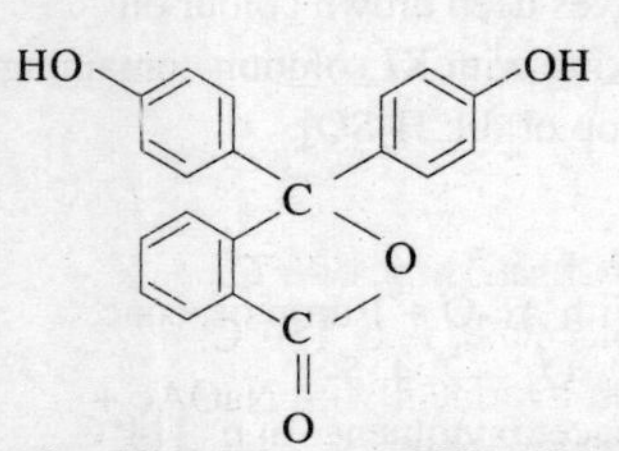 | 254 (261) | Colourless, insoluble in $H_2O$, soluble in alcohol | 169 (di) | — | — | — | - With NaOH solution gives red colour.The colour is discharged by addition of excess conc. NaOH solution.<br>- Diacetyl derivative, m.p. 143°C. |
| Alizarin | 289 | See quinones (Table 6.5) | | | | | |

**Table 6.5** Quinones (Solids)

| *Name and structure* | *M.P (°C)* | *Properties* | *Derivatives: 2,4-dinitro-phenyl-hydrazone* | *Derivatives: Semi-carba-zone* | *Derivatives: Oxime* | *Specific tests and derivatives* |
|---|---|---|---|---|---|---|
| Thymoquinone 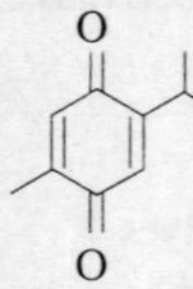 | 45 b.p. 232 | Yellow, steam volatile, sparingly soluble in $H_2O$ | 179 (mono) | 204 (mono) | 161 (mono) | - On reduction with aqueous $SO_2$, Gives hydrothymoquinone, m.p. 139°C.<br>- Forms adduct with hydroquinone, m.p. 137°C. |
| o-Benzoquinone 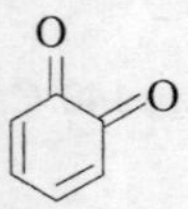 | 60–65 | Red, sparingly soluble in $H_2O$ | — | | 142 (di) | - It gives deep brown colour on shaking with KI solution containing a drop of dil. $H_2SO_4$. |
| Toluquinone 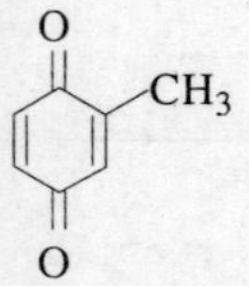 | 68 | Yellow, characteristic odour resembling that of benzoquinone, sparingly sol. in $H_2O$ | 269 (di) | 179 | 135 | - With $Ac_2O$ + 1 drop of conc. $H_2SO_4$—2, 4, 5-triacetoxytoluene. m.p. 114°C. |
| 2-Methyl-1, 4-naphthaquinone 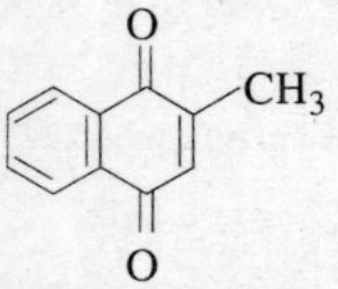 | 106 | Yellow, insoluble in $H_2O$ | — | 247 | 169 | - On reduction with $SO_2$ (aqueous) gives the corresponding quinol, m.p. 170°C. |
| p-Benzoquinone 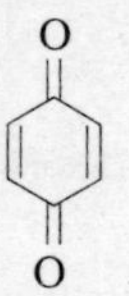 | 115 | Yellow, characteristic pungent odour, sparingly soluble in $H_2O$, soluble in dil. NaOH solution which turns brown on exposure to air. | — | 243 (di) | 240 (di) | - Reduces ammonical $AgNO_3$ solution in cold<br>- With $FeSO_4$ + dil. $H_2SO_4$ gives quinhydrone. m.p. 171°C.<br>- On warming with $Ac_2O$ + 1 drop conc. $H_2SO_4$, gives hydroxyhydro-quinone triacetate, m.p. 96°C.<br>- Liberates iodine from acidified KI solution. |
| 1, 2-Naphaquinone (β-Naphthoquinone) 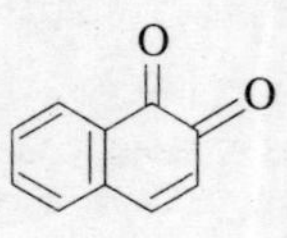 | 115-120 | Red, insoluble in $H_2O$ | — | 184 | 162 | - Gives green colour with conc. $H_2SO_4$.<br>- On oxidation with $KMnO_4$ gives phthalic acid, m.p. 195°C.<br>- With $Ac_2O$ + conc. $H_2SO_4$ Warm 1, 2, 4-triacetoxynaphthalene, m.p. 135°C.<br>- On warming with $NH_2OH$, HCl in alcohol gives β-nitroso-α-naphthol, m.p. 103°(d). |

| *Name and structure* | *M.P (°C)* | *Properties* | *Derivatives* | | | *Specific tests and derivatives* |
|---|---|---|---|---|---|---|
| | | | *2,4-dinitro phenyl-hydrazone* | *Semi-carba-zone* | *Oxime* | |
| 1, 4-Naphthaquinone ($\alpha$-Naphthaquinone) 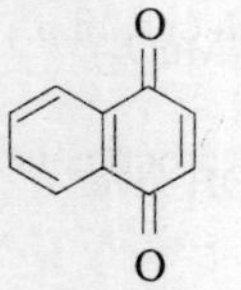 | 125 | Yellow, sparingly soluble in $H_2O$, soluble in dil. NaOH solution with red-brown colour, odour like quinone | | 247 | 198 | - With $Ac_2O$ + conc. $H_2SO_4$ warm 1, 2, 4-triacetoxynaphthalene, m.p. 135°C. |
| Quinhydrone 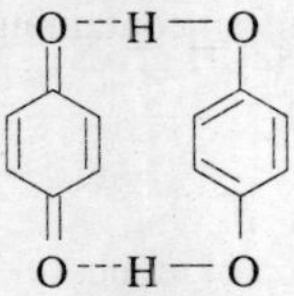 | 171 | Dark green, sparingly soluble in $H_2O$, soluble in alcohol and ether, giving yellow solution | — | — | — | - Reduces ammonical $AgNO_3$ sol.<br>- On reduction - Zn + HCl hydroquinone, m.p. 169<br>- On oxidn ($K_2Cr_2O_7$ + dil $H_2SO_4$), p-benzoquinone, m.p. 115. |
| Camphorquinone 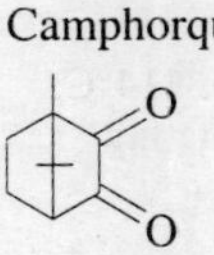 | 198 | Yellow, sparingly soluble in cold water, soluble in alcohol | 190 | 236 | 170 (di) | - On redn (Zn + HCl) - $\alpha$. hydroxy camphor, m.p. 203<br>- On long boiling with alc. KOH- camphoric acid m.p. 187.<br>- Monophenylhydrazone, m.p. 170 |
| Phenanthraquinone 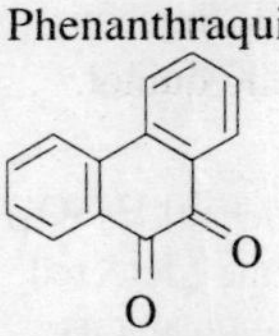 | 206 | Orange, sparingly soluble in $H_2O$ | — | 220 (di) | 162 | - On redn ($SO_2$ in warm alcohol)<br>- hydrophenanthraquinone, m.p. 147.<br>- On oxidn ($CrO_3$ or $KMnO_4$)<br>- diphenic acid, m.p. 228<br>- Di-p-nitrophenyhydrazone m.p. 245. |
| Anthraquinone 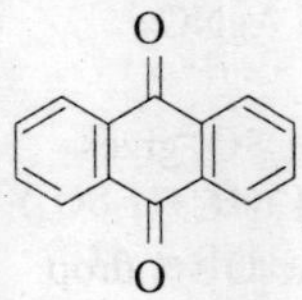 | 286 | Pale yellow, sparingly soluble in cold benzene and ether | — | — | 224 | - On heating with Zn dust yields anthracene. m.p. 216°C.<br>- On heating in alcohol with sodium hydrosulphite gives oxanthrol, m.p. 205°C(d)<br>- On warming with Zn dust $NH_4OH$ gives dihydroanthranol, m.p. 76°C.<br>- On heating with Zn dust and alkali a bright red colour is produced due to formation or anhydroanthraquinone; the colour disappears on shaking in presence of air due to regeneration of anthraquinone. |
| Alizarin (1, 2 Dihydroxy anthraquinone) 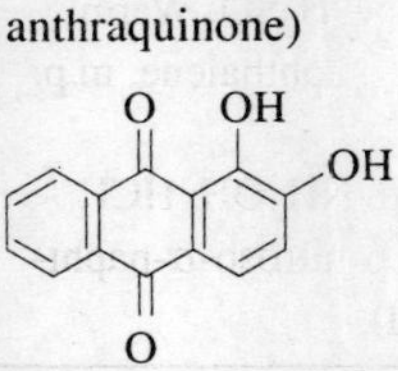 | 289 | | | | | - Sol. in conc. $H_2SO_4$ giving purple colour.<br>- $Ac_2O$+NaOAc yields mono acetyl, derivative, m.p. 201°C.<br>- Diacetate, m.p. 182°C. |

**Table 6.6 (a)** Alcohols (Liquids)

| *Name and structure* | *B.P. (°C)* | *Properties* | *3,5-dinitro benzoate* | *p-nitro-benzoate* | *Specific tests and derivatives* |
|---|---|---|---|---|---|
| Methyl alcohol $CH_3OH$ | 65 | Colourless, characteristic odour, soluble in $H_2O$ | 109 | 96 | - On oxidation (by a hot Cu wire) gives characteristic odour of formaldehyde. |
| Ethyl alcohol $C_2H_5OH$ | 78 | Colourless, sweet penetrating odour, soluble in $H_2O$ | 93 | 56 | - On oxidation (alkaline $KMnO_4$) gives acetic acid.<br>- With iodine and dil. NaOH gives iodoform, m.p. 119°C. |
| Isopropyl alcohol $(CH_3)_2CHOH$ | 82 | Colourless, alcoholic odour, soluble in $H_2O$. | 122 | 110 | - On oxidation ($K_2Cr_2O_7 + H_2SO_4$ gives acetone.<br>- With iodine and dil. NaCH gives iodoform, m.p. 119°C. |
| tert. Butyl alcohol $(CH_3)_3COH$ | 83 m.p.25 | Colourless, alcoholic odour, soluble in $H_2O$, alcohol | 142 | 116 | |
| Allyl alcohol $CH_2{=}CHCH_2OH$ | 97 | Colourless, irritating odour, soluble in alcohol | 50 | 28 | - Decolourises $Br_2$ water.<br>- On oxidation ($K_2Cr_2O_7$ + dil. $H_2SO_4$) gives pungent irritating odour of acrolein, b.p. 52°C. |
| n-Propyl alcohol $CH_3CH_2CH_2OH$ | 97 | Colourless, alcoholic odour, soluble in $H_2O$, alcohol | 74 | 35 | - On oxidation ($K_2Cr_2O_7$ + dil. $H_2SO_4$) —propionaldehyde<br>- On oxidation (alkaline $KMnO_4$)<br>- gives propionic acid. |
| sec. Butyl alcohol $CH_3CH(OH)CH_2CH_3$ | 99 | Colourless, alcoholic odour, soluble in $H_2O$, alcohol | 76 | 26 | - On oxidation ($K_2Cr_2O_7$ + dil $H_2SO_4$) gives ethyl methyl ketone (detected with sodium nitroprusside–NaOH solution—red colour) |
| tert. Amyl alcohol (2-Methylbutane-2-ol) $(CH_3)_2C(OH)CH_2CH_3$ | 102 | Colourless, alcoholic odour, sparingly soluble in $H_2O$, soluble in ether | 117 | 85 | |
| iso-Butyl alcohol $(CH_3)_2CHCH_2OH$ | 108 | Colourless, sparingly soluble in $H_2O$, soluble in alcohol | 88 | 69 | — On oxidation (dil. $K_2Cr_2O_7$+$H_2SO_4$) yields isobutyraldehyde, b.p. 63. |
| 3-Methylbutan-2-ol $(CH_3)_2CHCH\ (OH)CH_3$ | 112 | | 76 | — | — |
| Pentan-3-ol $CH_3CH_2CH(OH)$ $CH_2CH_3$ | 116 | | 100 | — | — |
| n-Butyl alcohol $CH_3CH_2CH_2CH_2OH$ | 118 | Colourless, characteristic odour, sparingly soluble in $H_2O$, soluble in alcohol, ether | 63 | 35 | — On oxidation (dil. $K_2Cr_2O_7$+$H_2SO_4$) yields n-butyraldehyde, b.p. 74°C. |
| 2, 3-Dimethyl-2-butanol | 118 | | 111 | 82 | |

| *Name and structure* | *B.P. (°C)* | *Properties* | *3,5-dinitro benzoate* | *p-nitro-benzoate* | *Specific tests and derivatives* |
|---|---|---|---|---|---|
| $(CH_3)_2C(OH)CH(CH_3)_2$ | | | | | |
| Sec-Amyl alcohol (Pentan-2-ol) $CH_3CH_2CH_2CHOHCH_3$ | 119 | | 62 | — | |
| 2-Methyl-2-pentanol $CH_3CH_2CH_2C(OH)(CH_3)_2$ | 121 | | 72 | 70 | |
| 2-Methoxyethanol (Ethylene glycol mono-methyl ether) $CH_3OCH_2CH_2OH$ | 124 | Soluble in $H_2O$ | — | 51 | |
| 3-Methyl-3-pentanol $CH_3CH_2C(OH)(CH_3)—CH_2CH_3$ | 128 | | 62 (79) | — | |
| 2-Methylbutan-1-ol (Active amyl alcohol) $CH_3CH_2CH(CH_3)CH_2OH$ | 129 | Sparingly soluble in $H_2O$ | 70 | — | |
| Ethylene chlorohydrin $ClCH_2CH_2OH$ | 129 | Colourless, soluble in $H_2O$, alcohol, ether | 92 | — | - On oxidation (alkaline $KMnO_4$) gives oxalic acid (identified by acidification and adding $CaCl_2$ solution-white ppt. of calcium oxalate is obtained. |
| iso-Amyl alcohol (3-Methylbutan-1-ol) $(CH_3)_2CH\ CH_2CH_2OH$ | 132 | Colourless, characteristic odour, sparingly soluble in $H_2O$, soluble in alcohol, ether | 62 | — | - On oxidation (dil $K_2Cr_2O_7$ + $H_2SO_4$) gives iso-valeraldehyde, b.p. 92°C. |
| 2-Ethoxy ethanol $C_2H_5OCH_2CH_2OH$ | 135 | Soluble in $H_2O$ | 76 | — | |
| n-Amyl alcohol (1-pentanol) $CH_3CH_2CH_2CH_2CH_2OH$ | 138 | Colourless, characteristic odour, insoluble in $H_2O$, soluble in ether, alcohol | 46 | | |
| 2-Hexanol $CH_3CH_2CH_2CH_2CH(OH)CH_3$ | 139 | | 39 | — | |
| Cyclopentanol 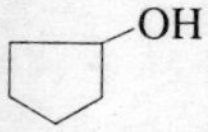 | 140 | | 115 | 62 | |
| Ethylene bromohydrin $BrCH_2CH_2OH$ | 149 | Colourless, soluble in $H_2O$, alcohol | 86 | — | |
| 4-Heptanol $CH_3CH_2CH_2CH(OH)CH_2CH_2CH_3$ | 156 | | 64 | 35 | |
| n-Hexyl alcohol | 157 | Colourless, almost | 58 | — | |

| *Name and structure* | *B.P. (°C)* | *Properties* | *3, 5-dinitro benzoate* | *p-nitro-benzoate* | *Specific tests and derivatives* |
|---|---|---|---|---|---|
| (1-Hexanol)<br>$CH_3(CH_2)_4CH_2OH$ | | insoluble in $H_2O$ | | | |
| Cyclohexanol<br>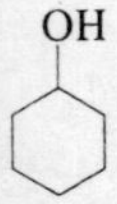 | 160<br>(m.p.<br>20) | Colourless, sparingly soluble in $H_2O$, soluble in alcohol, ether | 113 | 50 | - On oxidation (dil. $K_2Cr_2O_7 + H_2SO_4$) gives cyclohexanone b.p. 155°C.<br>- On oxidation (hot conc. $HNO_3$) gives adipic acid, m.p. 151°C. |
| Diacetone alcohol | 166 | See Carbonyl compounds (Table 6.3a) | | | |
| Furfuryl alcohol<br>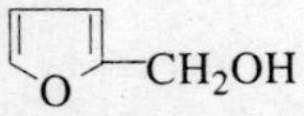 | 170 | Colourless, sparingly soluble in $H_2O$, soluble in alcohol | 80 | 76 | - On oxidation ($KMnO_4$) gives furoic acid, m.p. 133°C.<br>- Gives green colour with conc. HCl. |
| n-Heptanol<br>(1-Heptanol)<br>$CH_3(CH_2)_5CH_2OH$ | 176 | Colourless, insoluble in $H_2O$ | 48 | — | |
| Tetrahydrofurfuryl alcohol<br>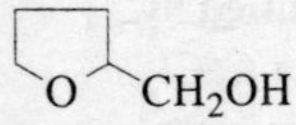 | 177 | Colourless, soluble in $H_2O$, alcohol | 84 | 47 | |
| Octan-2-ol<br>(capryl alcohol)<br>$CH_3(CH_2)_5$<br>$CHOH \cdot CH_3$ | 179 | Colourless, almost insoluble in $H_2O$ | 32 | 28 | - On oxidation (dil. $K_2Cr_2O_7$ + $H_2SO_4$) gives hexyl methyl ketone, b.p. 172°C. |
| Propylene glycol<br>(propan-1, 2 diol)<br>$CH_3CH(OH)CH_2OH$ | 188 | Colourless, soluble in $H_2O$ | 147<br>(di) | 127 | |
| n-Octanol<br>(1-Octanol)<br>$CH_3(CH_2)_6CH_2OH$ | 194 | Colourless, insoluble in $H_2O$ | 62 | — | |
| Ethylene glycol<br>$HOCH_2CH_2OH$ | 197 | Colourless, odourless, hygroscopic, soluble in $H_2O$, alcohol, insoluble in ether. | 169 | 140 | - On oxidation (conc. $HNO_3$) gives oxalic acid (tested by $CaCl_2$ solution<br>- white ppt.) |
| *l*-Linalool<br>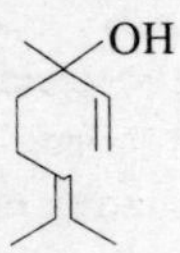 | 198 | Colourless, pleasant odour, insoluble in $H_2O$. | — | 70 | - Decolourises $KMnO_4$ solution. |
| Benzyl alcohol<br>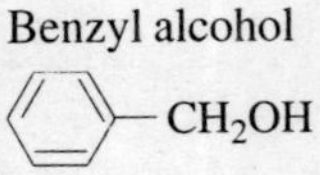 | 206 | Colourless, sparingly sol. in $H_2O$, soluble in alcohol, ether | 112 | 86 | - On oxidation ($KMnO_4$) gives benzoic acid, m.p. 122°C.<br>- On heating with dil. $HNO_3$, water bath, 2 min.) gives smell of bitter almonds (benzaldehyde). |

| *Name and structure* | *B.P. (°C)* | *Properties* | *3,5-dinitro benzoate* | *p-nitro- benzoate* | *Specific tests and derivatives* |
|---|---|---|---|---|---|
| Trimethylene glycol (propan-1, 3-diol) $CH_2OH\,CH_2CH_2OH$ | 216 | Colourless, soluble in $H_2O$ | 178 (di) | 119 | - Dibenzoate, m.p. 59°C. |
| β-Phenyl ethyl alcohol $C_6H_5CH_2CH_2OH$ | 219 | Colourless, soluble in $H_2O$, odour of rose oil | 108 | 62 | - On oxidation ($K_2Cr_2O_7$ + dil. $H_2SO_4$) gives phenyl acetic acid, m.p. 76°C. |
| (±) α-Terpineol 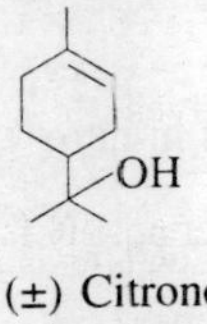 | 220 (m.p. 35°) | Colourless, pleasant odour of lime, insoluble in $H_2O$, soluble in alcohol, ether | 79 | 139 | - Decolourises $KMnO_4$ solution. |
| (±) Citronellol 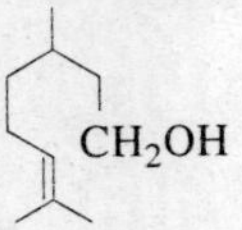 | 222 | Colourless, pleasant odour of lemon, insoluble in $H_2O$, soluble in alcohol, ether | — | — | - On oxidation first with alkaline $KMnO_4$ and then with $K_2Cr_2O_7$ + dil. $H_2SO_4$ yields β-methyladipic acid, m.p. 89°C.<br>- Decolourises $KMnO_4$ solution. |
| Geraniol 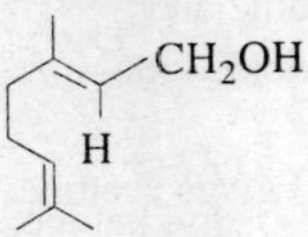 | 229 | Colourless, pleasant odour of rose and geranium, insoluble in $H_2O$, soluble in alcohol, ether | 63 | 35 | - Decolourises $KMnO_4$ solution.<br>- Reduces ammoniacaly $AgNO_3$ solution.<br>- On heating 2 drops with 2 drops of phenol and 2 drops of conc. $H_2SO_4$ at 120°, diluting and adding $NH_4OH$- red colour is formed. |
| n-Decanol $CH_3(CH_2)_8CH_2OH$ | 230 | | 57 | 30 | |
| γ-Phenyl propyl alcohol $C_6H_5CH_2CH_2\,CH_2OH$ | 237 | Colourless, sparingly soluble in $H_2O$ | 92 | 46 | |
| Diethylene glycol $(HOCH_2 \cdot CH_2)_2O$ | 245 | | 149 | | |
| Cinnamyl alcohol $C_6H_5$—CH= $CHCH_2OH$ | 257 (m.p. 33) | Colourless, insoluble in $H_2O$ soluble in alcohol | 121 | 78 | - Decolourlise $KMnO_4$ solution |
| Glycerol $CH_2OH$ \| $CHOH$ \| $CH_2OH$ | 290 (d) (m.p. 18) | Colourless, viscous, soluble in $H_2O$, alcohol, insoluble in ether | — | 188 | - On heating with $KHSO_4$ gives pungent irritating odour of acrolein.<br>- On heating with phenol and conc. $H_2SO_4$ (120°C) and then pouring in excess dil. $NH_4OH$ gives red colour.<br>- Reduces ammonical $AgNO_3$ solution. |

**Table 6.6(b)** Alcohols (Solids)

| *Name and structure* | *M.P. (°C)* | *Properties* | *3, 5-dinitro-benzoate* | *p-nitro-benzoate* | *Specific tests and derivatives* |
|---|---|---|---|---|---|
| Lactic acid | 18 b.p. 119/ 12 mm | See Carboxylic acids (Table 6.2b) | | | |
| Anisyl alcohol 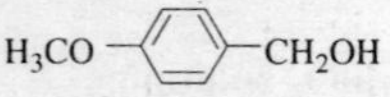  | 25 (b.p. 259) | | | 94 | - On oxidation ($KMnO_4$) gives anisic acid m.p. 184°C. |
| m-Nitrobenzyl alcohol 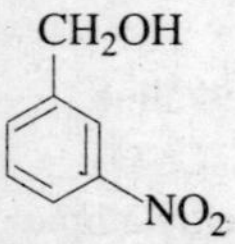  | 30 (b.p. 175-80 /30 mm) | Colourless | | | - Benzoate, m.p. 72°C.<br>- Oxidation ($KMnO_4$) gives m-nitro-benzoic acid, m.p. 141°C.<br>- On reduction (Zn dust + $CaCl_2$) solution gives m-amino benzyl alcohol, m.p. 97°C. |
| Cinnamyl alcohol | 33 | See liquids b.p. 257 | | | |
| (±) $\alpha$-Terpineol | 35 | See liquids, b.p. 220 | | | |
| Pinacol $(CH_3)_2C(OH)$—$C(OH)(CH_3)_2$ | 40 (b.p. 174) | Colourless, sparingly soluble in $H_2O$, readily soluble in hot $H_2O$, alcohol, ether | | | - On warming with iodine and dil. NaOH solution yields iodo-form, m.p. 119°C.<br>- On oxidation ($Cr_2O_3$) gives acetone, b.p. 56°C. |
| (–) Menthol 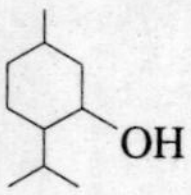  | 44 (b.p. 216) | Colourless, pepperamint odour, sparingly soluble in $H_2O$, soluble in alcohol, ether | 153 | 62 | - Benzoate, m.p. 54°C. |
| Cetyl alcohol $CH_3(CH_2)_{14}$ $CH_2OH$ | 50 (b.p. 189 15 mm) | | 66 | 52 | - On KOH fusion or on oxidation ($Cr_2O_3$ in acetic acid) yields palmitic acid, m.p. 62°C. |
| Diphenyl carbinol (Benzhydrol) $(C_6H_5)_2CHOH$ | 69 | | 141 | 131 | |
| o-Nitrobenzyl alcohol 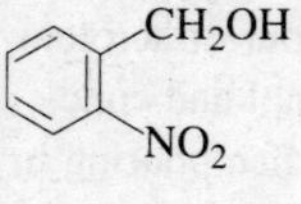  | 74 | Colourless, sparingly soluble in $H_2O$, soluble in alcohol, ether | | | - Benzoate, m.p. 101°C.<br>- On oxidation ($K_2Cr_2O_7$ + acid) gives o-nitrobenzoic acid, m.p. 144°C.<br>- On reduction (Zn-dust, $CaCl_2$ solution) gives o-amino benzyl alcohol m.p. 83°C. |

| *Name and structure* | *M.P.* | *Properties* | *3, 5-dinitro-benzoate* | *p-nitro-benzoate* | *Specific tests and derivatives* |
|---|---|---|---|---|---|
| Glycolic acid | 80 | See carboxylic acids (Table 6.2b) | | | |
| p-Nitrobenzyl alcohol 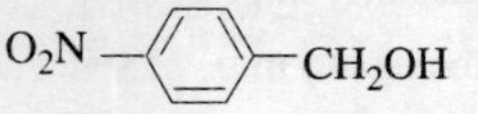 | 93 | Colourless, soluble in hot $H_2O$, alcohol, ether | | | - Benzoate, m.p. 94°C.<br>- Acetate, m.p. 78°C.<br>- On oxidation ($KMnO_4$) gives p-nitrobezoic acid, m.p. 240.<br>- On reduction (Zn dust + $CaCl_2$ solution) gives p-aminobenzyl alcohol, m.p. 65°C. |
| (–) Malic acid | 100 | See carboxylic acids (Table 6.2b) | | | |
| Citric acid (hyd.) | 100 | See carboxylic acids (Table 6.2b) | | | |
| (±) Mandelic acid | 118 | See carboxylic acids (Table 6.2b) | | | |
| Benzoin | 137 | See carbonyl compounds (Table 6.3b) | | | |
| (–) Cholesterol 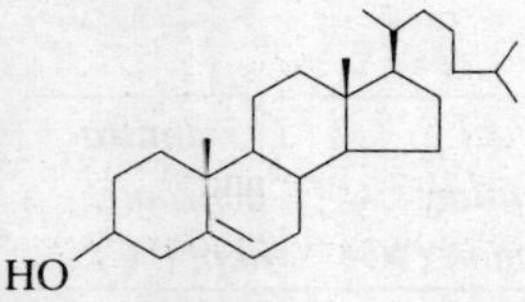 | 148 | Colourless, insoluble in $H_2O$, soluble in alcohol | | 190 | |
| Benzilic acid | 150 | See Carboxylic acids (Table 6.2b) | | | |
| Citric acid (anhyd.) | 153 | See Carboxylic acids (Table 6.2b) | | | |
| Triphenyl carbinol $(C_6H_5)_3COH$ | 162 (b.p. 380) | Colourless, insoluble in $H_2O$, soluble in alcohol, ether | | | - Acetate, m.p. 88°C.<br>- With conc. HCl gives chloride, m.p. 108°C.<br>- With conc. $H_2SO_4$ forms intense yellow solution. |
| Ergosterol 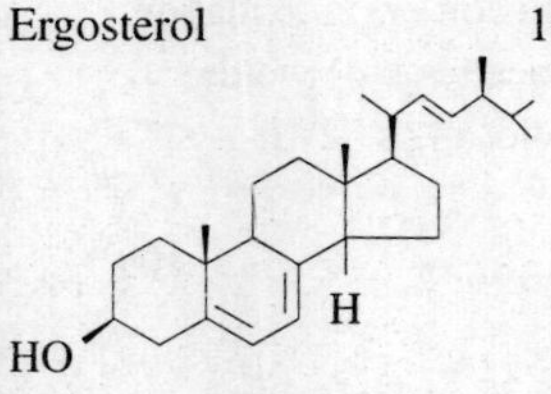 | 165 | Colourless, insoluble in $H_2O$, soluble in alcohol | 202 | | - Acetate, m.p. 180°C.<br>Benzoate, m.p. 168°C. |
| D(+)Mannitol $CH_2OH(CHOH)_4CH_2OH$ | 166 (d) | Colourless, soluble in $H_2O$, insoluble in ether, alcohol | | | - Hexaacetate, m.p. 120°C.<br>- Hexabenzoate, m.p. 149°C. |
| Tartaric acid | 170 | See Carboxylic acids (Table 6.2b) | | | |
| (±) Borneol 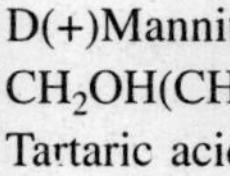 | 208 | Colourless, camphor like odour, insoluble in $H_2O$, soluble in alcohol, ether | 154 | 137 | - On heating with $HNO_3$ gives camphor m.p. 179°C. |

| *Name and structure* | *M.P.* | *Properties* | *3, 5-dinitro-benzoate* | *p-nitro-benzoate* | *Specific tests and derivatives* |
|---|---|---|---|---|---|
| Mucic acid | 213 | See Carboxylic acids (Table 6.2b) | | | |
| Inositol 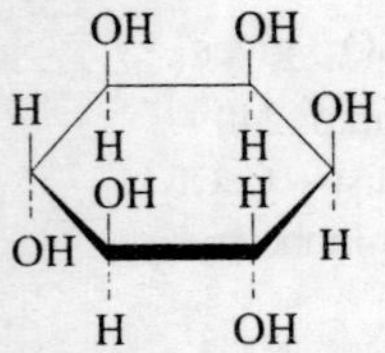 | 225 | Colourless, soluble in $H_2O$, sparingly soluble in alcohol | | | - Hexaacetate, m.p. 216°C.<br>- Hexabenzoate m.p. 258°C.<br>- On heating with $HNO_3$ gives oxalic acid. |
| Pentaerythritol $C(CH_2OH)_4$ | 269 | Colourless, sparingly soluble in $H_2O$, alcohol, insoluble in ether | | | - Tetraacetate, m.p. 83°C.<br>- Tetrabenzoate, m.p. 99°C. |

**Table 6.7(a)** Esters (Liquids)

*Note This table includes esters of all types of carboxylic acids, e.g. simple saturated, unsaturated, halo, keto and nitro substituted acids*

| *Name and structure* | *B.P. (°C)* | *Properties* | *Acid obtained by Saponification b.p. (°C)* | *m.p. (°C)* | *3, 5 dinitro-benzoate m.p. (°C)* |
|---|---|---|---|---|---|
| Methyl formate $HCOOCH_3$ | 32 | Colourless, soluble in $H_2O$ | 101 | — | 109 |
| Ethyl formate $HCOOC_2H_5$ | 54 | Colourless, sparingly soluble in alcohol, ether | 101 | — | 93 |
| Methyl acetate $CH_3COOCH_3$ | 57 | Colourless, sparingly soluble in $H_2O$, soluble in alcohol, ether | 118 | — | 109 |
| Isopropylformate $HCOOCH(CH_3)_2$ | 68 | Colourless, insoluble in $H_2O$, soluble in alcohol | 101 | — | 112 |
| Vinyl acetate $CH_3COOCH{=}CH_2$ | 72 | | 118 | — | — |
| Methyl chloroformate $ClCOOCH_3$ | 75 | Colourless, sharp lachymatory odour, decomposes in $H_2O$ (behaves as an acid chloride), solution in alcohol, benzene | — | — | 109 |
| Ethyl acetate $CH_3COOC_2H_5$ | 77 | Colourless, pleasant fruity odour, sparingly soluble in $H_2O$, soluble in alcohol, ether | 118 | — | 93 |
| Methyl propionate $CH_3CH_2COOCH_3$ | 79 | Colourless, sparingly soluble in $H_2O$ | 140 | — | 109 |
| *n*-Propyl formate $HCOOC_3H_7(n)$ | 81 | Colourless, sparingly soluble in $H_2O$ | 101 | — | 74 |

| *Name and structure* | *B.P. (°C)* | *Properties* | *Acid obtained by saponification b.p. (°C)* | *m.p. (°C)* | *3, 5 dinitro-benzoate m.p. (°C)* |
|---|---|---|---|---|---|
| Allyl formate $HCOOCH_2CH=CH_2$ | 83 | Colourless, characteristic odour, insoluble in $H_2O$, soluble in alcohol | 101 | — | 50 |
| Methyl acrylate $CH_2=CHCOOCH_3$ | 85 | Colourless, soluble in alcohol, ether, polymerises readily | 140 | — | 109 |
| Dimethylcarbonate $CH_3OCOOCH_3$ | 90 | | | | |
| Isopropyl acetate $CH_3COOCH(CH_3)_2$ | 91 | | 118 | — | 122 |
| Methyl isobutyrate $(CH_3)_2CHCOOCH_3$ | 92 | | 155 | — | 109 |
| Ethyl chloroformate $ClCOOC_2H_5$ | 93 | Colourless, characteristic odour, decomposes in $H_2O$, behaves as an acid chloride, soluble in alcohol, benzene | — | — | 93 |
| *t*-Butyl acetate $CH_3COOC(CH_3)_3$ | 98 | Colourless, soluble in alcohol, ether | 118 | — | 142 |
| iso-Butylformate $HCOOCH_2CH(CH_3)_2$ | 98 | | 101 | — | 88 |
| Ethylpropionate $CH_3CH_2COOC_2H_5$ | 98 | Colourless, sparingly soluble in $H_2O$ | 140 | — | 93 |
| Ethyl acrylate $CH_2=CHCOOC_2H_5$ | 101 | Colourless, sparingly soluble in $H_2O$, polymerises readily | 140 | — | 93 |
| Methyl orthoformate $CH(OCH_3)_3$ | 101 | | | | |
| n-Propyl acetate $CH_3COOC_3H_7(n)$ | 101 | Colourless, sparingly soluble in $H_2O$, soluble in alcohol | 118 | — | 122 |
| Methyl n-butyrate $CH_3CH_2CH_2COOCH_3$ | 102 | | 163 | — | 109 |
| Allyl acetate $CH_3COOCH_2CH=CH_2$ | 103 | | 118 | — | 50 |
| n-Butyl formate $HCOOC_4H_9$ (n) | 107 | Colourless, sparingly soluble in $H_2O$ | 101 | — | 63 |
| Ethyl isobutyrate $(CH_3)_2CHCOOC_2H_5$ | 110 | Colourless, insoluble in $H_2O$ | 155 | — | 93 |
| Sec-Butyl acetate $CH_3COOCH(CH_3)C_2H_5$ | 111 | | 118 | — | 76 |
| iso-Propyl priopionate $CH_3CH_2COOCH(CH_3)_2$ | 111 | | 140 | — | 122 |

| *Name and structure* | *B.P. (°C)* | *Properties* | *Acid obtained by saponification* | | *3, 5 dinitro-benzoate* |
|---|---|---|---|---|---|
| | | | *b.p. (°C)* | *m.p. (°C)* | *m.p. (°C)* |
| iso-Butyl acetate $CH_3COOCH_2CH(CH_3)_2$ | 116 | | 118 | — | 88 |
| Ethyl n-butyrate $CH_3CH_2CH_2COOC_2H_5$ | 120 | | 163 | — | 93 |
| *n*-Propyl propionate $CH_3CH_2COOC_3H_7$ (*n*) | 122 | | 140 | — | 74 |
| iso-Amyl formate $HCOO(CH_2)_2CH(CH_3)_2$ | 123 | | 101 | — | 62 |
| *n*-Butyl acetate $CH_3COOC_4H_9$(*n*) | 125 | | 118 | — | 63 |
| Diethyl carbonate $C_2H_5O\,COOC_2H_5$ | 126 | | — | — | — |
| Methyl n-valerate $CH_3CH_2CH_2CH_2COOCH_3$ | 128 | | 186 | — | 109 |
| iso-Propyl-n-butyrate $CH_3CH_2CH_2COOCH(CH_3)_2$ | 128 | | 163 | — | 122 |
| Ethyl glyoxalate $OHC\,COOC_2H_5$ | 130 | | — | 70–75 (Hemihydrate) | 93 |
| Methyl chloroacetate $ClCH_2COOCH_3$ | 130 | Colourless, slightly soluble in $H_2O$, soluble in alcohol | — | 63 | 109 |
| Ethyl isovalerate $(CH_3)_2CHCH_2COOC_2H_5$ | 134 | | 176 | — | 93 |
| Methyl pyruvate $CH_3,CO{\cdot}COOCH_3$ | 136 | Colourless, soluble in alcohol, ether | 165 | — | 109 |
| Iso-Butyl propionate $CH_3CH_2COOCH_2CH(CH_3)_2$ | 137 | | 140 | — | 88 |
| Ethyl crotonate $CH_3CH{=}CHCOOC_2H_5$ | 138 | | 165(d) | — | 93 |
| iso-Amyl acetate $CH_3COOCH_2CH_2CH(CH_3)_2$ | 142 | Colourless, banana like odour, insoluble in $H_2O$, soluble in alcohol, ether | 118 | — | 62 |
| Ethyl chloroacetate $ClCH_2COOC_2H_5$ | 144 | Colourless, insoluble in $H_2O$, soluble in alcohol, ether | — | 63 | 93 |
| Methyl bromoacetate $BrCH_2COOCH_3$ | 144 | Colourless, irritating lachy-matory odour | — | 50 | 109 |

| *Name and structure* | *B.P. (°C)* | *Properties* | *Acid obtained by saponification b.p. (°C)* | *m.p. (°C)* | *3, 5 dinitro-benzoate m.p. (°C)* |
|---|---|---|---|---|---|
| n-Propyl n-butyrate | 144 | | 163 | — | 122 |
| $CH_3CH_2CH_2COOC_3H_7(n)$ | | | | | |
| n-Butyl propionate | 144 | | 140 | — | 63 |
| $CH_3CH_2COO(CH_2)_3CH_3$ | | | | | |
| Ethyl n-valerate | 145 | | 186 | — | 93 |
| $CH_3CH_2CH_2CH_2COOC_2H_5$ | | | | | |
| Ethyl orthoformate | 145 | | — | — | — |
| $CH(OC_2H_5)_3$ | | | | | |
| Methyl lactate | 145 | | 119/12 mm | — | 109 |
| $CH_3CH(OH)COOCH_3$ | | | | | |
| *n*-Amyl acetate | 146 | Sparingly soluble in water, soluble in | 118 | — | 46 |
| $CH_3COO(CH_2)_4CH_3$ | | hot water, alcohol | | | |
| Ethyl-α-chloropropionate | 146 | | 186 | — | 93 |
| $CH_3CH(Cl)COOC_2H_5$ | | | | | |
| iso-Butyl isobutyrate | 147 | | 155 | — | 88 |
| $(CH_3)_2CHCOOCH_2CH(CH_3)_2$ | | | | | |
| Ethyl lactate | 152 | | 119/12 mm | — | 93 |
| $CH_3CH(OH)COOC_2H_5$ | | | | | |
| Ethyl pyruvate | 155 | Colourless, insoluble in $H_2O$, soluble in | 165 | — | 93 |
| $CH_3CO{\cdot}COOC_2H_5$ | | alcohol, ether | | | |
| iso-Butyl n-butyrate | 157 | | 163 | — | 88 |
| $CH_3CH_2CH_2COOCH_2CH(CH_3)_2$ | | | | | |
| Ethyl bromoacetate | 159 | | — | 50 | 93 |
| $BrCH_2COOC_2H_5$ | | | | | |
| iso-Amyl propionate | 160 | | 140 | — | 62 |
| $CH_3CH_2COO(CH_2)_2CH(CH_3)_2$ | | | | | |
| Ethyl glycollate | 160 | | — | 80 | 93 |
| $HOCH_2COOC_2H_5$ | | | | | |
| Cyclohexylformate | 162 | | 101 | — | 113 |
| HCOO 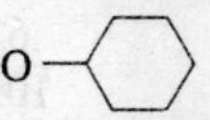 | | | | | |
| Ethyl-α-bromopropionate | 162 | | 205 | — | 93 |
| $CH_3CH(Br)COOC_2H_5$ | | | | | |
| *n*-Butyl *n*-butyrate | 165 | | 163 | — | 63 |
| $CH_3CH_2CH_2COO(CH_2)_3CH_3$ | | | | | |

| *Name and Structure* | *B.P. (°C)* | *Properties* | *Acid obtained by saponification b.p. (°C)* | *m.p. (°C)* | *3, 5 dinitro-benzoate m.p. (°C)* |
|---|---|---|---|---|---|
| Ethyl *n*-caproate $CH_3(CH_2)_4COOC_2H_5$ | 166 | | 205 | — | 93 |
| Ethyl trichloroacetate $Cl_3CCOOC_2H_5$ | 167 | | — | 58 | 93 |
| Isopropyl lactate $CH_3CH(OH)COOCH(CH_3)_2$ | 168 | | 119/12 mm | — | 122 |
| Methyl acetoacetate $CH_3COCH_2COOCH_3$ | 169 | Colourless, sparingly soluble in $H_2O$, soluble in alcohol, ether | — | — | 109 |
| Cyclohexylacetate $CH_3COOC_6H_{11}$ | 175 | | 118 | — | 113 |
| Isoamyl-n-butyrate $CH_3CH_2CH_2COOCH_2CH_2CH(CH_3)_2$ | 178 | | 163 | — | 62 |
| Ethyl-$\beta$-bromopropionate $BrCH_2CH_2COOC_2H_5$ | 179 | | — | 62 | 93 |
| Dimethyl malonate $CH_2(COOCH_3)_2$ | 181 | Colourless, insoluble in $H_2O$, soluble in alcohol, ether | — | 133(d) | 109 |
| Ethyl acetoacetate $CH_3COCH_2COOC_2H_5$ | 181 | Colourless, pleasant odour, sparingly soluble in $H_2O$, soluble in alcohol, ether, chloroform | — | — | 93 |
| Ethylene glycol monoacetate $HOCH_2CH_2OOCCH_3$ | 183 | | 118 | — | 169 |
| Diethyl oxalate $(COOC_2H_5)_2$ | 186 | Colourless, sweet smelling, slightly soluble in alcohol, ether | — | 101 | 93 |
| Ethyl methyl acetoacetate $CH_3COCH(CH_3)COOC_2H_5$ | 187 | | — | — | 93 |
| Di-isopropyl oxalate $[COOCH(CH_3)_2]_2$ | 190 | | — | 101 | 122 |
| Ethylene glycol diacetate $(CH_2OOCCH_3)_2$ | 190 | Colourless, sparingly soluble in $H_2O$, soluble in alcohol, ether. | 118 | — | 169 |
| Dimethyl succinate $(CH_2COOCH_3)_2$ | 195 | Colourless, soluble in $H_2O$ | — | 185 | 109 |
| Phenyl acetate $CH_3COOC_6H_5$ | 196 | Colourless, sweet smelling sparingly soluble in $H_2O$, soluble in alcohol, ether, chloroform. | 118 | — | 145 |
| Diethyl malonate $CH_2(COOC_2H_5)_2$ | 198 | Colourless, faint odour, insoluble in $H_2O$, soluble in alcohol, ether | — | 133(d) | 93 |

| *Name and structure* | *B.P. (°C)* | *Properties* | *Acid obtained by saponification* b.p. (°C) | m.p. (°C) | *3, 5 dinitro-benzoate* m.p. (°C) |
|---|---|---|---|---|---|
| Methyl benzoate $C_6H_5COOCH_3$ | 198 | Colourless, insoluble in $H_2O$ and soluble in hot alcohol, ether | — | 122 | 109 |
| Benzyl formate $HCOOCH_2C_6H_5$ | 203 | | 101 | — | 112 |
| Ethyl levulinate $CH_3COCH_2CH_2COOC_2H_5$ | 205 | Soluble in $H_2O$ | 245 | — | 93 |
| Ethyl cyanoacetate $NCCH_2COOC_2H_5$ | 207 | See Nitriles (Table 6.19a) | — | 66 | 93 |
| o-Cresyl acetate 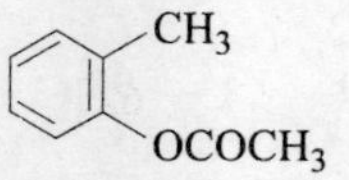 | 208 | | 118 | — | 138 |
| Phenyl propionate $CH_3CH_2COOC_6H_5$ | 211 | | 140 | — | 145 |
| *m*-Cresyl acetate 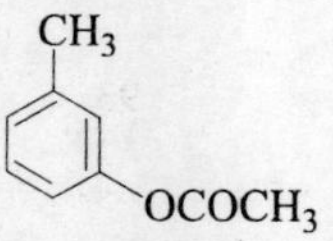 | 212 | | 118 | — | 165 |
| p-Cresyl acetate 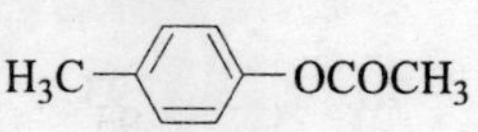 | 212 | | 118 | — | 188 |
| Di-n-propyl oxalate $[COOC_3H_7(n)]_2$ | 213 | | — | 101 | 74 |
| Ethyl benzoate 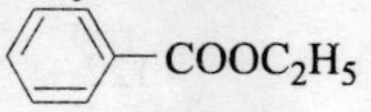 | 213 | Colourless, sweet smelling, in soluble in $H_2O$, soluble in alcohol, ether, chloroform | — | 122 | 93 |
| Methyl pelargonate $CH_3(CH_2)_7COOCH_3$ | 213 | | 253 | — | 109 |
| Methyl-o-toluate 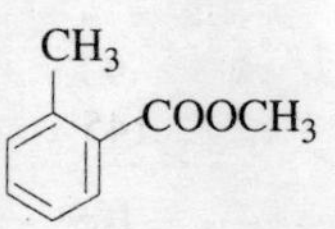 | 213 | | — | 104 | 109 |
| Methyl m-toluate 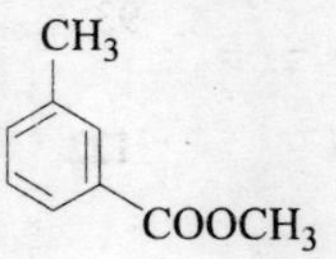 | 215 | | — | 110 | 109 |

| *Name and structure* | *B.P. (°C)* | *Properties* | *Acid obtained by saponification* | | *3, 5 dinitro-benzoate* |
|---|---|---|---|---|---|
| | | | *b.p. (°C)* | *m.p. (°C)* | *m.p. (°C)* |
| Benzyl acetate $CH_3COOCH_2C_6H_5$ | 216 | Colourless, insoluble in $H_2O$, soluble in alcohol, ether | 118 | — | 112 |
| Diethyl succinate $(CH_2COOC_2H_5)_2$ | 216 | Colourless, insoluble in $H_2O$, soluble in alcohol, ether | — | 185 | 93 |
| Methyl phenyl acetate $C_6H_5CH_2COOCH_3$ | 218 | Colourless, insoluble in $H_2O$, soluble in alcohol, ether | — | 76 | 109 |
| Isopropyl benzoate $C_6H_5COOCH(CH_3)_2$ | 218 | | — | 122 | 122 |
| Benzyl propionate $CH_3CH_2COOCH_2C_6H_5$ | 222 | | 140 | — | 112 |
| Diethyl maleate $CHCOOC_2H_5$ ‖ $CHCOOC_2H_5$ | 224 | Colourless, insoluble in $H_2O$, soluble in alcohol, ether | — | 130 | 93 |
| Methyl salicylate 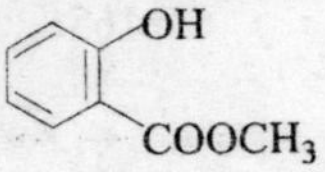  | 224 | Colourless, characteristic smell like oil of wintergreen, insoluble in $H_2O$, soluble in alcohol, ether, chloroform | — | 159 | 109 |
| Phenyl n-butyrate $CH_3CH_2CH_2COOPh$ | 227 | | 163 | — | 145 |
| Ethyl pelargonate $CH_3(CH_2)_7\ COOC_2H_5$ | 227 | | 253 | — | 93 |
| Ethyl phenyl acetate $C_6H_5CH_2COOC_2H_5$ | 229 | Colourless, insoluble in $H_2O$, soluble in alcohol, ether | — | 76 | 93 |
| *n*-Propyl benzoate $C_6H_5COOCH_2CH_2CH_3$ | 230 | | — | 122 | 74 |
| Methyl n-caprate $CH_3(CH_2)_8COOCH_3$ | 232 | | 269 | — | 109 |
| Ethyl salicylate 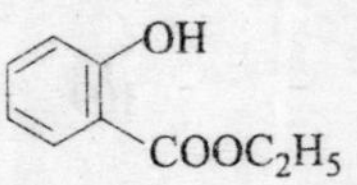  | 234 | Colourless, insoluble in $H_2O$, solution alcohol, ether | — | 159 | 93 |
| Ethyl glutarate $CH_2(CH_2COOC_2H_5)_2$ | 234 | | — | 98 | 93 |
| Methyl o-chlorobenzoate 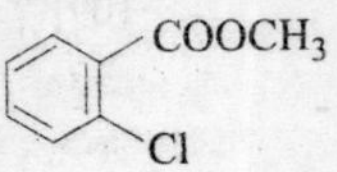  | 234 | | — | 142 | 109 |
| Isopropyl salicylate 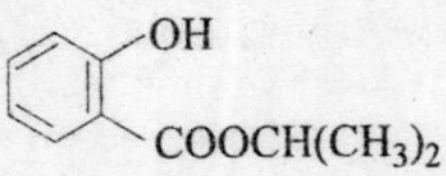  | 237 | | — | 159 | 122 |

| *Name and structure* | *B.P. (°C)* | *Properties* | *Acid obtained by saponification* | | *3, 5 dinitro-benzoate* |
|---|---|---|---|---|---|
| | | | *b.p. (°C)* | *m.p. (°C)* | *m.p. (°C)* |
| Benzyl n-butyrate<br>$CH_3CH_2CH_2COOCH_2C_6H_5$ | 238 | | 163 | — | 112 |
| Ethyl p-chlorobenzoate<br>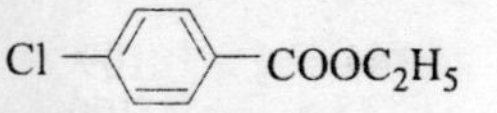 | 238 | | — | 243 | 93 |
| *n*-Propyl salicylate<br>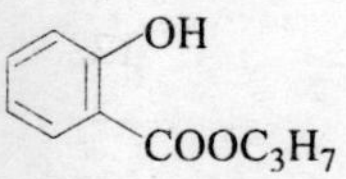 | 238 | | — | 159 | 74 |
| Di-n-butyl oxalate<br>$[COOC_4H_9(n)]_2$ | 242 | | — | 101 | 63 |
| Isobutyl benzoate<br>$C_6H_5COOCH_2CH(CH_3)_2$ | 242 | | — | 122 | 88 |
| Ethyl o-Chlorobenzoate<br>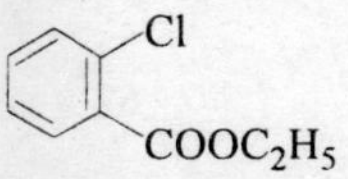 | 243 | | — | 142 | 93 |
| Methyl o-bromobenzoate<br>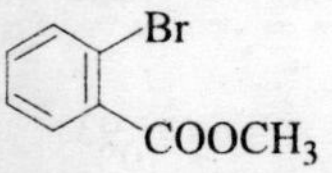 | 244 | | — | 150 | 109 |
| Diethyl adipate<br>$(CH_2CH_2COOC_2H_5)_2$ | 245 | | — | 151 | 93 |
| Ethyl n-caprate<br>$CH_3(CH_2)_8COOC_2H_5$ | 245 | | 269 | — | 93 |
| Ethyl m-chlorobenzoate<br>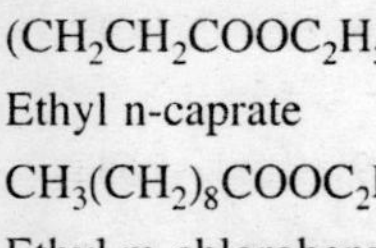 | 245 | | — | 158 | 93 |
| Di-n-propylsuccinate<br>$(CH_2COOC_3H_7(n))_2$ | 246 | | — | 185 | 74 |
| *n*-Butyl benzoate<br>$C_6H_5COO(CH_2)_3CH_3$ | 248 | | — | 122 | 63 |
| Diethyl acetone dicarboxylate<br>$CO(CH_2COOC_2H_5)_2$ | 250 | Colourless, sparingly soluble in $H_2O$ | — | 135 (d) | 93 |
| Diethyl (–) maliate<br>$CH_2COOC_2H_5$<br>\|<br>$CH(OH)COOC_2H_5$ | 253 | | — | 100 | 93 |

| *Name and structure* | *B.P. (°C)* | *Properties* | *Acid obtained by saponification* | | *3, 5 dinitro-benzoate* |
|---|---|---|---|---|---|
| | | | *b.p. (°C)* | *m.p. (°C)* | *m.p. (°C)* |
| Ethyl o-bromobenzoate 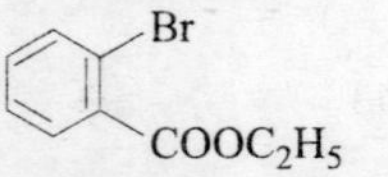  | 254 | | — | 150 | 93 |
| Methyl anthranilate 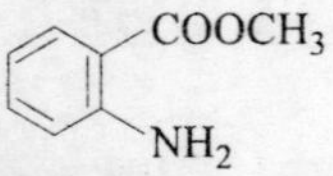  | 255 | | — | 144 | 109 |
| Glyceryl triacetate (triacetin) $CH_3COOCH_2CH(OCOCH_3)CH_2OCOCH_3$ | 258 | Colourless, sparingly soluble in $H_2O$, soluble alcohol, ether | 118 | — | — |
| Ethyl m-bromobenzoate 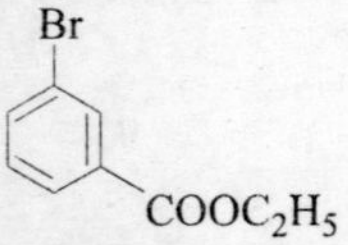  | 259 | | — | 155 | 93 |
| Isoamyl benzoate $C_6H_5COO\ (CH_2)_2CH(CH_3)_2$ | 262 | | — | 122 | 62 |
| Ethyl p-bromobenzoate 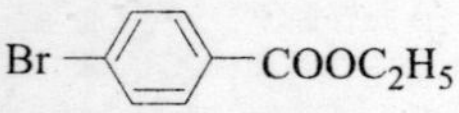  | 263 | | — | 251 | 93 |
| Ethyl anthranilate 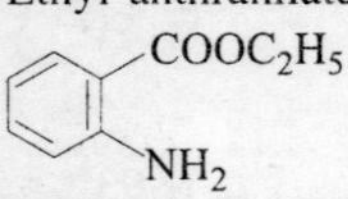  | 265 | | — | 144 | 93 |
| n-Butyl salicylate 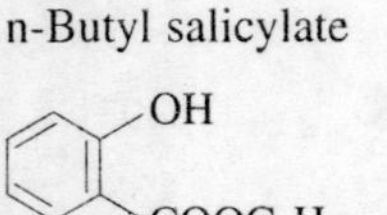  | 268 | | — | 159 | 63 |
| Ethyl-p-anisate 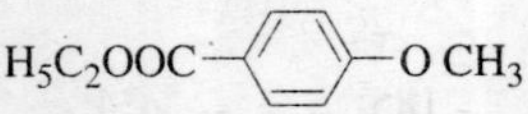  | 269 | | — | 184 | 93 |
| Ethyl benzoyl acetate $C_6H_5COCH_2COOC_2H_5$ | 270 | See Carbonyl compounds (Table 6.3b) | — | — | 93 |
| Ethyl cinnamate 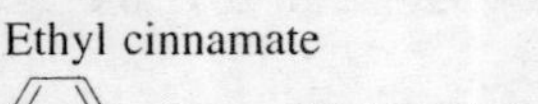  | 271 | Colourless, insoluble in $H_2O$, soluble in alcohol, ether | — | 133 | 93 |
| Methyl o-nitrobenzoate 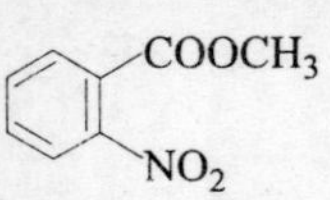  | 275 | | — | 146 | 109 |

| *Name and structure* | *B.P. (°C)* | *Properties* | *Acid obtained by saponification b.p. (°C)* | *Acid obtained by saponification m.p. (°C)* | *3, 5 dinitro-benzoate m.p. (°C)* |
|---|---|---|---|---|---|
| Resorcinol diacetate 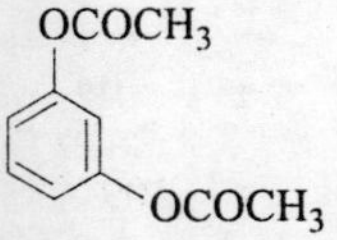 | 277 | Colourless liquid | 118 | — | 201 |
| Diethyl (±) tartrate $C_2H_5OOC\ (CHOH)_2COOC_2H_5$ | 280 | Colourless, soluble in $H_2O$ | — | 206 | 93 |
| Dimethyl phthalate 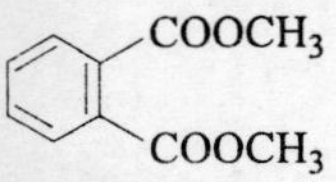 | 282 | Colourless, insoluble in $H_2O$ | — | 195 | 109 |
| Triethyl citrate $CH_2COOC_2H_5$ \| $C(OH)COOC_2H_5$ \| $CH_2COOC_2H_5$ | 294 | | — | 100 | 93 |
| Ethyl m-aminobenzoate 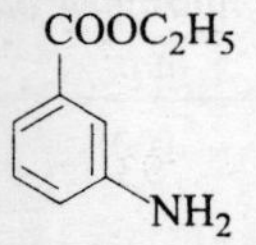 | 294 | | — | 174 | 93 |
| Diethylphthalate 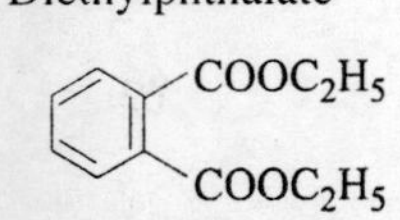 | 298 | Colourless, insoluble in $H_2O$, soluble in alcohol, ether | — | 195 | 93 |
| o-Cresyl benzoate 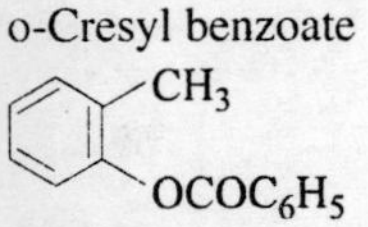 | 307 | Colourless, insoluble in $H_2O$, soluble in alcohol, ether | — | 122 | 138 |
| Ethyl 1-naphthoate 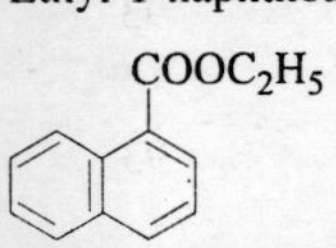 | 309 | | — | 162 | 93 |
| Benzyl phenyl acetate $C_6H_5CH_2COOCH_2C_6H_5$ | 317 | | — | 76 | 112 |
| Benzyl salicylate 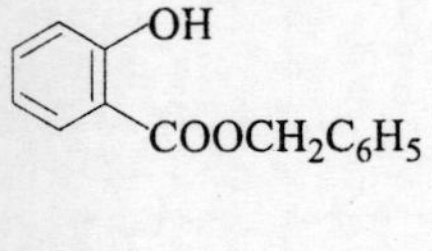 | 320 | | — | 159 | 112 |

| *Name and structure* | *B.P. (°C)* | *Properties* | *Acid obtained by saponification b.p. (°C)* | *m.p. (°C)* | *3, 5 dinitro-benzoate m.p. (°C)* |
|---|---|---|---|---|---|
| Benzyl benzoate $C_6H_5COOCH_2C_6H_5$ | 323 m.p. 21 | Colourless, insoluble in $H_2O$, soluble in alcohol, ether | — | 122 | 112 |
| Di-*n*-butylphthalate 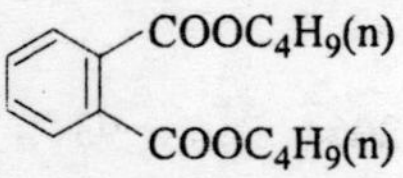 | 338 | | — | 195 | 63 |
| Ethyl cyclohexanone-2-carboxylate 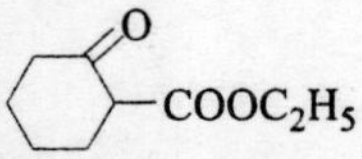 | 107/ 11 mm | | — | 81 (d) | 93 |
| Ethyl cyclopentanone-2-carboxylate 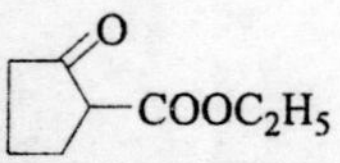 | 117/ 26 mm | | — | — | 93 |

**Table 6.7(b)** Esters (Solids)

| *Name and structure* | *M.P. (°C)* | *Properties* | *Acid obtained by saponification b.p.* | *m.p.* | *3, 5-dinitrobenzoate m.p.* |
|---|---|---|---|---|---|
| Methyl m-chlorobenzoate 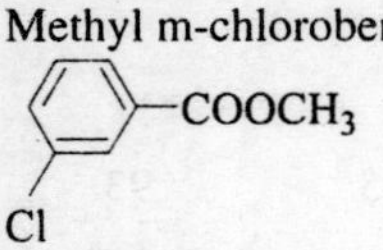 | 21 (b.p. 231) | | — | 158 | 109 |
| Cetyl acetate $CH_3COOCH_2(CH_2)_{14}CH_3$ | 22 | | 118 | — | 66 |
| Ethyl palmitate $CH_3(CH_2)_{14}COOC_2H_5$ | 24 b.p. 185/10 mm | | — | 62 | 93 |
| Methyl anthranilate 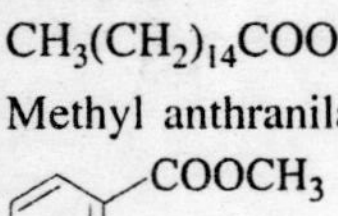 | 25 (b.p. 255) | | — | 144 | 109 |
| Methyl m-bromobenzoate 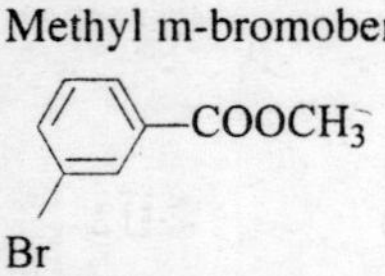 | 29 | | — | 155 | 109 |
| Bornyl acetate 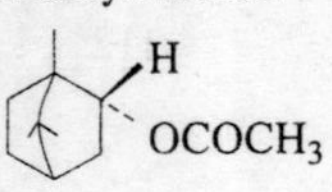 | 30 b.p. 221 | | 118 | — | 154 |

| *Name and structrue* | *M.P. (°C)* | *Properties* | *Acid obtained by saponification b.p.* | *m.p.* | *3, 5-dinitrobenzoate m.p.* |
|---|---|---|---|---|---|
| Ethyl o-nitrobenzoate<br>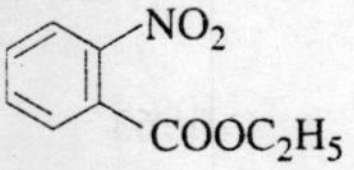 | 30 | | — | 146 | 93 |
| Ethyl 2-naphthoate<br>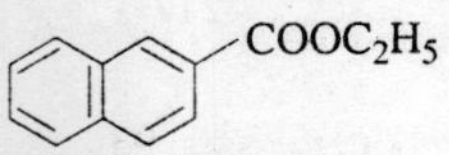 | 32 | | — | 184 | 93 |
| Ethyl stearate<br>$CH_3(CH_2)_{16}COOC_2H_5$ | 33 | | — | 69 | 93 |
| Methyl p-toluate<br>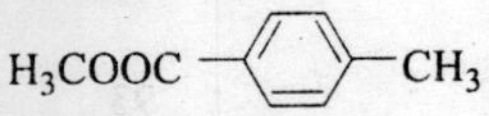 | 33<br>(b.p. 217) | | — | 178 | 109 |
| Methyl cinnamate<br>$C_6H_5CH{=}CHCOOCH_3$ | 36<br>(b.p. 262) | Colourless, insoluble in $H_2O$, soluble in alcohol, ether | — | 133 | 109 |
| Ethyl mandalate (±)<br>$C_6H_5CH(OH)COOC_2H_5$ | 37 | | — | 118 | 93 |
| Benzyl cinnamate<br>$C_6H_5CH{=}CHCOOCH_2C_6H_5$ | 39 | | — | 133 | 112 |
| Ethyl m-nitrobenzoate<br>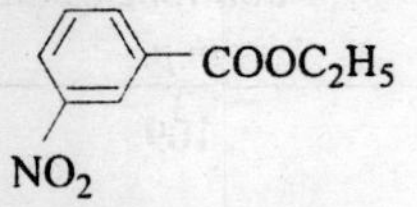 | 41 | | — | 141 | 93 |
| Dibenzyl phthalate<br>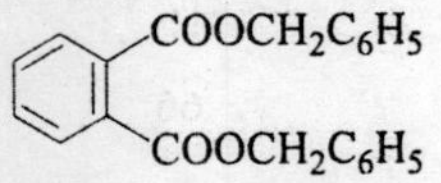 | 42 | | — | 195 | 112 |
| Dibenzyl succinate<br>$(CH_2COOCH_2C_6H_5)_2$ | 42 | | — | 185 | 112 |
| Phenyl salicylate (salol)<br>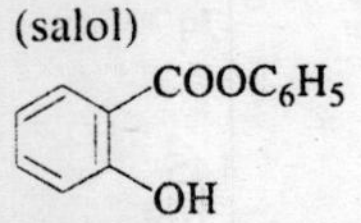 | 42<br>(b.p. 173/12 mm) | Colourless, insoluble in $H_2O$, soluble in hot alcohol, ether, benzene | — | 159 | 145 |
| Methyl p-chlorobenzoate<br>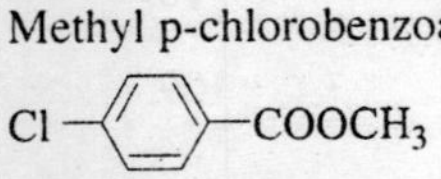 | 44 | | — | 243 | 109 |
| Cinnamyl cinnamate<br>$C_6H_5CH{=}CHCOOCH_2CH{=}CHC_6H_5$ | 44 | | — | 133 | 121 |
| Diethyl terephthalate<br>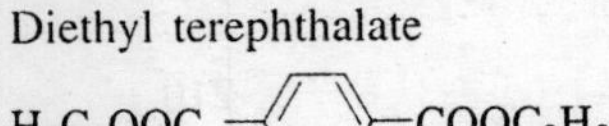 | 44 | | — | > 300 | 93 |

| *Name and structure* | *M.P. (°C)* | *Properties* | *Acid obtained by saponification b.p.* | *m.p.* | *3, 5-dinitrobenzoate m.p.* |
|---|---|---|---|---|---|
| Methyl-p-anisate 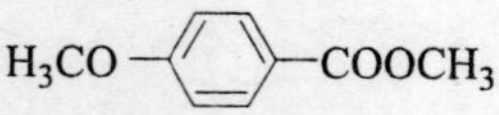 | 45 | | — | 184 | 109 |
| Dimethyl (±) tartrate $CH_3OOC{\cdot}[CH(OH)]_2\ COOCH_3$ | 48 | | — | 206 | 109 |
| 1-Naphthyl acetate 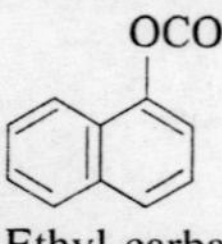 | 49 | | 118 | — | 217 |
| Ethyl carbamate (Urethane) $NH_2COOC_2H_5$ | 50 | Colourless, soluble in $H_2O$, alcohol, ether | — | — | 93 |
| Methyl carbamate $H_2NCOOCH_3$ | 52 | Soluble in $H_2O$, alcohol, ether | — | — | 109 |
| Dimethyl oxalate $(COOCH_3)_2$ | 53 | Colourless, sparingly soluble in alcohol, ether | — | 101 | 109 |
| m-Cresyl benzoate 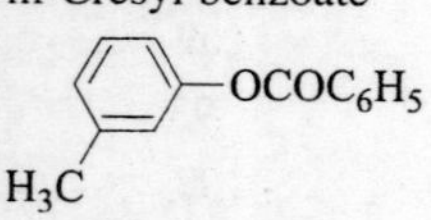 | 54 | | — | 122 | 165 |
| Ethyl p-nitrobenzoate 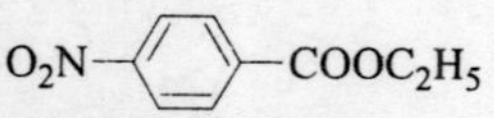 | 56 | | — | 240 | 93 |
| 1-Naphthyl benzoate 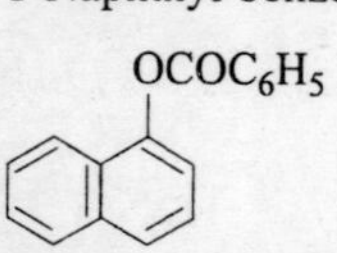 | 56 | | — | 122 | 217 |
| Methyl mandelate (±) $C_6H_5CH(OH)COOCH_3$ | 58 | | — | 118 | 109 |
| *n*-Propyl carbamate $NH_2COOC_3H_7(n)$ | 60 | Colourless, soluble in $H_2O$, alcohol, ether | | — | 74 |
| Phenyl benzoate $C_6H_5COOC_6H_5$ | 68 | Colourless, insoluble in $H_2O$, soluble in hot alcohol, warm ether | — | 122 | 145 |
| Methyl m-hydroxybenzoate 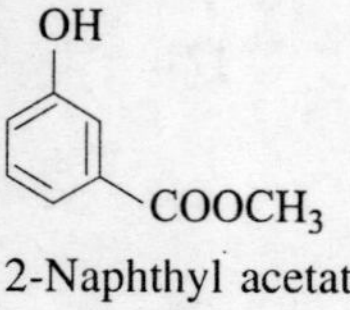 | 70 | Colourless solid | — | 201 | 109 |
| 2-Naphthyl acetate 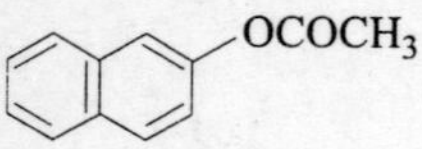 | 70 | | 118 | — | 210 |

| Name and structure | M.P. (°C) | Properties | Acid obtained by saponification b.p. | m.p. | 3, 5-dinitrobenzoate m.p. |
|---|---|---|---|---|---|
| Di-Phenyl phthalate | 70 | | — | 195 | 145 |
| 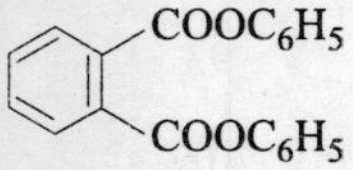 | | | | | |
| p-Cresyl benzoate | 71 | | — | 122 | 188 |
| 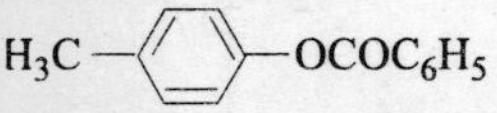 | | | | | |
| Ethyl m-hydroxybenzoate | 72 | Colourless solid | — | 201 | 93 |
| 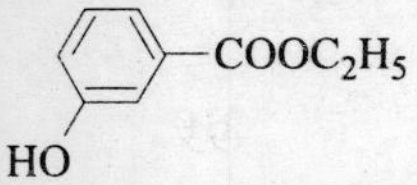 | | | | | |
| Methyl o-nitrocinnamate | 72 | | — | 240 | 109 |
| 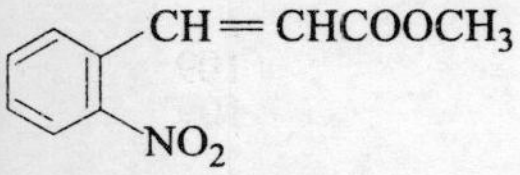 | | | | | |
| Phenyl cinnamate | 72 | Colourless, insoluble in $H_2O$ | — | 133 | 145 |
| 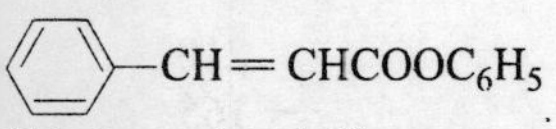 | | | | | |
| Ethylene glycol dibenzoate $(CH_2OCOC_6H_5)_2$ | 73 | | | 122 | 169 |
| Methyl 2-naphthoate | 77 | | — | 184 | 109 |
| 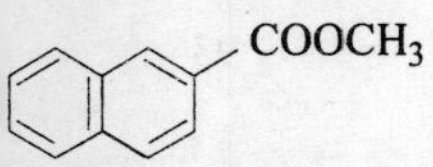 | | | | | |
| Diphenyl carbonate $C_6H_5OCOOC_6H_5$ | 78 | | — | — | 145 |
| Ethyl m-nitrocinnamate | 78 | | — | 200 | 93 |
| 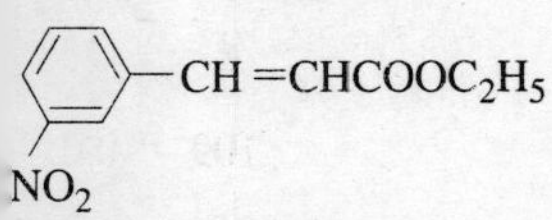 | | | | | |
| Methyl m-nitrobenzoate | 78 | | — | 141 | 109 |
| 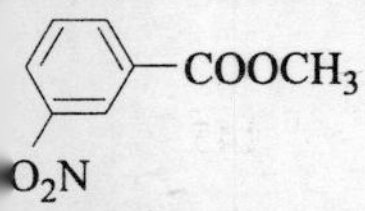 | | | | | |
| Trimethyl citrate | 78 | | — | 100 | 109 |
| $CH_2COOCH_3$<br>\|<br>$C(OH)COOCH_3$<br>\|<br>$CH_2COOCH_3$ | | | | | |
| Dibenzyl oxalate $C_6H_5CH_2OOC–COOCH_2C_6H_5$ | 80 | | — | 101 | 112 |
| Methyl p-bromobenzoate | 81 | | — | 251 | 109 |

| *Name and structure* | *M.P. (°C)* | *Properties* | *Acid obtained by saponification b.p.* | *m.p.* | *3, 5-dinitrobenzoate m.p.* |
|---|---|---|---|---|---|
| 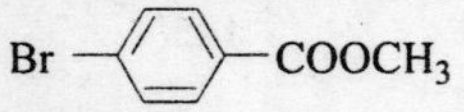 | | | | | |
| Ethyl p-aminobenzoate | 92 | | — | 186 | 93 |
| 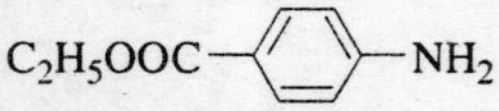 | | | | | |
| Ethyl 3, 5-dinitrobenzoate | 94 | | — | 205 | 93 |
| 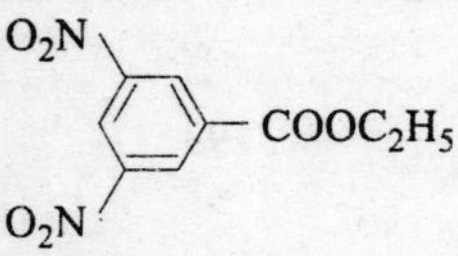 | | | | | |
| β-Naphthyl salicylate | 95 | | — | 159 | 210 |
| 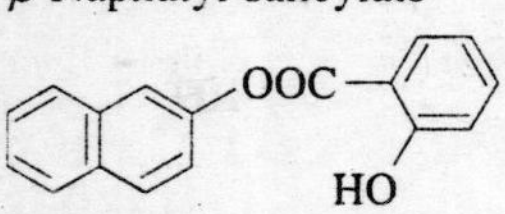 | | | | | |
| Methyl p-nitrobenzoate | 96 | | — | 240 | 109 |
| 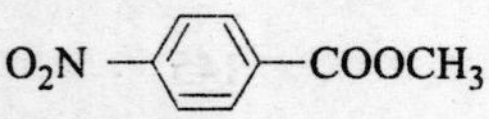 | | | | | |
| β-Naphthyl benzoate | 107 | | — | 122 | 210 |
| 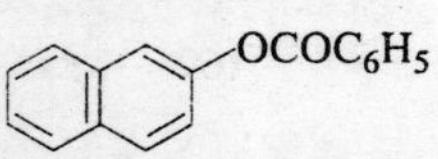 | | | | | |
| Methyl 3, 5-dinitrobenzoate | 108 | | — | 205 | 109 |
| 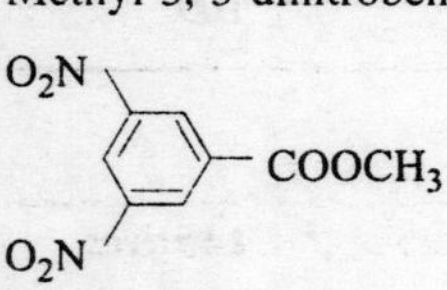 | | | | | |
| Ethyl oxamate $NH_2COCOOC_2H_5$ | 114 | Soluble in hot $H_2O$, colourless | — | | 93 |
| Ethyl p-hydroxybenzoate 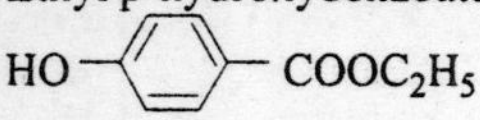 | 116 | Colourless, insoluble in $H_2O$, soluble in alcohol, ether | — | 213 | 93 |
| Resorcinol dibenzoate 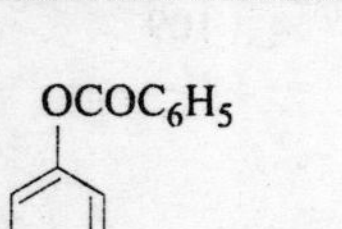 | 117 | | — | 122 | 201 |
| Diphenyl succinate $(CH_2COOC_6H_5)_2$ | 121 | | — | 185 | 145 |
| Hydroquinone diacetate 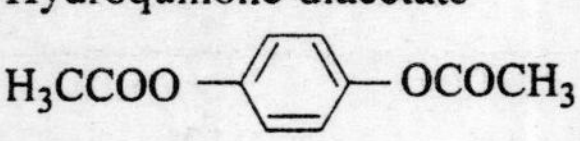 | 123 | Colourless, sparingly soluble in $H_2O$, soluble in alcohol, ether | 118 | — | 317 |
| Methyl m-nitrocinnamate | 124 | | — | 200 | 109 |

| *Name and structure* | *M.P. (°C)* | *Properties* | *Acid obtained by saponification b.p.* | *m.p.* | *3, 5-dinitrobenzoate m.p.* |
|---|---|---|---|---|---|
| 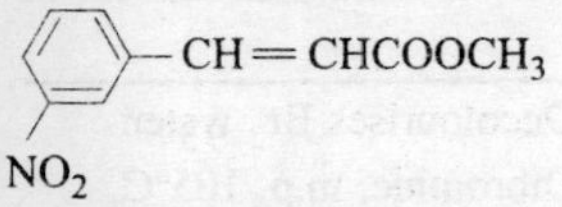 | | | | | |
| Methyl p-hydroxybenzoate 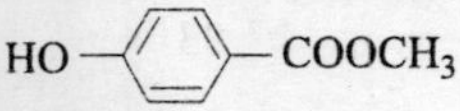 | 131 | Colourless, insoluble in $H_2O$, soluble in alcohol, ether | — | 213 | 109 |
| Diphenyl oxalate $(COOC_6H_5)_2$ | 136 | | — | 101 | 145 |
| Ethyl p-nitrocinnamate 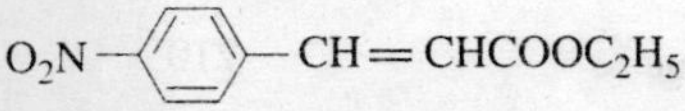 | 137 | | — | 285 | 93 |
| Dimethyl terephthalate 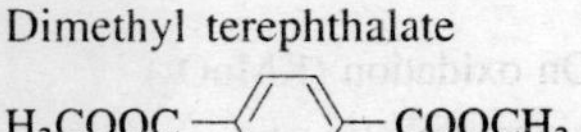 | 140 | | — | >300 | 109 |
| Methyl p-nitrocinnamate 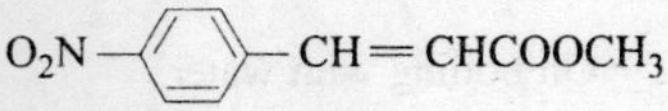 | 161 | | — | 285 | 109 |
| Pyrogallol triacetate 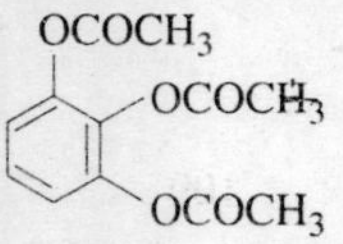 | 161 | Colourless, insoluble in $H_2O$, soluble in alcohol | 118 | — | 205 |

## Table 6.8(a) Lactones (Liquids)

| *Name and structure* | *B.P. (°C)* | *Properties* | *Derivatives* | *Specific tests and derivatives* |
|---|---|---|---|---|
| β-Propiolactone 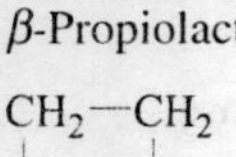 | *51/10 mm* | | | |
| γ-Butyrolactone 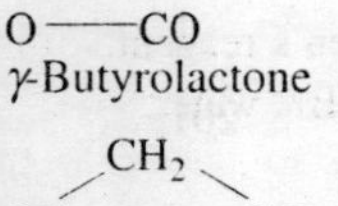 | 204 | | | |
| δ-Valerolactone 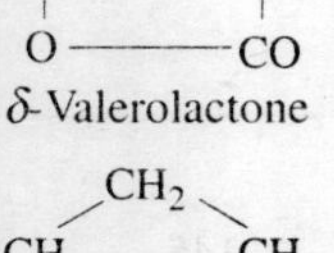 | 207 | | | |

**Table 6.8(b)** Lactones (Solids)

| *Name and structure* | *M.P. (°C)* | *Properties* | *Derivatives* | *Specific tests and derivatives* |
|---|---|---|---|---|
| Coumarin 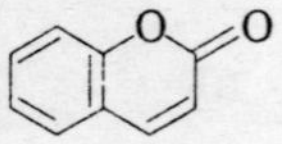 | 67° | White, fairly soluble in $H_2O$, characteristics odour, soluble in warm NaOH solution and on acidification is regenerated | On methylation with dimethyl sulphate and NaOH solution (shaking 20 min.) and acidification gives cis o-methoxycinnamic acid, m.p. 88°C: on long boiling with conc. KOH solution (20 min) followed by acidification gives coumaric acid, m.p. 207 °C. | -Decolourises $Br_2$ water.<br>-Dibromide, m.p. 105°C. |
| Phthalide 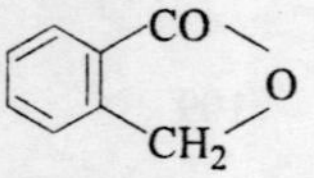 | 73 | Colourless, insoluble in $H_2O$ | On boiling with alkali yields o-hydroxymethyl benzoic acid, m.p. 120°C. | On oxidation ($KMnO_4$) gives phthalic acid, m.p. 195°C. |
| Glycollide 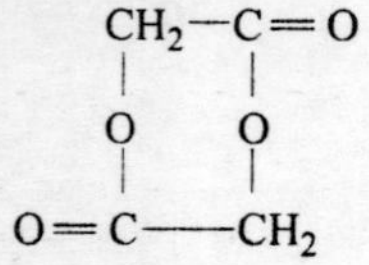 | 84 | Colourless, insoluble in $H_2O$ | Glycolamide, m.p. 120°C<br>Glycol anilide, m.p. 96°C | On boiling with water gives glycollic acid. m.p. 80°C. |
| Lactide 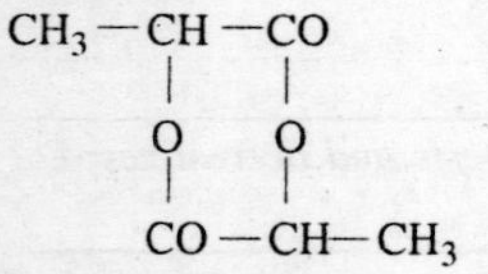 | 124 | Colourless, insoluble in $H_2O$ | Lactamide, m.p. 74°C,<br>Lactanilide, m.p. 59°C | - On boiling with $H_2O$ gives lactic acid. |
| L-Ascorbic acid 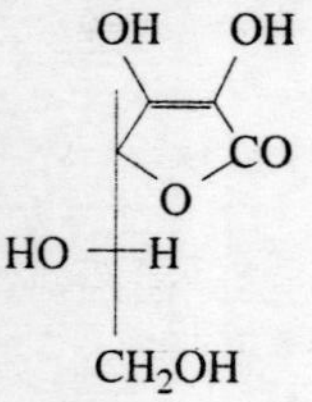 | 190 (d) | Colourless, soluble in $H_2O$, sparingly soluble in alcohol | | - With neutral ferric chloride gives intense violet colour.<br>- Reduces Tollen's reagent.<br>- Decolourises $Br_2$ water. |
| Phenolphthalein | 261 | See phenols, Table 6.4b | | |

**Table 6.9(a)** Acid anhydrides (Liquids)

| *Name and strusture* | *B.P.* | *Properties* | *Derivatives* Anilide | *p-Totuidide* | *Amide* | *Specific tests and derivatives* |
|---|---|---|---|---|---|---|
| Acetic anhydride $(CH_3CO)_2O$ | 138 | Colourless, pungent irritating odour, sparingly soluble in $H_2O$ | 114 | 148 | 82 | - On boiling with $H_2O$ gives acetic acid. The aqueous solution is made neutral by $NH_4OH$ and boiling off excess $NH_3$, on adding $FeCl_3$ gives red colour. |
| Propionic anhydride $(CH_3CH_2CO)_2O$ | 168 | Colourless, pungent odour, sparingly soluble in $H_2O$ | 104 | 124 | 79 | - On boiling with water gives propionic acid. |
| *n*-Butyric anhydride $(CH_3CH_2CH_2CO)_2O$ | 191 | Colourless, pungent odour, sparingly soluble in $H_2O$ | 96 | 72 | 114 | |
| Citraconic anhydride 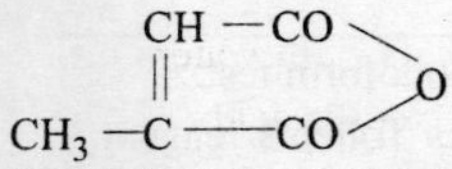 | 213 | — | 175 | — | 187 | |

**Table 6.9(b)** Acid anhydrides (Solids)

| *Name and structure* | *M.P. (°C)* | *Properties* | *Derivatives* Anilide | *p-Toluidide* | *Amide* | *Specific tests and derivatives* |
|---|---|---|---|---|---|---|
| Benzoic anhydride $(C_6H_5CO)_2O$ | 42 (b.p. 360°) | Colourless, insoluble in $H_2O$, soluble in alcohol, ether | 164 | 158 | 128 | - On heating with NaOH solution and acidification gives benzoic acid, m.p. 122°C. |
| Glutaric anhydride 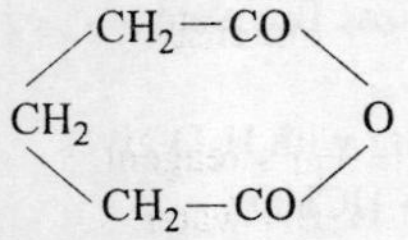 | 56 | | 224 | 217 | 174 | - On heating with NaOH solution and acidification gives glutaric acid, m.p. 98°C. |
| Maleic anhydride 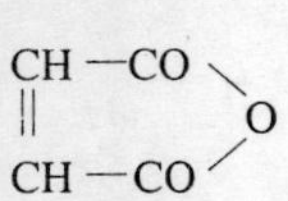 | 56 (b.p. 202°) | Colourless soluble in $H_2O$ | 198 (mono) 187 (di) | 142 | 153 | On heating with NaOH and acidification gives maleic acid, m.p. 130°C. - Decolourises $Br_2$ water on warming. |
| Succinic anhydride 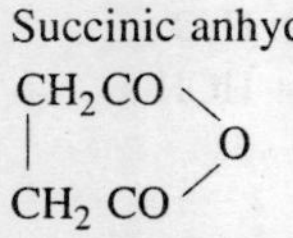 | 119 (b.p. 250°) | Colourless characteristic and peneterating odour | 228 (di) 145 (mono) | 255 (di) | 260 (di) | - On heating with NaOH gives succinic acid, m.p. 185°C. |
| Phthalic anhydride 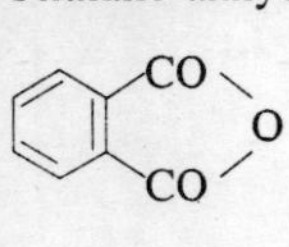 | 131 (b.p. 284°) | Colourless, insoluble in cold $H_2O$, soluble in hot $H_2O$ (with hydrolysis), soluble in alcohol | 250 (di) 170 (mono) | 201 (di) | 149 (mono) 220 (di) | - On heating with NaOH solution and acidification gives phthalic acid, m.p. 195°C. |

| | | | | | | |
|---|---|---|---|---|---|---|
| 3-Nitrophthalic anhydride 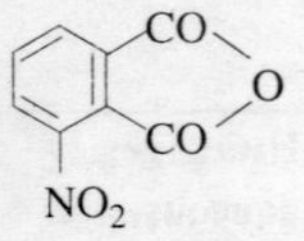 | 162 | Light yellow colour, insoluble in cold $H_2O$, soluble in alcohol | 234 (di) 181 (mono) | 224 | 201 | - On heating with NaOH solution and acidification gives 3-nitrophthalic acid, m.p. 218°C. |
| (±) Camphoric anhydride 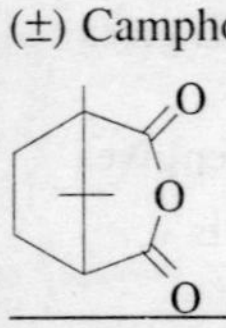 | 221 | | 226 | — | 192 | - On boiling with water gives cis camphoric acid, m.p. 187°C. |

## Table 6.10(a) Ethers (liquids)

| *Name and structure* | *B.P. (°C)* | *Properties* | *3, 5-dinitro benzoate* | *Picrate* | *Nitro derivative* | *Specific tests and derivatives* |
|---|---|---|---|---|---|---|
| Ethylene oxide (Oxirane) 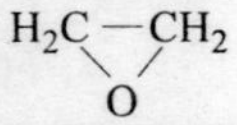 | 11 | Soluble in $H_2O$, alcohol, ether | | | | - Gives iodoform test. - Reduces Tollen's reagent. |
| Diethyl ether $C_2H_5OC_2H_5$ | 35 | Colourless, characteristic odour, very sparingly soluble in $H_2O$, soluble in alcohol, chloroform | | | | - Soluble in cold conc. $H_2SO_4$ and is recovered unchanged on dilution. |
| Propylene oxide 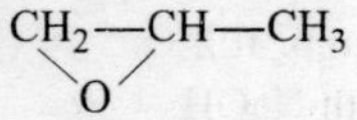 | 35 | Soluble in $H_2O$, alcohol, ether | | | | - With hot water gives propylene glycol. |
| Ethylvinyl ether $CH_2{=}CH\,O\,CH_2CH_3$ | 36 | | | | | - Decolourises $Br_2$ water. |
| Chloromethyl methyl ether $CH_3{\cdot}O{\cdot}CH_2Cl$ | 59 | Colourless liquid | — | 163 | — | - On shaking with $H_2O$ gives $CH_3OH + HCHO$. |
| Tetrahydrofuran 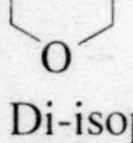 | 65 | Colourless, characteristic odour, soluble in $H_2O$ | | | | |
| Di-isopropyl ether $(CH_3)_2CHOCH(CH_3)_2$ | 68 | Colourless, insoluble in $H_2O$, soluble in alcohol, ether | 121 | | | |
| Chloromethyl ethyl ether $C_2H_5OCH_2Cl$ | 83 | | | | | - On shaking with $H_2O$ gives $C_2H_5OH + HCHO$. |
| Ethylene glycol dimethyl ether $CH_3OCH_2CH_2OCH_3$ | 85 | Colourless, soluble in $H_2O$. | | | | |
| Tetrahydropyran 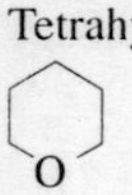 | 85 | Colourless, soluble in $H_2O$ | | | | |

| *Name and structure* | *B.P. (°C)* | *Properties* | *3, 5-dinitro-benzoate* | *Picrate* | *Nitro derivative* | *Specific tests and derivatives* |
|---|---|---|---|---|---|---|
| Di-n-propyl ether<br>n)$C_3H_7—O—C_3H_7$ | 90 | Colourless, sparingly soluble in $H_2O$, soluble in alcohol, ether | 74 | | | |
| x-Chloroethyl ethyl ether<br>$CH_3CHCl·O·C_2H_5$ | 98 | Colourless | | | | - On shaking with $H_2O$ gives $C_2H_5OH + CH_3CHO$. |
| Dioxan<br>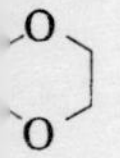 | 102 | Colourless, soluble in $H_2O$, alcohol, ether | | | | - After heating with alkaline $KMnO_4$, the filtrate is treated with $H_2O_2$ (to destroy $KMnO_4$). To the residual solution is added acetic acid and $CaCl_2$ solution and heated a white ppt. of Ca oxalate is obtained. |
| Epichlorohydrin<br>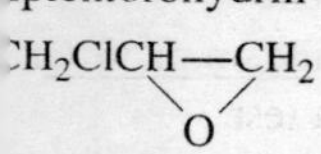 | 117 | Colourless, insoluble in $H_2O$, soluble in alcohol | | | | |
| Ethylene glycol diethyl ether<br>$C_2H_5OCH_2CH_2OC_2H_5$ | 121 | | | | | |
| Ethylene glycol monoethyl ether<br>$HOCH_2CH_2OC_2H_5$ | 135 | Colourless, soluble in $H_2O$, alcohol, ether | 75 | | | |
| Di-n-butyl ether<br>-$C_4H_9OC_4H_9(n)$ | 140 | Colourless, insoluble in $H_2O$, soluble in alcohol, ether | 63 | | | |
| Cyclohexyl ethyl ether<br>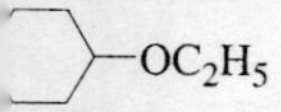 | 149 | | | | | |
| Anisole<br>Phenyl methyl ether)<br>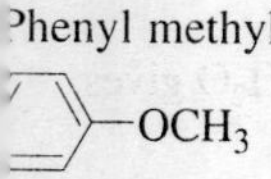 | 153 | Colourless, pleasant smell | — | 81 | 87 (di) | - The compound in conc. $H_2SO_4$ is treated with conc. $HNO_3$ + conc. $H_2SO_4$ (1:1) at room temp. (30°). After 10 min, the mixture is poured on ice to give dinitro derivative, m.p. 87°C. |
| Diethylene glycol dimethyl ether (Diglyme)<br>$H_3OCH_2CH_2OCH_2CH_2OCH_3$ | 162 | Colourless, soluble in $H_2O$ | | | | |
| Methyl benzyl ether<br>$C_6H_5CH_2OCH_3$ | 169 | | | 116 | | - On oxidation (alkaline $KMnO_4$) gives benzoic acid. |
| Di-isoamyl ether<br>$[(CH_3)_2CHCH_2CH_2]_2O$ | 171 | Colourless, insoluble in $H_2O$, soluble in alcohol, ether | 62 | — | — | |
| Cresyl methyl ether<br>-methoxytoluene) | 171 | Colourless, insoluble in $H_2O$, soluble in alcohol, | — | 119 | 64 (mono) | - Bromo derivative, m.p. 63°C. |

| *Name and structure* | *B.P. (°C)* | *Properties* | *3, 5-dinitro-benzoate* | *Picrate* | *Nitro derivative* | *Specific tests and derivatives* |
|---|---|---|---|---|---|---|
| 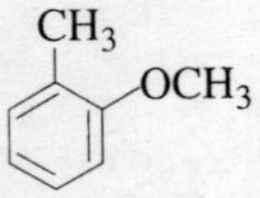 | | ether | | | | - On nitration with conc. $H_2SO_4$ + conc. $HNO_3$ gives nitro derivative, m.p. 64°C.<br>- On oxidation (alkaline $KMnO_4$) gives o-methoxy-benzoic acid, m.p. 100°C. |
| Phenetole (Phenyl ethyl ether) 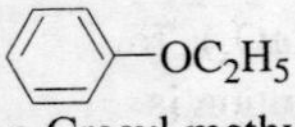 | 172 | Colourless, insoluble in $H_2O$ | | 92 | 58 | |
| p-Cresyl methyl ether (p-Methoxytoluene) 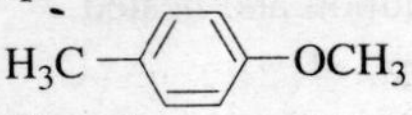 | 176 | Colourless, insoluble in $H_2O$. | — | 89 | 122 (di) | - Oxidation ($KMnO_4$) gives anisic acid, m.p. 184°C. |
| Cineole  | 176 | | | | | - With dry HCl at 40-50° yields dipentene dihydrochloride, m.p. 50°C. |
| *m*-Cresyl methyl ether (m-Methoxytoluene) 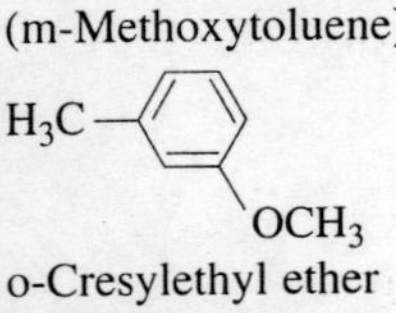 | 177 | | — | 114 | 91 (tri) | - Oxidation ($KMnO_4$) gives m-methoxybenzoic acid, m.p. 100°C. |
| o-Cresylethyl ether (o-Ethoxytoluene) 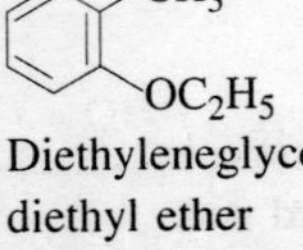 | 184 | | — | 118 | 51 (di) | |
| Diethyleneglycol diethyl ether $(C_2H_5OCH_2CH_2)_2O$ | 188 | | | | | |
| p-Cresylethyl ether (p-Ethoxytoluene) 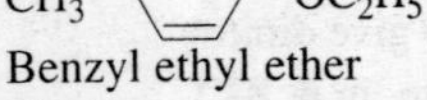 | 189 | | — | 111 | — | |
| Benzyl ethyl ether 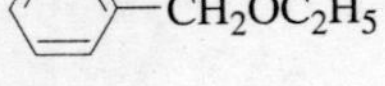 | 189 | Colourless, insoluble in $H_2O$, soluble in alcohol, ether | | | | |
| m-Ethoxytoluene m-Cresylethyl ether 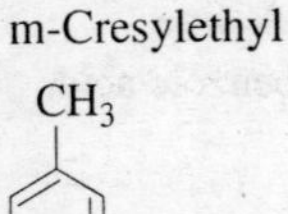 | 192 | | — | 115 | — | |
| o-Chloroanisole | 195 | Colourless, insoluble in $H_2O$, | — | 57 | 95 (mono) | - With fuming $HNO_3$ gives |

| *Name and structure* | *B.P. (°C)* | *Properties* | *3, 5-dinitro-benzoate* | *Picrate* | *Nitro derivative* | *Specific tests and derivatives* |
|---|---|---|---|---|---|---|
| 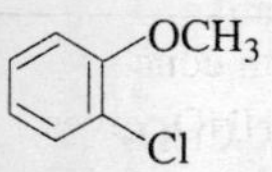  | | soluble in alcohol, ether | | | | mono nitro derivative, m.p. 95°C. |
| p-Chloroanisole 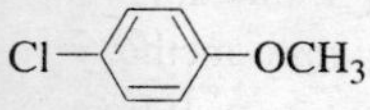  | 200 | Colourless, insoluble in $H_2O$, soluble in alcohol, ether | — | | 98 (mono) | - With fuming $HNO_3$ gives mono nitro derivative, m.p. 98°C. |
| Guaiacol | 205 | See phenols, Table 6.4a | | | | |
| Veratrole (Catechol dimethyl ether) 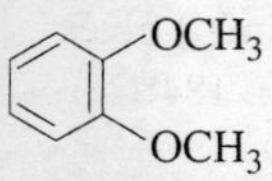  | 206 (m.p. 22) | Colourless, insoluble in $H_2O$, soluble in alcohol, ether | — | 57 | 95 (mono) | - Dibromo derivative, m.p. 92°C. - With fuming $HNO_3$ gives mononitro derivative, m.p. 95°C. |
| o-Chlorophenetole 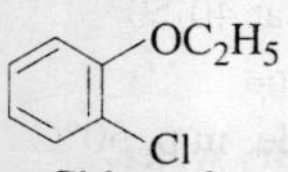  | 208 | Colourless, insoluble in $H_2O$, soluble in alcohol, ether | — | | 82 (mono) | - With fuming $HNO_3$ gives 4-nitro derivative, m.p. 82°C. |
| p-Chlorophenetole 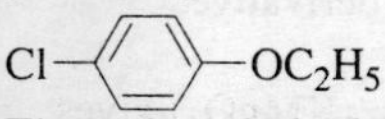  | 212 (m.p. 21) | Colourless, insoluble in $H_2O$, soluble in alcohol, ether | — | | 61 (mono) 54 (di) | |
| Thymol methyl ether 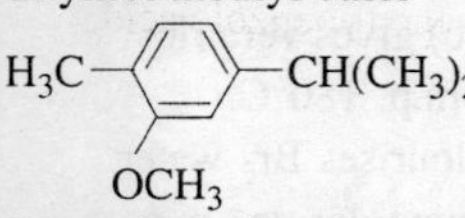  | 212 | | — | | 92 (trinitro) | - On boiling with 48% HBr yields thymol, m.p. 50°C. |
| Resorcinol dimethyl ether 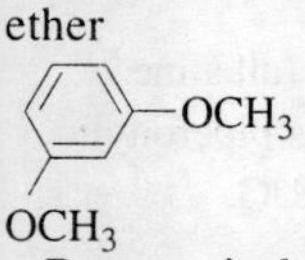  | 214 | | — | 58 | 124 (tri.) | - Dibromo derivative, m.p. 140. |
| o-Bromoanisole 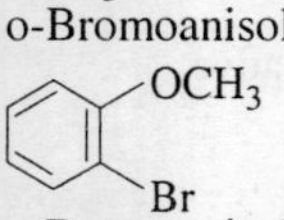  | 218 | Colourless, soluble in alcohol | | | 106 (mono) | - With fuming $HNO_3$ + glacial acetic acid gives 4-nitro derivative m.p. 106°C. |
| p-Bromoanisole 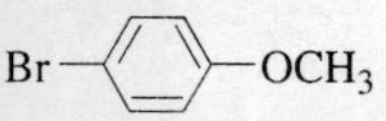  | 222 (m.p. 13) | Colourless, soluble in alcohol | — | — | 88 (mono) | - On nitration (fuming $HNO_3$ + glacial acetic acid) gives 2-nitro derivative, m.p. 88°C. |
| o-Bromophenetole 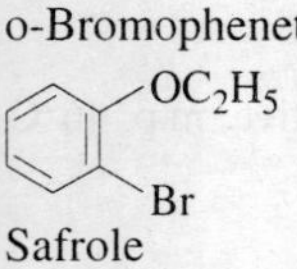  | 224 | Colourless, soluble in alcohol ether | — | — | 98 (mono) | - With fuming $HNO_3$ + gives 4-nitro derivative m.p. 98°C. |
| Safrole 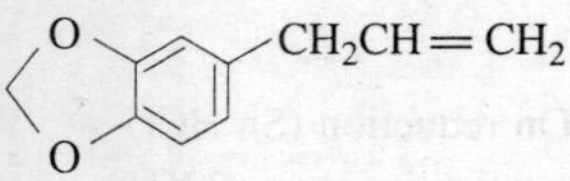  | 232 | | | | | - Decolourises $Br_2$ water. - Penta bromo derivative, m.p. 169°C. - Oxidation (alkaline $KMnO_4$) gives piperonylic acid, m.p. 228°C. |

| *Name and structure* | *B.P. (°C)* | *Properties* | *3, 5-dinitro-benzoate* | *Picrate* | *Nitro derivative* | *Specific tests and derivatives* |
|---|---|---|---|---|---|---|
| Anethole (p-propenylanisole) 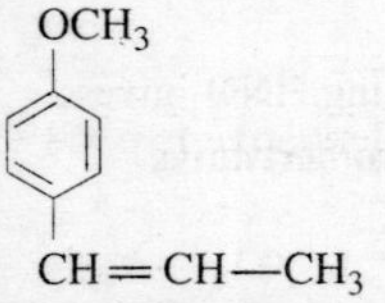 | 232 (m.p. 22) | Colourless, characteristic odour, insoluble in $H_2O$, soluble in alcohol, ether | — | 70 | — | - Tribromo derivative m.p. 108°C.<br>- On oxidation ($K_2Cr_2O_7$ + $H_2SO_4$) yields anisic acid, m.p. 184°C. |
| Resorcinol diethyl ether 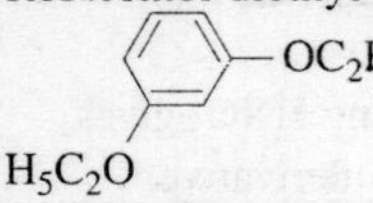 | 235 | Colourless, insoluble in $H_2O$, soluble in alcohol, ether | — | 58 | — | |
| m-Methoxyphenol (Resorcinol monomethyl ether) 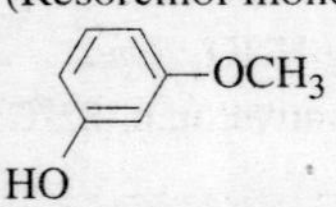 | 243 | See Phenols (Table 6.4a) | | | | |
| Eugenol methyl ether 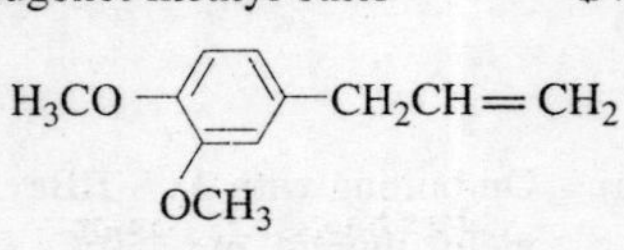 | 244 | | — | 114 | — | - Decolourises $Br_2$ water.<br>- Dibromo derivative, m.p. 77°C.<br>- On oxidation ($K_2Cr_2O_7$ in HOAc) gives veratric acid, m.p. 180°C. |
| Isosafrole 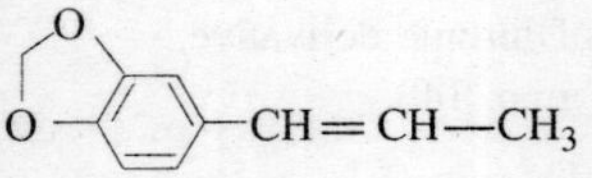 | 246 | | — | 75 | — | - Decolourises $Br_2$ water.<br>- Tribromo derive., m.p. 109°C.<br>- On oxidation (alkaline $KMnO_4$) gives piperonylic acid, m.p. 228°C. |
| Anisaldehyde | 247 | See carbonyl compounds (Table 6.3a) | | | | |
| Diphenyl ether 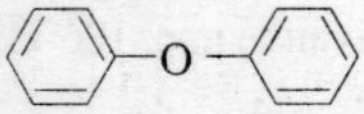 | 252 m.p. 28 | Colourless, Germanium like odour, insoluble in $H_2O$, soluble in alcohol, ether | — | 110 | 135 (di) | - Dibromo derivative, m.p. 58°C. |
| Iso-eugenol methyl ether 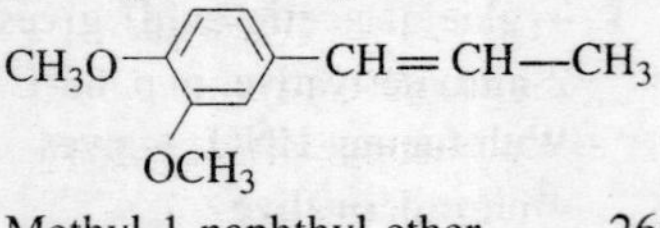 | 264 | | | | | |
| Methyl-1-naphthyl ether 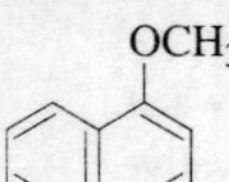 | 265 | Colourless, insoluble in $H_2O$, soluble in alcohol, ether | — | 113 | — | - Bromo derivative, m.p. 46°C. |
| o-Nitroanisole 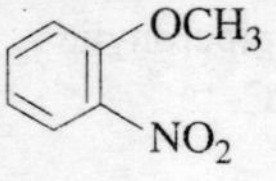 | 272 (m.p. 10) | Colourless, insoluble in $H_2O$, soluble in alcohol, ether | — | — | 95 | - On reduction (Sn HCl)—o-anisidine, m.p. 225°C.<br>- Nitration (50°; 15 min) (conc. $H_2SO_4$ + conc. $HNO_3$) gives nitro derivative, m.p. 95°C. |

| *Name and structure* | *B.P. (°C)* | *Properties* | *3, 5-dinitro-benzoate* | *Picrate* | *Nitro derivative* | *Specific tests and derivatives* |
|---|---|---|---|---|---|---|
| o-Methoxydiphenyl 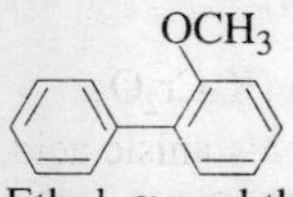 | 274 | | — | — | 95 | |
| Ethyl $\alpha$-naphthyl ether 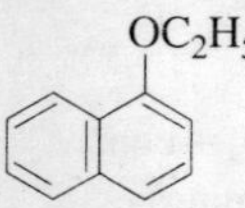 | 276 | Colourless, insoluble in $H_2O$, soluble in alcohol, ether | — | 119 | — | - Bromo derivative m.p. 48°C. |
| Dibenzyl ether $C_6H_5CH_2OCH_2C_6H_5$ | 298 | Colourless, insoluble in $H_2O$, soluble in hot alcohol, ether | 112 | 78 | — | |

**Table 6.10(b)** Ethers (Solids)

| *Name and structure* | *M.P. (°C)* | *Properties* | *3, 5-dinitro-benzoate* | *Picrate* | *Nitro derivative* | *Specific tests and derivatives* |
|---|---|---|---|---|---|---|
| p-Chlorophenetol | 21 (b.p. 212) | See liquids, b.p. 212. | | | — | |
| Diphenyl ether | 28 (b.p. 252) | See liquids, b.p. 252 | — | — | — | |
| Apiole 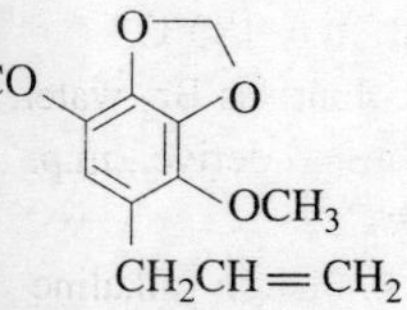 | 29 | | | | | - Gives blood red colour in conc. $H_2SO_4$. |
| Ethyl $\beta$-naphthyl ether 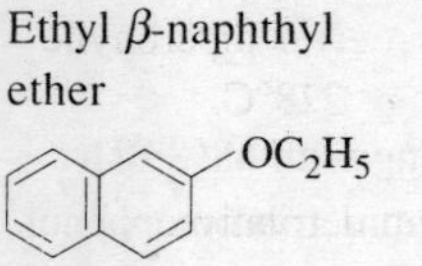 | 37 (b.p. 282) | Colourless, floral odour, insoluble in $H_2O$, soluble in alcohol, ether | — | 44 | — | - Bromo derivative, m.p. 66°C. |
| m-Nitroanisole 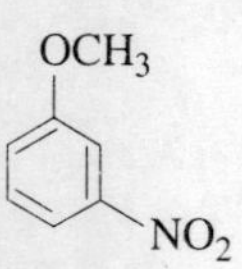 | 39 (b.p. 258) | Colourless, insoluble in $H_2O$, soluble in alcohol | — | — | — | - On reduction (Sn + HCl) yields m-anisidine, b.p. 251°C. (Picrate, m.p. 169°C (d)); - On heating with 48% HBr gives m-nitrophenol, m.p. 97°C. |
| Pyrogallol triethyl ether 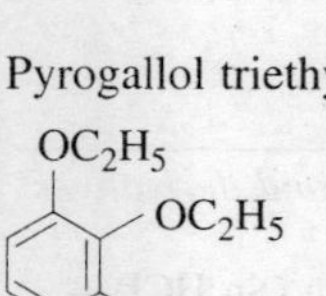 | 39 | | — | — | — | |
| Catechol diethyl ether 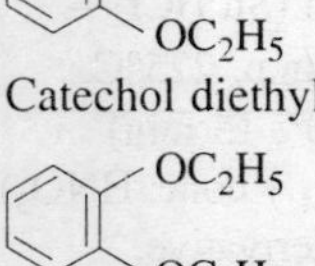 | 43 | | — | 71 | 122 (tri) | |
| Pyrogallol trimethyl ether | 47 | | — | 81 | — | |

| *Name and structure* | *M.P. (°C)* | *Properties* | *3, 5-dinitro-benzoate* | *Picrate* | *Nitro derivative* | *Specific tests and derivatives* |
|---|---|---|---|---|---|---|
| 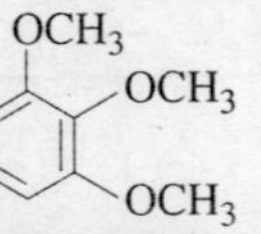 | | | | | | |
| Hydroquinone dimethyl ether 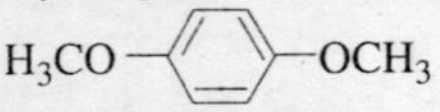 | 56 | | — | 119 | — | - Bromo derivative, m.p. 142°C. |
| Symm. trichloroanisole 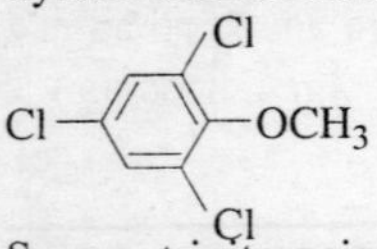 | 62 | Colourless, insoluble in $H_2O$ | — | — | 95 (di) | - With conc. $H_2SO_4$ + conc. $HNO_3$ (long warming) gives dinitroderivative, m.p. 95°C. |
| Symm. trinitroanisole 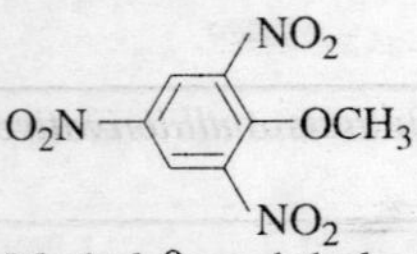 | 68 | Colourless, insoluble in $H_2O$, soluble in alcohol | — | — | — | - On heating with aq. NaOH gives picric acid, m.p. 122°C. |
| Methyl-β-naphthyl ether 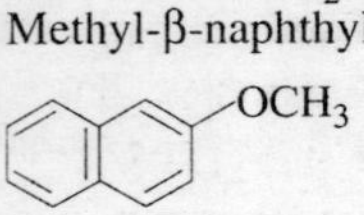 | 72 | Colourless, orange blossom odour, insoluble in $H_2O$, sparingly soluble in alcohol, soluble in ether | — | 108 | — | - Bromo derivative, m.p. 86°C. |
| Vanillin | 81 | See Carbonyl compounds (Table 6.3b) | | | | |
| p-Methoxydiphenyl 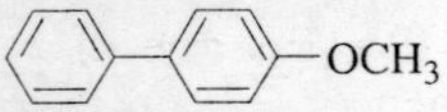 | 85 | | | | | |
| Dibenzofuran (Diphenylene oxide) 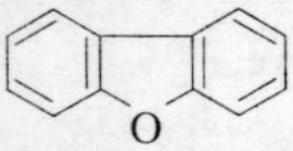 | 86 | | — | 94 | 182 | |
| Symm. tribromoanisole 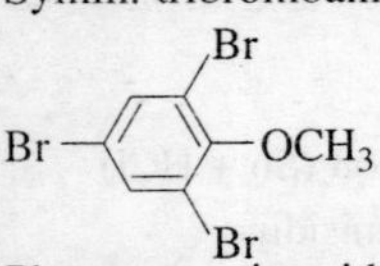 | 88 | Colourless, insoluble in $H_2O$ | — | — | — | - On heating with 48% BHr yields symm. tribromophenol, m.p. 95°C. |
| Phenoxyacetic acid | 98 | See carboxylic acids (Table 6.2b) | | | | |
| o-Methoxybenzoic acid | 100 | See carboxylic acids (Table 6.2b) | | | | |
| p-Methoxybenzoic acid (Anisic acid) | 184 | See carboxylic acids (Table 6.2b) | | | | |

**Table 6.11(a)** Hydrocarbons (Liquids)

| *Name and structure* | *B.P. (°C)* | *Properties* | *Pic-rate* | *Nitro- M.P.* | *derivative (Position of $NO_2$ group)* | *Pro-cedure* | *Specific tests and derivatives* |
|---|---|---|---|---|---|---|---|
| Isopentane $(CH_3)_2CHCH_2CH_3$ | 30 | | | | | | |
| Isoprene | 34 | | | | | | - Decolourises $Br_2$ water. |

| *Name and structure* | *B.P. (°C)* | *Properties* | *Picrate* | *Nitro-derivative M.P.* | *(Position of $NO_2$ group)* | *Procedure* | *Specific tests and derivatives* |
|---|---|---|---|---|---|---|---|
| $CH_2{=}C(CH_3){-}CH{=}CH_2$ | | | | | | | |
| *n*-Pentane $CH_3(CH_2)_3CH_3$ | 36 | | | | | | |
| Trimethylethylene (Amylene) $(CH_3)_2C{=}CHCH_3$ | 38 | | | | | | - Decolourless $Br_2$ water.<br>- On dilution and distillation yields tert. amyl alcohol, b.p. 102°C. |
| *n*-Hexane $CH_3(CH_2)_4CH_3$ | 68 | Colourless, petrol odour insoluble in $H_2O$, sparingly soluble in alcohol, ether | | | | | |
| Benzene  | 80 | Colourless, characteristic odour (caution: carcinogenic), soluble in alcohol | 84 | 90 | (1, 3) | 1 | |
| Cyclohexane 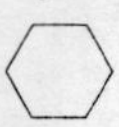 | 80 (m.p. 4°) | Colourless, petrol odour insoluble in $H_2O$, soluble in alcohol, ether | | | | | - On oxidation (on boiling with fuming $HNO_3$) yields adipic acid, m.p. 151°C. |
| Cyclohexene  | 83 | Colourless, insoluble in $H_2O$, soluble in alcohol, ether | | | | | - Decolourises $Br_2$ water<br>- On oxidation ($HNO_3$) yields adipic acid m.p. 151°C (See procedure A). page 195) |
| *n*-Heptane $CH_3(CH_2)_5CH_3$ | 98 | Colourless, petrol odour, insoluble in $H_2O$, soluble in alcohol, ether | | | | | |
| Toluene 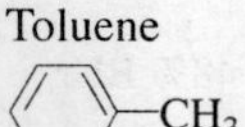 | 110 | Colourless, characteristic odour, insoluble in $H_2O$, soluble in alcohol, ether | 88 | 70 | (2, 4) | 2 | - On oxidation (alkaline $KMnO_4$) gives benzoic acid, m.p. 122°C. |
| *n*-Octane $CH_3(CH_2)_6CH_3$ | 125 | Colourless, petrol odour, insoluble in $H_2O$, soluble in ether, alcohol | | | | | |
| Ethyl benzene 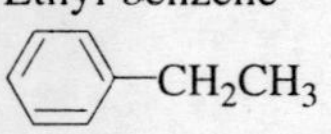 | 136 | Colourless, characteristic odour, insoluble in $H_2O$, soluble in alcohol, ether | 96 | 37 | (2, 4, 6) | 1 | - On oxidation ($KMnO_4$) benzoic acid m.p. 122°C. |
| *p*-Xylene 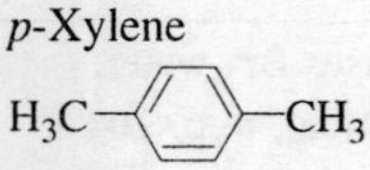 | 137 (m.p. 15) | Colourless, characteristic odour, insoluble in $H_2O$, soluble in alcohol, ether | 90 | 137 | (2, 3, 5) | 1 | - On oxidn. ($KMnO_4$)— terephthalic acid, m.p. > 300. |
| m-Xylene 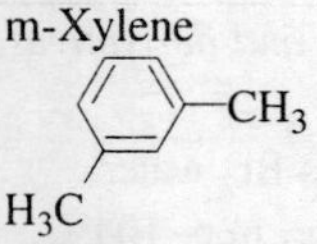 | 139 | Colourless, characteristic odour, insoluble in $H_2O$, soluble in alcohol, ether | 91 | 182<br>83 | (2, 4, 6)<br>(2, 4) | 1 | - On oxidn. ($KMnO_4$)— isophthalic acid, m.p. > 300. |
| o-Xylene 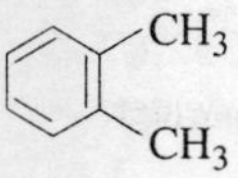 | 142 | Colourless, characteristic odour, insoluble in $H_2O$, soluble in alcohol, ether | 88 | 71<br>178 | (4, 5)<br>(tri) | | - On oxidn. ($KMnO_4$)— phthalic acid, m.p. 195.<br>- With $Br_2$ (130°) yields dibromide, m.p. 93. |

| *Name and structure* | *B.P. (°C)* | *Properties* | *Pic-rate* | *Nitro- derivative M.P.* | *(position of $NO_2$ group)* | *Procedure* | *Specific tests and derivatives* |
|---|---|---|---|---|---|---|---|
| Styrene 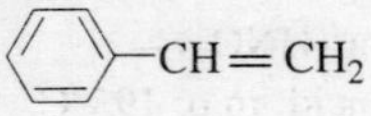 | 146 | Colourless, sparingly soluble in $H_2O$, soluble in alcohol, ether, pleasant smell | | | | | - Decolourises $Br_2$ water<br>- On oxidn. (alk. $KMnO_4$) — Benzoic acid m.p. 122.<br>- Dibromide, m.p. 73.<br>- On oxidn. (acidic $KMnO_4$)—Benzaldehyde, b.p. 179. |
| Cumene (Isopropylbenzene) 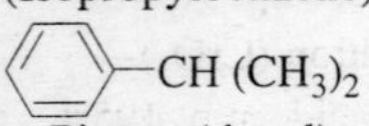 | 152 | Colourless, insoluble in $H_2O$, soluble in alcohol, ether | — | 109 | (2, 4, 6) | | - On oxidation ($KMnO_4$)—benzoic acid, m.p. 122 |
| α-Pinene (d or l)  | 156 | Colourless, turpentine odour, insoluble in $H_2O$, soluble in alcohol, ether | | | | | - Decolourises $Br_2$ water.<br>- Dibromide derivative, m.p. 164°C. |
| n-Propylbenzene 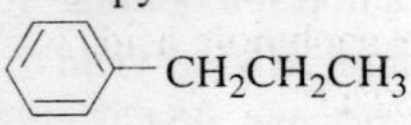 | 159 | Colourless, insoluble in $H_2O$ soluble in alcohol, ether | 103 | — | — | — | - On oxidation ($KMnO_4$)—benzoic acid, m.p. 122°C. |
| Mesitylene 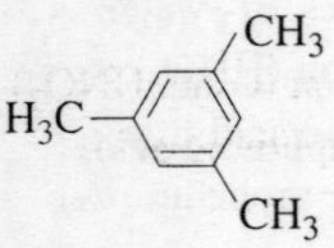 | 164 | Colourless, insoluble in $H_2O$, soluble in alcohol, ether | 97 | 235<br>86 | (2, 4, 6)<br>(2, 4) | 2 | |
| tert-Butyl benzene 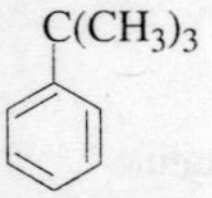 | 169 | | — | 62 | (2,4) | | |
| pseudo cumene 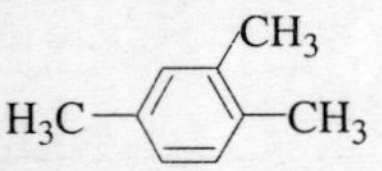 | 169 | | 97 | 185 | (3, 5, 6) | | |
| p-Cymene 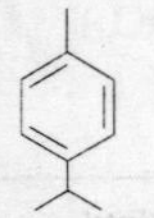 | 176 | Colourless, eucalyptus odour, insoluble in $H_2O$, soluble in alcohol | —<br>— | 118<br>54 | (2, 3, 6)<br>(2, 6) | 1 | |
| Limonene 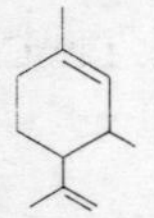 | 176 | Colourless, lemon odour, insoluble in $H_2O$, soluble in alcohol, ether $[\alpha]_p^{20°}$ + 127. | | | | | - Decolourises $Br_2$ water.<br>- Tetrabromide., m.p. 104°C |
| Dipentene (dl-limonene) | 176 | Colourless, insoluble in $H_2O$, soluble in alcohol | | | | | - Decolourises $Br_2$ water.<br>- Tetrabromide, m.p. 104°C. |
| Indene 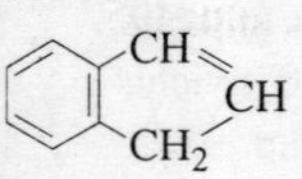 | 180 | Colourless, insoluble in $H_2O$,soluble in alcohol, ether | 98 (explosive) | — | — | — | - Decolourises $Br_2$ water.<br>- Dibromide, m.p. 32°C.<br>- Gradually polymerises |

| *Name and structure* | *B.P. (°C)* | *Properties* | *Pic-rate* | *Nitro-derivative M.P.* | *(position of $NO_2$ group)* | *Pro-cedure* | *Specific tests and derivatives* |
|---|---|---|---|---|---|---|---|
| | | | | | | | on heating.<br>- Oxidation ($HNO_3$)— phthalic acid, m.p. 195°C. |
| Decalin (Trans)<br>cis<br>(Decahydronaphthalene)<br>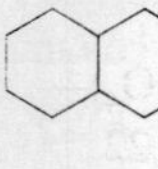 | 187<br>195 | Colourless, insoluble in $H_2O$, soluble in alcohol, ether | — | 164 | — | — | - On boiling with 35% $HNO_3$ gives a nitro derivative (insoluble in alcoholic KOH), b.p. 96/2 mm and a dinitro derivative, m.p. 164°C. |
| Tetralin (Tetrahydronaphthalene)<br>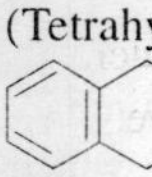 | 206 | Colourless, insoluble in $H_2O$, soluble in alcohol, ether | — | 95 | (5, 7) | 2 | - On oxidation ($CrO_3$)— phthalic acid, m.p. 195°C..<br>- Bromination (2 moles $Br_2$) yields dibromo derivative, m.p. 70°C. |
| $\alpha$-Methylnaphthalene<br>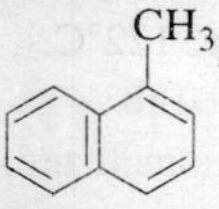 | 241 | Colourless, insoluble in $H_2O$, soluble in alcohol, ether | 141 | 71 | (4) | 3 | - On oxidation (dil. $HNO_3$) gives $\alpha$-naphthoic acid, m.p. 162°C. |
| $\beta$-Methylnaphthalene<br>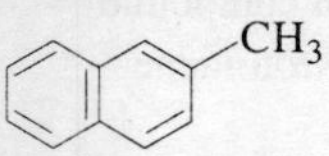 | 241 (m.p. 32) | Colourless, insoluble in $H_2O$, soluble in alcohol, ether | 115 | 81 (1) | 3 | | - On oxidation (conc. $HNO_3$) gives $\beta$-Naphthoic acid, m.p. 184°C. |
| Diphenylmethane<br>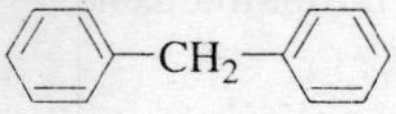 | 264 (m.p. 26) | Colourless, orange like odour, insoluble in $H_2O$, soluble in alcohol, ether | — | 172 | (2, 2′, 4, 4′) | 1 | |

**Table 6.11(b)** Hydrocarbons (Solids)

| *Name and structure* | *B.P. (°C)* | *Properties* | *Pic-rate* | *Nitro derivative M.P.* | *(position of $NO_2$ group)* | *Pro-cedure* | *Specific tests and derivatives* |
|---|---|---|---|---|---|---|---|
| (–) Camphene<br>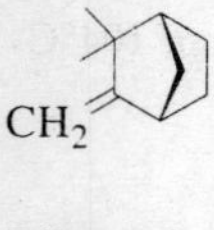 | 52 (b.p. 158) | Colourless, turpentine and camphor like odour, insoluble in $H_2O$, soluble in ether, alcohol | | | | | - Decolourises $Br_2$ water.<br>- Dibromide, m.p. 89°C.<br>- Hydrochloride, m.p. 149°C. |
| Dibenzyl ($(C_6H_5CH_2)_2$) | 53 | Colourless, insoluble in $H_2O$, soluble in alcohol, ether | | 180 | (4, 4′) | 4 | - On oxidation ($KMnO_4$)— benzoic acid, m.p. 122°C.<br>- On boiling with $KClO_3$ in HCl yields stilbene, m.p. 125°C. |

| *Name and structure* | *M.P. (°C)* | *Properties* | *Picrate* | *Nitro- derivative M.P. (position of $NO_2$ group)* | *Procedure* | *Specific tests and derivatives* |
|---|---|---|---|---|---|---|
| Diphenyl 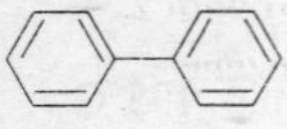 | 70 | Colourless, characteristic odour, insoluble in $H_2O$, soluble in alcohol, ether | | 234 (4, 4′) | 5 | - On oxidation ($CrO_3$)—benzoic acid, m.p. 122°C.<br>- With $Br_2$ in presence of Fe yields *p, p′*-dibromo derivative, m.p. 169°C. |
| Naphthalene 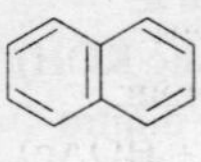 | 80 | Colourless, characteristic unpleasant penetrating odour, insoluble in $H_2O$, solution in alcohol, ether | 150 | 61 (1) | 3 | - Addition of anhydrous $AlCl_3$ to chloroform solution of the compound gives green colour.<br>- On oxidation ($KMnO_4$) gives phthalic acid, m.p. 195°C. |
| Triphenylmethane $(C_6H_5)_3CH$ | 92 | Colourless, insoluble in $H_2O$, soluble in hot alcohol, ether | | 206 (4, 4′, 4″) | 4 | - Reduction of trinitro derivative (Zn dust + $CH_3COOH$) gives triamino derivative, m.p. 208°C, which on oxidation ($PbO_2$) gives intense red colour of rosaniline.<br>- Bromination ($Br_2$ at 130°) gives triphenylbromomethane, m.p. 152°C. |
| Acenaphthene 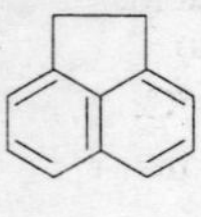 | 95 | Colourless, insoluble in $H_2O$, soluble in hot alcohol, ether | 161 | 101 (5) | 6 | - Gives addition compound with 2, 4-dinitrotoluene, m.p. 60°C.<br>- Gives addition compound with 2, 4, 6-trinitrotoluene, m.p. 109°C. |
| Retene (8-Methyl-2-isopropyl phenanthrene) 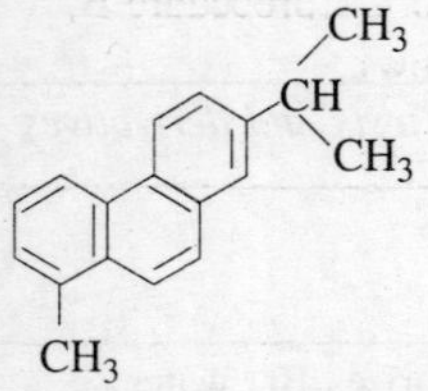 | 98 | | 124 (unstable) | — | — | - On oxidation ($K_2Cr_2O_7$ + $H_2SO_4$) yields retenequinone, m.p. 197°C.<br>- With $Br_2$ at 90-100° yields dibromo derivative, m.p. 180°C. |
| Azulene 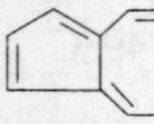 | 99 | Deep blue colour, insoluble in $H_2O$, soluble in alcohol, dissolves in conc. acids | | | | |
| Phenanthrene 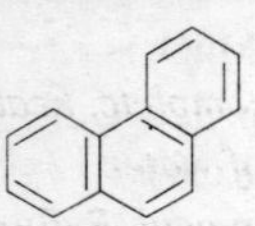 | 100 | Colourless, insoluble in $H_2O$, soluble in ether | 143 | | | - On oxidation gives phenanthraquinone, m.p. 206° and diphinic acid, m.p. 228°C. |

| *Name and structure* | *M.P. (°C)* | *Properties* | *Picrate* | *Nitro- derivative M.P. (position of $NO_2$ group)* | *Procedure* | *Specific tests and derivatives* |
|---|---|---|---|---|---|---|
| | | | | | | - Gives compound with 2, 4-dinitrochlorobenzene, m.p. 44°C.<br>- Gives compound with 2, 4, 6-trinitrotoluene, m.p. 87°C. |
| Fluorene<br>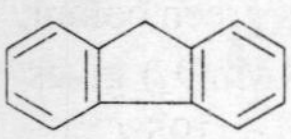 | 114 | Colourless insoluble in $H_2O$, soluble in hot alcohol | 79 | 199 (2, 7) | | - On warming with conc. $H_2SO_4$ gives blue colour.<br>- On oxidation ($CrO_3$ + HOAc) gives fluorenone, m.p. 83°C. (procedure B, See below).<br>- On oxidation (alkaline $KMnO_4$) gives phthalic acid, m.p. 195°C. |
| Stilbene (trans)<br>(cis)<br>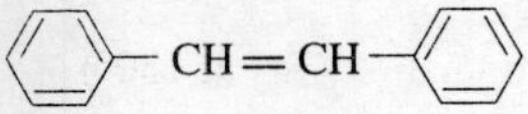 | 125<br>(141/12) | Colourless, insoluble in $H_2O$, soluble in alcohol, ether | | | | - Decolourises $Br_2$ water on warming.<br>- Dibromide, m.p. 237°C.<br>- On oxidation ($KMnO_4$ + dil. $H_2SO_4$) gives smell of bitter almonds. |
| Anthracene<br>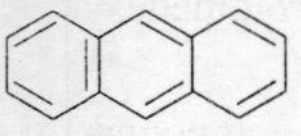 | 216 | Colourless or pale yellow, insoluble in $H_2O$, soluble in alcohol. | 138 | | | - On bromination ($Br_2$ in $CCl_4$) yields dibromo derivative, m.p. 221°C.<br>- On oxidation ($CrO_3$ in $CH_3COOH$) gives anthraquinone, m.p. 280°C. (procedure B, see below). |
| Chrysene<br>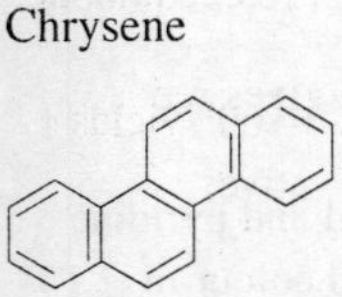 | 255 | | | 273 (unstable) | | - On oxidation ($CrO_3$ in HOAc) gives chrysaquinone, m.p. 240°C. (procedure B, see below). |

*Procedure for nitration:*

1. *Fuming $HNO_3$ with conc. $H_2SO_4$.*
2. *Mixed conc. $HNO_3$ and $H_2SO_4$.*
3. *Cold conc. $HNO_3$, 10 min.*
4. *Fuming $HNO_3$ at room temp., 10 min.*
5. *Conc. $HNO_3$ at 100°, 10 min.*
6. *Add solution of fuming $HNO_3$ in glacial acetic acid dropwise to ice-cold solution of hydrocarbon in glacial acetic acid. Keep at room temp. for 10 min. Do not heat.*

*Procedure A: Add 1 g. cyclohexene in small portions to 5 ml of warm conc. $HNO_3$ (caution). When the addition is complete, heat under reflux for 15 min. Cool, filter the separated adipic acid and recrystallise from small volume of water.*

*Procedure B: Add a solution of 1 g of $CrO_3$ in water (2 ml) to a solution of the hydrocarbon (0.5 g) in glacial acetic acid. Reflux for 30 min. cool and add water. Filter the formed quinone and crystallise from glacial acetic acid.*

**Table 6.12(a)** Halohydrocarbons (Alkyl and Aryl halides) Liquids

| *Name and structure* | *B.P.(°C)* | *Derivatives and tests* |
|---|---|---|
| Methyl chloride<br>$CH_3Cl$ | Gas | - Alkyl-$\beta$-naphthyl ether, m.p. 72°C.<br>- S-Alkylisothiouronium picrate, m.p. 224°C. |
| Ethyl chloride<br>$C_2H_5Cl$ | 12 | S-Alkylisothiouronium picrate, m.p. 188°C. |
| Isopropyl chloride<br>$(CH_3)_2CHCl$ | 36 | - Alkyl-$\beta$-naphthyl ether, m.p. 41°C.<br>- S-Alkylisothiouronium picrate, m.p. 196°C (2 hr heating) |
| Ethyl bromide<br>$C_2H_5Br$ | 38 | - S-Alkylsothiouronium picrate, m.p. 188°C. |
| Methylene chloride<br>(Dichloromethane)<br>$CH_2Cl_2$ | 40 | $\beta$-Naphthyl ether, m.p. 133°C. |
| Methyl iodide<br>$CH_3I$ | 43 | - Gives yellow ppt. with alcoholic $AgNO_3$ in cold.<br>- $\beta$-Naphthyl ether m.p. 70°C.<br>- Quatenary salt (with quinoline, m.p. 72°), (with pyridine m.p. 117°C). |
| n-Propyl chloride<br>$CH_3CH_2CH_2Cl$ | 46 | - $\beta$-Naphthyl ether, m.p. 39°C.<br>- S-Alkylisothiouronium picrate, m.p. 177°C (unstable; 3 hr heating). |
| Allyl chloride<br>$CH_2{=}CHCH_2Cl$ | 46 | - Decolourises $Br_2$ water.<br>- S Alkylisothiouronium picrate, m.p. 154°C. |
| Dichloroethylene<br>$CHCl{=}CHCl$ (trans) | 48 | - Decolourises bromine water.<br>- Dibromide m.p. 192°C. |
| tert. Butyl chloride<br>$(CH_3)_3CCl$ | 52 | - On hydrolysis with cold $H_2O$ gives tert. butyl alcohol.<br>- S-Alkylisothiouronium picrate, m.p. 151°C. |
| Ethylidene chloride<br>(1, 1-Dichloroethane)<br>$CH_3CHCl_2$ | 57 | - $\beta$-Naphthyl ether, m.p. 200°C(di).<br>- Reduces Fehling solution. |
| Dichloroethylene<br>$ClCH{=}CHCl$ (cis) | 60 | - Decolourises $Br_2$ water.<br>- Dibromide, m.p. 192°C. |
| Chloroform<br>$CHCl_3$ | 61 | - On warming with aniline and alcoholic KOH yields obnoxious odour of phenylisocyanide.<br>- On warming with resorcinol and dil. NaOH solution gives red colour.<br>- Treatment of NaOH solution (20%) of the compound and pyridine (sufficient to form a layer) with chloroform gives red colour in pyridine layer.<br>- Reduces fehling solution on heating. |
| sec. Butyl chloride<br>$CH_3CH_2CHClCH_3$ | 67 | - Picrate of $\beta$-naphthyl ether, m.p. 85°C.<br>- S-Alkylisothiouronium picrate, m.p. 166°C. |
| Isobutyl chloride<br>$(CH_3)_2CHCH_2Cl$ | 68 | - $\beta$-Naphthyl ether, m.p. 33°C (forms picrate, m.p. 84°C).<br>- S-Alkyl isothiouronium picrate m.p. 167°C. |
| Allyl bromide<br>$CH_2{=}CHCH_2Br$ | 70 | - Decolourises $Br_2$ water.<br>- Picrate of $\beta$-naphthyl ether, m.p. 99°C. |
| Ethyl iodide<br>$C_2H_5I$ | 72 | - S-Alkyisothiouronium picrate, m.p. 188°C.<br>- Picrate of $\beta$-naphthyl ether, mp. 102°C.<br>- Addition product with pyridine, m.p. 90°C. |
| Carbon tetrachloride<br>$CCl_4$ | 77 | - On warming with aniline and alcoholic KOH gives obnoxious smell of phenylisocyanate, the reaction is slower than with chloroform. |

| *Name and structure* | *B.P.(°C)* | *Derivatives and tests* |
|---|---|---|
| *n*-Butyl chloride<br>$CH_3CH_2CH_2CH_2Cl$ | 77 | - Pictrate of $\beta$-naphthyl ether, m.p. 67°C.<br>- S-Alkylisothiouronium picrate, m.p. 177°C. |
| Ethylene chloride<br>$ClCH_2CH_2Cl$ | 84 | - Alkyl-2-Naphthyl ether, m.p. 217°C.<br>- Does not reduce fehling solutions. |
| Trichloroethylene<br>$CHCl{=}CCl_2$ | 87 | |
| Methylene bromide<br>$CH_2Br_2$ | 98 | - Alkyl-$\beta$-naphthyl ether m.p. 133°C. |
| Isoamyl chloride<br>$(CH_3)_2CH{\cdot}CH_2CH_2Cl$ | 100 | - Picrate of $\beta$-naphthyl ether, m.p. 94°C. |
| Allyl iodide<br>$CH_2{=}CHCH_2I$ | 101 | - Picrate of $\beta$-naphthyl ether, m.p. 99°C.<br>- Forms addition product with dimethyl aniline, m.p. 86°C.<br>- Yellow ppt. with alcoholic $AgNO_3$ solution. |
| *n*-Amyl chloride<br>$CH_3(CH_2)_4\ Cl$ | 106 | - Picrate, of Alkyl-$\beta$-naphthyl ether m.p. 67°C.<br>- S-Alkylisothiouronium picrate, m.p. 154°C. |
| Ethylene bromide<br>$CH_2BrCH_2Br$ | 132<br>(m.p. 10) | - Addition product with pyridine, m.p. 295°C (decompose).<br>- S-Alkylthiouronium picrate, m.p. 260°C. |
| Chlorobenzene<br>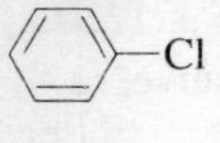 | 132 | - Nitration (conc. $HNO_3$ + conc. $H_2SO_4$ warming 80–90°) yields 2, 4-dinitro derivative m.p. 51°C.<br>- Mono nitro (p -) m.p. 84°C.<br>- On adding to large excess of chlorosulphonic acid gives p-chlorobenzene sulphonyl chloride m.p. 53°, which on treatment with $NH_4OH$ yields the sulphonamide, m.p. 144°C. |
| *n*-Hexyl chloride<br>$CH_3(CH_2)_4CH_2Cl$ | 134 | |
| Chlorocyclohexane<br>(Cyclohexyl chloride)<br>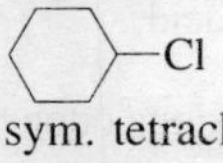 | 142 | - Alkyl $\beta$-naphthyl ether, m.p. 116°C. |
| sym. tetrachloroethane<br>$Cl_2CH{-}CHCl_2$ | 147 | |
| Bromoform<br>$CHBr_3$ | 151<br>(m.p. 9) | - Similar to chloroform |
| Bromobenzene<br>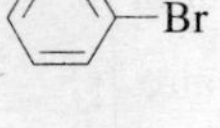 | 156 | - On nitration (conc. $HNO_3$ + conc. $H_2SO_4$ at 80–90°) yields 2, 4-dinitro deriv., m.p. 75°C.<br>- p-nitro bormobenzene m.p. 124°C.<br>- On adding to large excess of chlorosulphonic acid yields *p*-bromobenzenesulphonyl chloride, m.p. 75°, which on treatment with $NH_3$ gives the sulphonomaide, m.p. 160° C. |
| o-Chlorotoluene<br>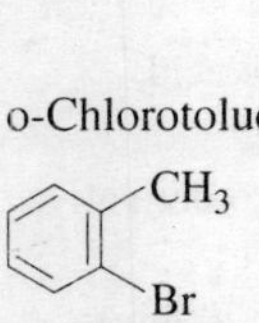 | 159 | - On oxidation (alkaline $KMnO_4$) gives *o*-chlorobenzoic acid, m.p. 142°C.<br>- On nitration (conc. $HNO_3$ + conc. $H_2SO_4$ at 80–90°) yields, 3, 5-dinitro derivative, m.p. 63°C. |
| *p*-Chlorotoluene<br>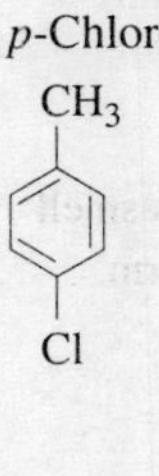 | 162 | - Oxidation (alkaline $KMnO_4$) gives *p*-chlorobenzoic acid, m.p. 243°C.<br>- On nitration (conc. $HNO_3$ + $H_2SO_4$) gives 2-nitro derivative m p. 38°C (temp. 40–50°, 20 min). |

| *Name and structure* | *B.P.(°C)* | *Derivatives and tests* |
|---|---|---|
| m-Chlorotoluene 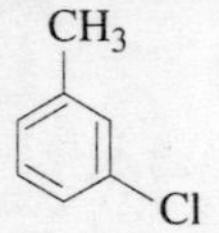  | 162 | - On oxidation (alkaline $KMnO_4$) gives m-chlorobenzoic acid, m.p. 158°C.<br>- On nitration (conc. $HNO_3$ + conc. $H_2SO_4$) gives 4, 6-dinitroderivative, m.p. 91°C. |
| Bromocyclohexane (Cyclohexyl bromide) 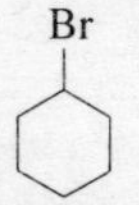  | 164 | - β-Naphthyl ether, m.p. 116°C. |
| *m*-Dichlorobenzene 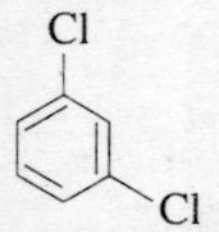  | 172 | - On nitration (fuming $HNO_3$) gives 4, 6-dinitro derivative, m.p. 103°C. |
| Benzyl chloride 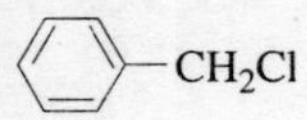  | 179 | - On boiling with lead nitrate solution gives smell of bitter almonds (benzaldehyde, b.p. 179°C).<br>- On oxidation ($K_2Cr_2O_7$ + $H_2SO_4$) gives benzoic acid, m.p. 122°C.<br>- On nitration (fuming $HNO_3$ in cold) gives p-nitro derivative, m.p. 71°C.<br>- Forms quartenary salt with with quinoline, m.p. 170°C.<br>- β-Naphthyl ether, m.p. 101°C (Picrate m.p. 123°). |
| o-Dichlorobenzene 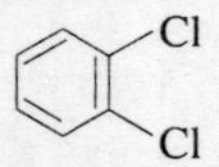  | 179 | - On nitration gives 4, 5-dinitro derivative, m.p. 110°C. |
| o-Bromotoluene 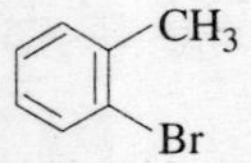  | 181 | - On oxidation (alkaline $KMnO_4$) gives o-bromobenzoic acid, m.p. 150°C.<br>- On bromination (2 moles $Br_2$) gives 2, 4, 5-tribromotoluene, m.p. 113°C.<br>- On nitration (conc. $HNO_3$ + conc. $H_2SO_4$-warm) gives dinitro derivative, m.p. 82°C. |
| *m*-Bromotoluene 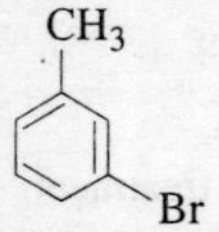  | 183 | - On oxidation (alkaline $KMnO_4$) gives m-bromobenzoic acid, m.p. 155°C.<br>- On nitration (conc. $HNO_3$ + conc. $H_2SO_4$) gives 4, 6-dinitro derivative, m.p. 103°C (temp. 40-50° for 20 min). |
| *p*-Bromotoluene 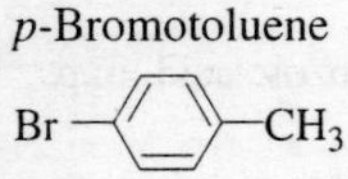  | 185<br>m.p. 27 | - On oxidation (alkaline $KMnO_4$) gives p-bromobenzoic acid, m.p. 251°C.<br>- On nitration (conc. $HNO_3$ + conc. $H_2SO_4$) gives 2-nitro derivative m.p. 47°C. |
| Iodobenzene 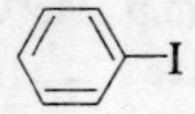  | 188 | - On nitration (conc. $HNO_3$ + conc. $H_2SO_4$) (40-50° for 20 min) gives 4-nitro derivative, m.p. 171°C.<br>- On bromination with bromine in glacial acetic acid gives p-bromoiodobenzene, m.p. 91°C. |

| *Name and structure* | *B.P.(°C)* | *Derivatives and tests* |
|---|---|---|
| β-Phenylethyl chloride 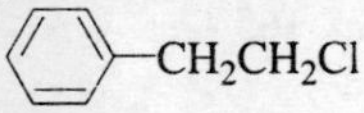  | 190 | - β-Naphthyl ether, m.p. 70°C.<br>- Anilide m.p. 97°C. |
| Benzyl bromide 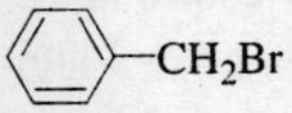  | 198 | - On oxidation give benzoic acid, m.p. 122°C. |
| m-Iodotoluene 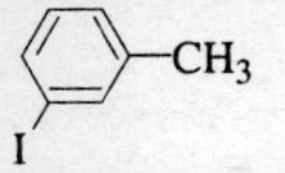  | 204 | - On oxidation gives m-iodobenzoic acid, m.p. 186°C.<br>- On nitration (conc. $HNO_3$ + conc. $H_2SO_4$) gives nitro derivative, m.p. 108°C. |
| o-Iodotoluene 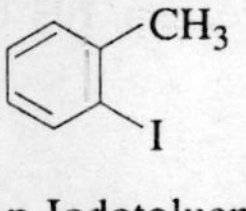  | 211 | - On oxidation (dil. $HNO_3$) gives o-iodobenzoic acid, m.p. 162°C.<br>- On nitration (fuming $HNO_3$ in cold) gives 6-nitro derivative, m.p. 103°C. |
| p-Iodotoluene 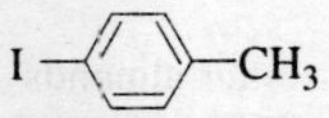  | 211<br>(m.p. 35) | - On oxidation ($CrO_3$) gives p-iodobenzoic acid, m.p. 265°C.<br>- On heating (210–250°) with copper powder yields di-p-toyl, m.p. 125°C. |
| Benzal chloride 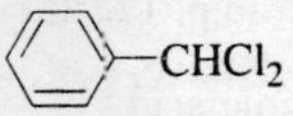  | 212 | - On refluxing with excess of phenyl hydrazine in ethanol yields benzaldehyde phenylhydrazone, m.p. 158°C.<br>On boiling with $Na_2CO_3$ solution yields benzaldehyde, b.p. 179°C. |
| m-Dibromobenzene 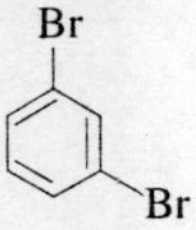  | 219 | - On nitration (conc. $HNO_3$ + conc. $H_2SO_4$) gives 4-nitroderivative, m.p. 61°C.<br>- With excess chlorosulphonic acid yields sulphonyl chloride, m.p. 79°, which with $NH_4OH$ yields the sulphonamide, m.p. 190°C. |
| Benzotrichloride<br>$C_6H_5CCl_3$ | 220 | - On heating with $Na_2CO_3$ solution gives benoic acid, m.p. 122°C.<br>- On boiling with alcohol yields HCl and ethyl benzoate, b.p. 213°C. |
| o-Dibromobenzene 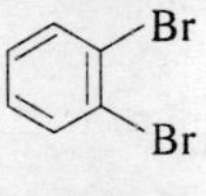  | 224 | - On nitration (conc. $HNO_3$ + conc. $H_2SO_4$) gives 4, 5-dinitro derivative, m.p. 114°C.<br>- With excess chlorosoluphonic acid yields a sulphonyl chloride, m.p. 34°, which with $NH_4OH$ gives the sulphonamide, m.p. 175°C. |
| α-Chloronaphthalene 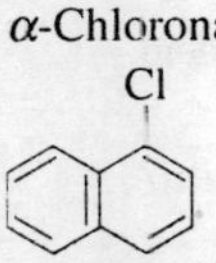  | 259 | - On nitration (fuming $HNO_3$) yields 4-5-dinitro derivative, m.p. 180°C.<br>- On heating with conc. $H_2SO_4$ gives 4-sulphonic acid, m.p. 130°C.<br>- Picrate, m.p. 137°C. |
| α-Bromonaphthalene 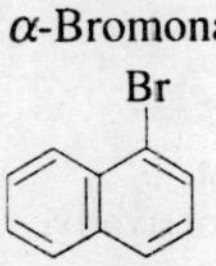  | 281 | - On oxidation ($Cr_2O_3$-acetic acid) yields phathalic acid, m.p. 195°C.<br>- On nitration (conc. $HNO_3$) gives 4-nitro derivative, m.p. 85°C.<br>- Pictrate, m.p. 134°C. |
| α-Iodonaphthalene 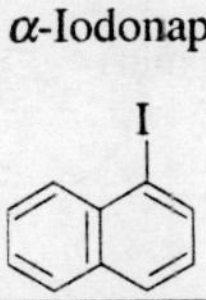  | 305 | - Picrate, m.p. 128°C. |

**Table 6.12b** Halohydrocarbons (Alkyl and Aryl halides) Solids

| *Name and structure* | *M.P.(°C)* | *Derivatives and tests* |
|---|---|---|
| Benzyl iodide 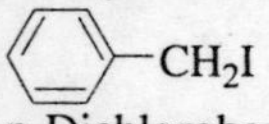 | 24 | - Forms quaternary salt with dimethylaniline, m.p. 165°C. |
| p-Dichlorobenzene 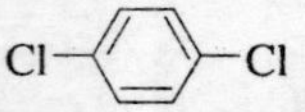 | 53 | - On nitration (fuming $HNO_3$; the mixture is initially cooled in ice and then allowed to stand at room temp. for 10 min) gives 2-nitro derivative, m.p. 54°C. |
| β-Iodonaphthalene 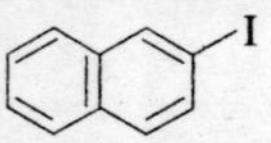 | 55<br>b.p. 309 | - Picrate, m.p. 95°C.<br>- Anilide m.p. 170°C. |
| β-Chloronaphthalene 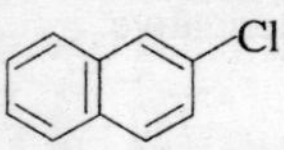 | 56 | - On nitration (conc. $HNO_3$ + conc. $H_2SO_4$; warming for 6-8 hrs) gives 1, 8-dinitro derivative, m.p. 175°C.<br>- Picrate m.p. 81°C. |
| β-Bromonaphthalene 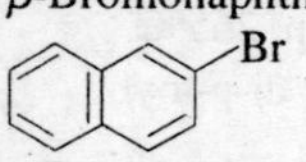 | 59 | - Picrate m.p. 86°C.<br>- Anilide m.p. 171°C. |
| p-Bromochlorobenzene 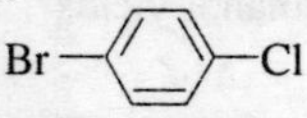 | 66 | - On nitration (conc. $HNO_3$ + conc. $H_2SO_4$) gives 2-nitro derivative, m.p. 72°C. |
| p-Chlorodiphenyl 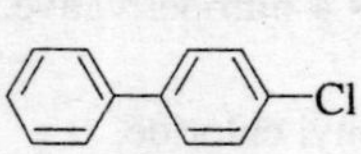 | 77 | - On oxidation (alkaline $KMnO_4$) gives p-chlorobenzoic acid, m.p. 240°C. |
| Ethylene iodide<br>$CH_2ICH_2I$ | 81 | - With $Br_2$ yields ethylene bromide, b.p. 132°C. |
| p-Dibromobenzene  | 89 | - On nitration (fuming $HNO_3$; the mixture initially cooled in ice and then left for 10 min. at room temp) gives 2, 5-dinitro derivative, m.p. 84°C. |
| p-Bromo biphenyl 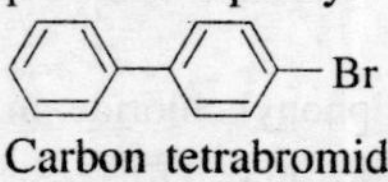 | 89 | - On oxidation (alkaline $KMnO_4$) gives p-bromobenzoic acid, m.p. 251°C. |
| Carbon tetrabromide<br>$CBr_4$ | 92 | - With alcoholic $AgNO_3$ gives yellow precipitate of AgBr. |
| Triphenyl methyl chloride<br>$(C_6H_5)_3CCl$ | 113 | - On boiling with $H_2O$ for 5 min gives triphenyl carbinol, m.p. 162°C. |
| p-Iodo biphenyl 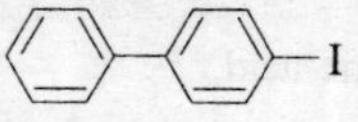 | 114 | |
| Iodoform<br>$CHI_3$ | 119 | - With alcoholic $AgNO_3$ gives yellow ppt. of AgI.<br>- On warming with small quantity of phenol or resorcinol in dil. alcoholic NaOH gives red colour. |
| 1, 3, 5-Tribromobenzene<br>(sym. Tribromobenzene) 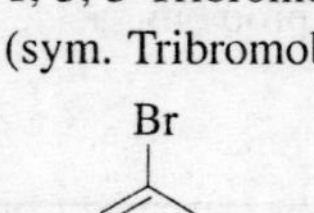 | 120 | - Forms addition product (in dry ether) with quinoline, m.p. 65°C. |

| *Name and structure* | *M.P. (°C)* | *Derivatives and tests* |
|---|---|---|
| p-Diiodobenzene  | 129 | - 2, 5-Dinitro derivative, m.p. 171°C. |
| 1, 2, 4, 5-Tetrachlorobenzene 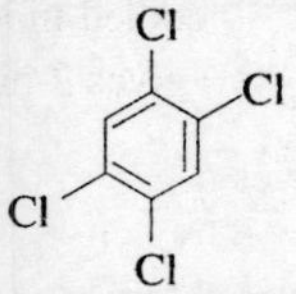 | 139 | - 3-Nitro derivative, m.p. 99°C. |
| p, p′-Dichlorodiphenyl 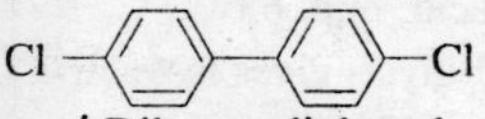 | 149 | - On oxidation ($CrO_3$) gives p-chlorobenzoic acid, m.p. 240°C. |
| p, p′-Dibromodiphenyl 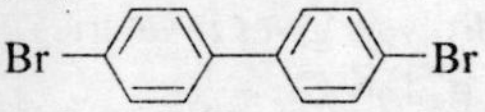 | 164 | - On oxidation ($CrO_3$) gives p-bromobenzoic acid, m.p. 251°C. |
| 1, 2, 4, 5-Tetrabromobenzene 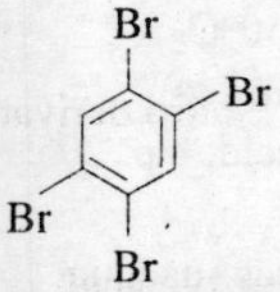 | 181 | - 3-Nitro derivative, m.p. 168°C. |
| Hexa-chloroethane $Cl_3C—CCl_3$ | 187 | - With Zn dust + dil. $H_2SO_4$ yields tetrachloro ethylene, b.p. 121°C. |
| Hexachlorobenzene 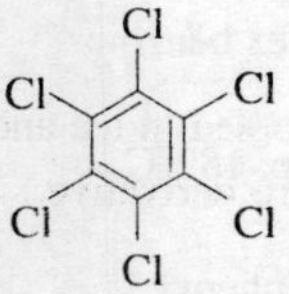 | 228 | - On boiling with fuming $HNO_3$ yields chloranil, m.p. 290°C. |

**Table 6.13(a)** Carboxylic acid halides (Liquids)

| *Name and structure* | *B.P. (°C)* | *Properties* | *Derivatives* | | | *Specific tests and derivatives* |
|---|---|---|---|---|---|---|
| | | | *Anilide* | *p-Toluidide* | *Amide* | |
| Acetyl chloride $CH_3COCl$ | 55 | Colourless, pungent irritating odour, decomposes in $H_2O$, alcohol, soluble in ether, chloroform | 114 | 148 | 82 | - The solution in $H_2O$ (Obtained on heating) is made alkaline line ($NH_4OH$) and excess $NH_3$ boiled off. Addition of $FeCl_3$ solution gives wine red colour. |
| Oxalyl chloride $(COCl)_2$ | 64 | | 246 (di) | 267 (di) | 419 (di) | |
| Propionyl chloride $CH_3CH_2COCl$ | 80 | Colourless, pungent irritating odour | 104 | 124 | 79 | - On hydrolysis (heating with water) gives propionic acid, b.p. 140°C. |
| Acetyl bromide $CH_3COBr$ | 80 | Colourless, pungent odour, decomposes in $H_2O$, soluble in benzene, ether | 114 | 148 | 82 | - Same test as in case of acetyl chloride. On hydrolysis gives acetic acid, b.p. 118°C. |

| *Name and structure* | *B.P. (°C)* | *Properties* | *Derivatives* | | | *Specific tests and derivatives* |
|---|---|---|---|---|---|---|
| | | | *Anilide* | *p-Toluidide* | *Amide* | |
| Iso-Butyryl chloride $(CH_3)_2CHCOCl$ | 92 | Colourless, pungent odour, decomposes in $H_2O$, soluble in benzene, ether | 105 | 104 | 129 | - On hydrolysis (NaOH solution) gives isobutyric acid, b.p. 155°C. |
| n-Butyryl chloride $CH_3CH_2CH_2COCl$ | 100 | -do- | 96 | 72 | 114 | - On hydrolysis gives n-butyric acid, b.p. 163°C |
| Chloroacetyl chloride $ClCH_2COCl$ | 105 | -do- | 137 | 162 | 120 | - On hydrolysis gives chloro-acetic acid, m.p. 63°C. |
| iso-Valeryl chloride $(CH_3)_2CHCH_2COCl$ | 115 | -do- | 110 | 109 | 135 | - On hydrolysis gives isovaleric acid, b.p. 176°C. |
| n-Valeryl chloride $CH_3CH_2CH_2CH_2COCl$ | 127 | -do- | 63 | 70 | 104 (114) | - On hydrolysis gives n-valeric acid, b.p. 186°C. |
| Bromoacetyl bromide $BrCH_2COBr$ | 149 | -do- | 130 | 91 | 91 | - On hydrolysis gives bromo-acetic acid, m.p. 50°C. |
| $\alpha$-Bromopropionyl bromide $CH_3CHBrCOBr$ | 153 | -do- | 99 | 125 | 123 | - On hydrolysis gives $\alpha$-bromopropionic acid, b.p. 205°. |
| Succinoyl chloride $CH_2COCl$ \| $CH_2COCl$ | 190 (d) (m.p. 16) | -do- | 228 (di) (148 mono) | 255 (di) | 260 (di) | - On hydrolysis gives succinic acid, m.p.185°C. |
| Benzoyl chloride 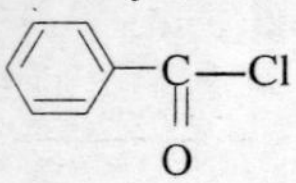 | 197 | Colourless, pungent irritating odour, decomposes in $H_2O$, soluble in ether, benzene | 164 | 158 | 128 | - On heating with NaOH solution and acidification gives benzoic acid, m.p. 122°C.<br>- Benzoyl glycine, m.p. 187°C. |
| Phenylacetyl chloride 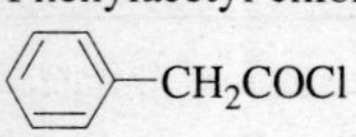 | 210 94/12 mm (m.p. 20) | Colourless, pungent, irritating odour, decomposes in $H_2O$, sol. in ether, benzene | 118 | 136 | 154 | - On hydrolysis gives phenyl acetic acid, m.p. 76°C. |
| Anisoyl chloride 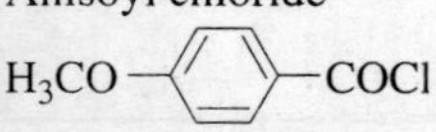 | 262 145/14 mm (m.p. 22) | -do- | 168 | 186 | 162 | - On hydrolysis yields anisic acid, m.p. 184°C. |
| Phthaloyl chloride 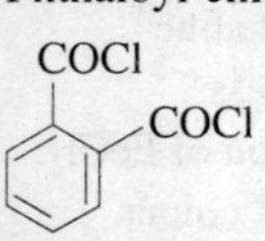 | 276 (m.p. 12) | -do- | 250 (di) 169 (mono) | 201 | 220 (di) 149 (mono) | - On hydrolysis yields phthalic acid, m.p. 195°C. |
| Cinnamoyl bromide 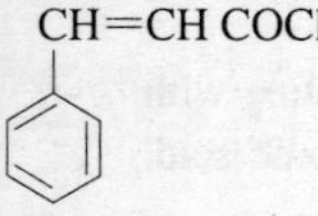 | 257 (trans) (m.p. 35) | -do- | 151 | 168 | 147 | - On hydrolysis yields cinnamic acid, m.p. 133°C. |

**Table 6.13(b)** Carboxylic acid halides (solids)

| *Name and structure* | *M.P. (°C)* | *Properties* | *Derivatives* | | | *Specific tests and derivatives* |
|---|---|---|---|---|---|---|
| | | | *Anilide* | *p-Toluidide* | *Amide* | |
| p-Bromobenzoyl chloride 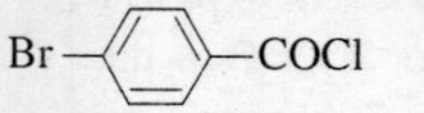 | 42 (b.p. 245) | | 197 | – | 189 | - On hydrolysis (dil. NaOH solution) yields p-bromobenzoic acid, m.p. 251°C. |
| 2-Naphthoyl chloride 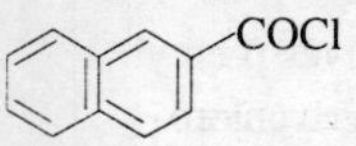 | 51 | | 170 | 191 | 192 | - On hydrolysis (dil. NaOH) yields 2-naphthoic acid, m.p. 184°C. |
| 3, 5-Dinitrobenzoyl chloride 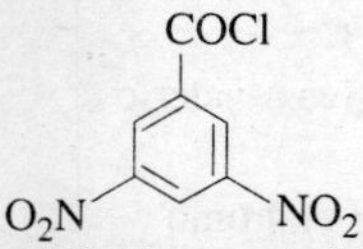 | 74 | Colourless or light brown, decomposes in $H_2O$, soluble in ether | 234 | – | 183 | - On hydrolysis (dil. NaOH solution) gives 3, 5-dinitrobenzoic acid, m.p. 205°C. |
| p-Nitrobenzoyl chloride 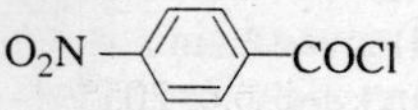 | 75 | Yellow, decomposes in $H_2O$, soluble in ether | 204 (211) | 203 | 201 | - On hydrolysis (dil. NaOH) yields p-nitrobenzoic acid, m.p. 240°C. |
| Terephthaloyl chloride  | 83 (b.p. 263) | | 337 | – | – | - On hydrolysis (dil. NaOH solution) yields terephthalic acid, m.p. > 300°C. |

**Table 6.14(a)** Nitro compounds (Liquids)

***Note*** This Table includes following types of nitro substituted compounds, e.g., Hydrocarbons, Phenols, Carboxylic acids, Alcohols and Carbonyl compounds

| *Name and structure* | *B.P. (°C)* | *Properties* | *Derivatives and tests* |
|---|---|---|---|
| Nitromethane $CH_3NO_2$ | 101 | | |
| Nitrobenzene 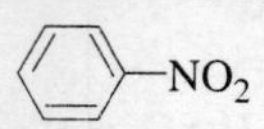 | 209 (m.p. 5°) | Pale yellow, bitter almond odour, insoluble in $H_2O$, soluble in alcohol, ether | - On reduction (Sn + HCl, heat) gives aniline, b.p. 183°C (characterised by dye test).<br>- On boiling with dil. Na arsenite yields azoxybenzene, m.p. 36°C.<br>- On warming with fuming $HNO_3$+$H_2SO_4$ yields m-dinitrobenzene, m.p. 90°C.<br>- On warming with $Br_2$ + $FeCl_3$ (or $FeBr_3$) yields m-bromo derive., m.p. 54°C. |
| o-Nitrotoluene 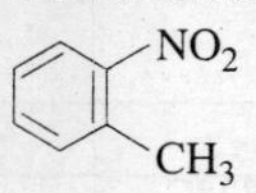 | 224 | Very pale yellow, odour of bitter almonds, insoluble in $H_2O$, soluble in alcohol, ether | - On reduction (Sn + HCl) gives o-toluidine, b.p. 199°C.<br>- On oxidation ($KMnO_4$, heat) gives o-nitrobenzoic acid, m.p. 146°C.<br>- On nitration (conc. $HNO_3$ + $H_2SO_4$) at 100° for 3 min. yields 2, 4-dinitrotoluene, m.p. 70°C |

| *Name and structure* | *B.P. (°C)* | *Properties* | *Derivatives and tests* |
|---|---|---|---|
| Phenylnitromethane 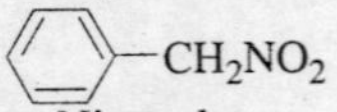  | 226 (d) | | |
| m-Nitrotoluene 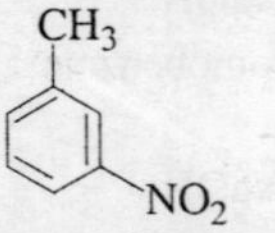  | 230 (m.p. 16) | Very pale yellow, odour of bitter almonds, insoluble in $H_2O$, soluble in alcohol, ether | - On reduction (Sn + HCl) gives m-toluidine, b.p. 202°C. - On oxidation ($KMnO_4$ or $K_2Cr_2O_7$ + dil. $H_2SO_4$ boiling) gives m-nitrobenzoic acid, m.p. 141°C. |
| p-Nitroethylbenzene 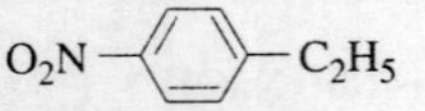  | 243 | | - On reduction (Sn + HCl) gives p-ethyl aniline, b.p. 214, (acetyl derivative, m.p. 94°) |
| p-Nitro-m-xylene 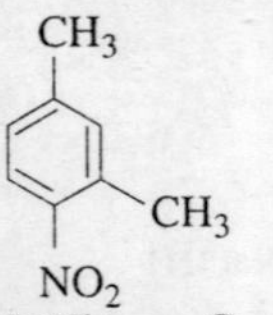  | 244 | Pale yellow, insoluble in $H_2O$, sol. in alcohol, ether | - On oxidation ($KMnO_4$) gives 4-nitro-isophthalic acid, m.p. 258°C. - On reduction (Sn + HCl) gives 4-m-xylidine, b.p 212°C. - With fuming $HNO_3$ in cold yields 4, 6-dinitro-m-xylene, m.p. 93°C. |
| 2-Nitro-p-Cymene 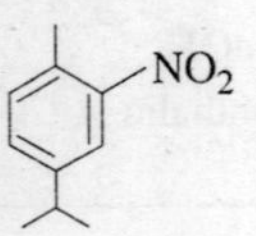  | 264 | Insoluble in $H_2O$, soluble in alcohol, ether | - On reduction (Sn + HCl) gives 2-amino-p-cymene, b.p. 241 (acetyl deriv., m.p. 115°C; benzoyl derivative, m.p. 102°) - On oxidation (alkaline $KMnO_4$) yields 2-nitro-4-isopropylbenzoic acid, m.p. 168°C. - On warming with conc. $HNO_3$ + $H_2SO_4$ yields 2, 6-nitro-p-cymene, m.p. 54°C. |
| o-Nitrophenetole 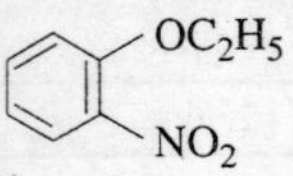  | 267 | Insoluble in $H_2O$, soluble in alcohol, ether | - On redn. (Sn + HCl) yields o-phenetidine b.p. 229°C. - With cold $HNO_3$ + $H_2SO_4$ yields 2, 4-dinitrophenetole, m.p. 86°C. - With $HNO_3$ + $H_2SO_4$ on warming yields trinitrophenetole, m.p. 78°C. - On boiling with 48% BHr yields o-nitrophenol, m.p. 45°C. |
| o-Nitroanisole 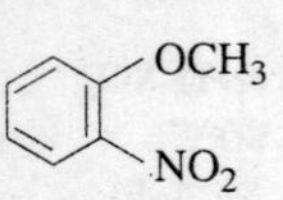  | 272 | See ethers (Table 6.10a) | - On nitration (conc.$HNO_3$+conc.$H_2SO_4$ at 0°C for 15 min.) gives nitro deriv. m.p. 95°C. |

**Table 6.14(b)** Nitro compounds (solids)

| *Name and structure* | *M.P. (°C)* | *Properties* | *Derivatives and tests* |
|---|---|---|---|
| m-Nitrobenzyl alcohol 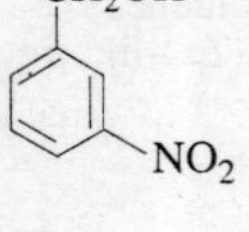  | 27 | See alcohols (Table 6.6b) | |

| *Name and structure* | *M.P. (°C)* | *Properties* | *Derivatives and tests* |
|---|---|---|---|
| 4-Nitro-o-xylene 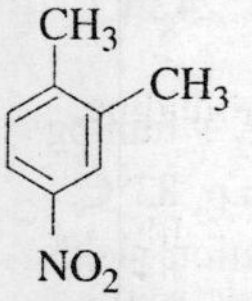 | 30 | Insoluble in $H_2O$, soluble in alcohol, ether | - On redn. (Sn + HCl) gives 3, 4-Dimethyl aniline m.p. 50 (acetyl derive., m.p. 99°)<br>- On nitration—dinitro deriv., m.p. 82°C. |
| o-Chloronitrobenzene 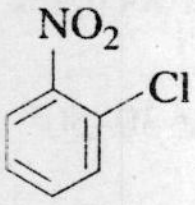 | 32 | | - On nitration (conc. $HNO_3$ + conc. $H_2SO_4$) (30-50°, 10-15 min. or at 100° for 2 min) gives 2, 4-dinitro chlorobenzene, m.p. 51°C. |
| o-Nitrobiphenyl 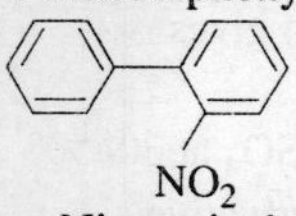 | 37 | | - On reduction gives o-aminobiphenyl, m.p. 49°C (acetyl derivative, m.p. 121°C) |
| m-Nitroanisole 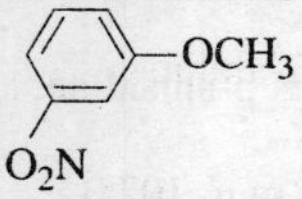 | 39 | See ethers (Table 6.10b) | |
| o-Bromonitrobenzene 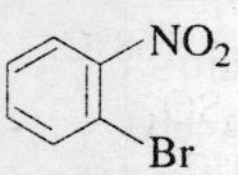 | 41 (b.p. 261) | Pale yellow, insoluble in $H_2O$, soluble in alcohol, ether | - With conc. $HNO_3$ and conc. $H_2SO_4$ at 30-50° for 10-15 min. gives 2, 4-dinitro derivative, m.p. 74°C. |
| Nitromesitylene 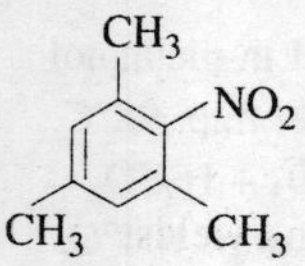 | 44 (b.P. 255) | | - On redn. (Sn + HCl) gives mesidine, b.p. 232°C (acetyl deriv., m.p. 216°C; benzoyl derivative, m.p. 204°C)<br>- On warming with fuming $HNO_3$ yields dinitromesitylene, m.p. 86°C.<br>- With conc. $HNO_3$ + $H_2SO_4$ at 100° for 5 min gives trinitromesitylene, m.p. 235°C. |
| o-Nitrobenzaldehyde | 44 | See carbonyl compounds (Table 6.3b) | |
| o-Nitrophenol | 45 | See phenols, (Table 6.4b) | |
| m-Chloronitrobenzene 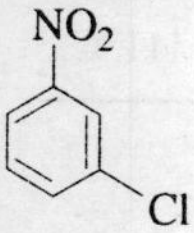 | 45 (b.p. 236) | Pale yellow, insoluble in $H_2O$, soluble in hot alcohol, ether, chloroform | - With fuming $HNO_3$ + conc. $H_2SO_4$ at 100°C for 2 min, gives 3, 4-dinitro derivative, m.p. 36°C.<br>- On boiling with KOH and $Na_2S$ in methanol yields 3, 3′-dichloroazoxybenzene m.p. 97°C. |
| o-Nitrobenzyl chloride 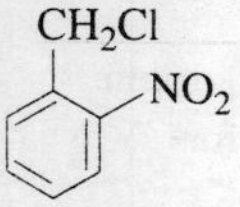 | 48 | Colourless, insoluble in $H_2O$, soluble in alcohol, hot ether | - On adding $SnCl_2$ +HCl at 40-50° and then adding Zn dust yields o-toluidine, b.p. 199°C.<br>- With $KMnO_4$ soln. yields o-nitrobenzoic acid, m.p. 146°C.<br>- With conc. $HNO_3$+$H_2SO_4$ yields 2, 4-dinitro derivative, m.p. 34°C.<br>- On boiling with KCNS in 90% alcohol yields the thiocyanate, m.p. 75°C. |

| *Name and structure* | *M.P. (°C)* | *Properties* | *Derivatives and tests* |
|---|---|---|---|
| 2, 4-Dinitrochlorobenzene 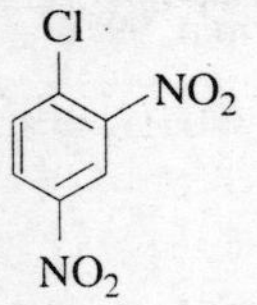 | 51 | Pale yellow, insoluble in $H_2O$, soluble in hot alcohol, ether | - With Sn + HCl yields chloro-m-phenylene diamine, m.p. 91 (diacetyl, m.p. 242°C; dibenzoyl derive., m.p. 178°C).<br>- On warming with fuming $HNO_3$ + fuming $H_2SO_4$ yields picryl chloride, m.p. 83°C.<br>- On boiling with dil. NaOH solution yields 2, 4-dinitrophenol m.p. 114°C, |
| p-Nitrotoluene 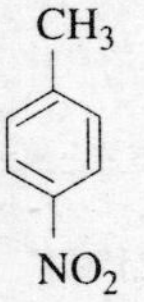 | 53 | Light pale yellow, odour of bitter almonds, insoluble in $H_2O$, soluble in alcohol, ether | - On oxidation ($KMnO_4$) gives p-nitrobenzoic acid, m.p. 240°C.<br>- With conc. $HNO_3$ + conc. $H_2SO_4$ at 30–50° for 10–15 min gives nitro deriv., m.p.70°C.<br>- On reduction (Sn +HCl, heat) gives p-toluidine, m.p. 44°C. |
| p-Nitroanisole 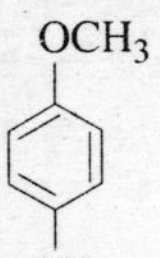 | 54 | Insoluble in $H_2O$ | - With conc. $HNO_3$ + conc. $H_2SO_4$ at 30–50° for 10–15 min. gives nitro derivative m.p. 88°C.<br>- On reduction (Sn +HCl) gives p-anisidine, m.p. 57°C.<br>- On boiling with 48% HBr gives p-nitrophenol, m.p. 114°C. |
| 2-Nitro-1, 4-dichlorobenzene 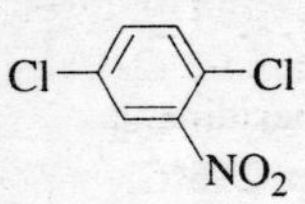 | 55 | Pale yellow, soluble in alcohol | - With Sn + HCl yields 2, 5-dichloroaniline, m.p. 50°C b.p. 251°C (Acetyl derivative, m.p. 133°C, benzoyl derivative m.p. 120°C).<br>- On boiling with 1 mole KOH in methanol yields 2-nitro-4-chloroanisole, m.p. 98°C.<br>- On boiling with fuming $HNO_3$ + $H_2SO_4$ (caution) yields, nitro deriv., m.p. 104°C. |
| m-Nitrobromobenzene 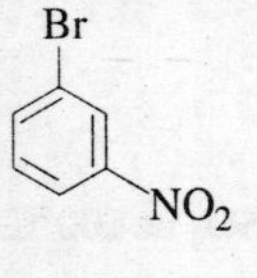 | 56 | Pale yellow, soluble in alcohol | - With Sn + HCl yields m-bromoaniline, b.p. 251°C.<br>- On heating with excess conc. $HNO_3$ + $H_2SO_4$ yields nitro derivative, m.p. 59°C.<br>- On boiling with methanolic KOH yields 3, 3′-dibromoazoxy benzene, m.p. 111°C. |
| m-Nitrobenzaldehyde 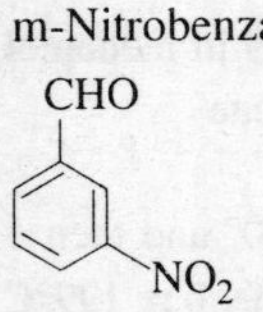 | 58 | See carbonyl compounds (Table 6.3b) | |
| p-Nitrophenetole 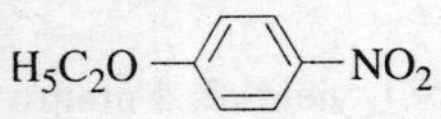 | 59 | Insoluble in $H_2O$ | - With conc. $HNO_3$ + conc. $H_2SO_4$, room temp. (ice cooling if necessary) for 10 min. yields nitro derivative, m.p. 86°C.<br>- With Sn + HCl gives p-phenetidine, b.p. 254°C. (acetyl derivative, m.p. 134°C). |

| *Name and structure* | *M.P. (°C)* | *Properties* | *Derivatives and tests* |
|---|---|---|---|
| 1-Nitronaphthalene 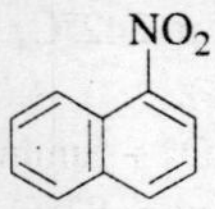 | 60 | Yellow, insoluble in $H_2O$, soluble in alcohol, ether | - On nitration with excess conc. $HNO_3$ + $H_2SO_4$ gives 1, 3, 8-trinitro naphthalene, m.p. 218°C.<br>- On oxidation ($CrO_3$—$CH_3COOH$) gives 3-nitrophthalic acid, m.p. 218°C. |
| 2, 6-Dinitrotoluene 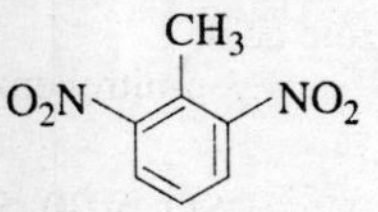 | 66 | | - On reduction (Sn + HCl) gives 2, 6-diaminotolune, m.p. 105°C.<br>- On oxidation ($KMnO_4$) gives 2, 6-dinitrobenzoic acid m.p. 202°C. |
| 2, 4, 6-Trinitroanisole | 68 | See ethers (Table 6.10b) | |
| 2, 4-Dinitrotoluene 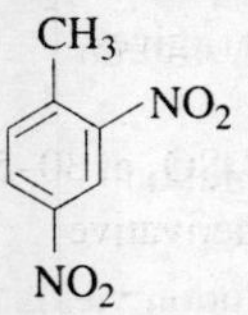 | 70 | Colourless or light yellow, insoluble in $H_2O$, sparingly soluble in alcohol, ether | - On reduction with excess Sn + HCl gives m-toluylene diamine, m.p. 99°C.<br>- On oxidation ($KMnO_4$) gives 2, 4, dinitrobenzoic acid, m.p. 183°C.<br>- With acetone + KOH gives blue colour changing to violet by addition of acetic acid.<br>- With benzaldehyde + little piperidine yields dinitrostilbene, m.p. 139°C. |
| o-Nitroaniline | 71 | See primary amines (Table 6.15b) | |
| p-Nitrobenzyl chloride 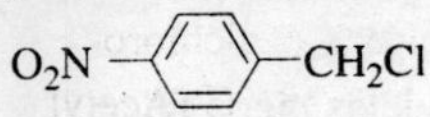 | 71 | Very pale yellow, insoluble in $H_2O$, sparingly soluble in alcohol, soluble in ether | - On oxidation ($KMnO_4$) yields p-nitro-benzoic acid, m.p. 240°C.<br>- On heating with $Na_2CO_3$ solution gives p-Nitrobenzyl alcohol, m.p. 93°C. |
| 2, 4-Dinitrobromobenzene 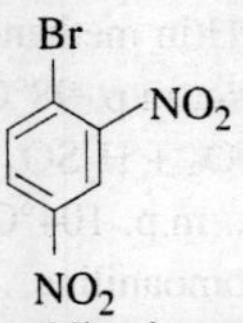 | 74 | Pale yellow, soluble in hot alcohol | - With Sn + HCl yields m-phenylene-diamine, m.p. 63°C.<br>- On heating with $Na_2CO_3$ soln. yields 2, 4-dinitrophenol, m.p. 114°C. |
| o-Nitrobenzyl alcohol | 74 | See alcohols (Table 6.6b) | |
| 2, 4-Dimethyl-6-nitroaniline 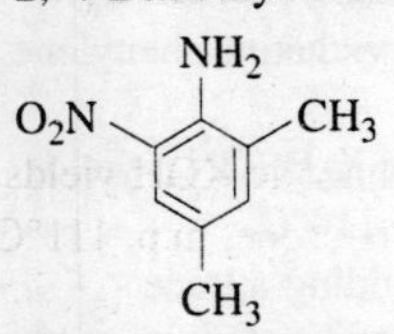 | 76 | Yellow, insoluble in water | - Acetyl, m.p. 176°C.<br>- Benzoyl, m.p. 185°C. |
| 2-Nitro-p-toluidine | 77 | See Amines (Table 6.15b) | |
| 2, 4, 6-Trinitrophenetole 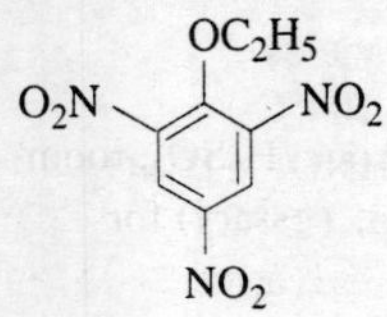 | 78 | Insoluble in $H_2O$, soluble in ether | - On boiling with aq. KOH yields picric acid, m.p. 122°C. |

| *Name and structure* | *M.P. (°C)* | *Properties* | *Derivatives and tests* |
|---|---|---|---|
| m-Nitroacetophenone 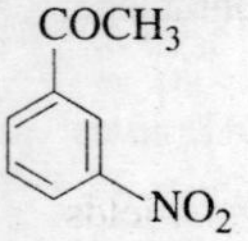 | 81 | See carbonyl compounds (Table 6.3b) | |
| 2, 4, 6-Trinitrotoluene 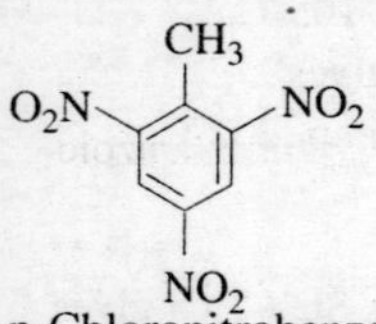 | 82 | Pale yellow, insoluble in $H_2O$, sparingly soluble in alcohol, ether | - On oxidation ($CrO_3$ + conc. $H_2SO_4$) yields 2, 4, 6-trinitrobenzoic acid, m.p. 220°C. |
| p-Chloronitrobenzene 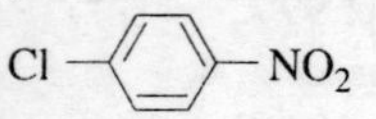 | 83 | Pale yellow, insoluble in $H_2O$, soluble in hot alcohol, ether | - With conc. $HNO_3$ + conc. $H_2SO_4$ at 30–50° for 10–15 min. yields 2, 4-dinitro derivative m.p. 51°C.<br>- On boiling with conc. NaOH solution yields p-nitrophenol, m.p. 114°C. |
| Picryl chloride (2, 4, 6-Trinitrocholorobenzene) 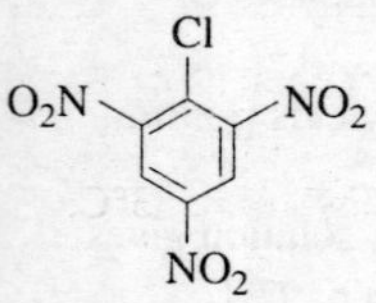 | 83 | Yellow, sparingly soluble in ether, soluble in alcohol. | - On boiling with dill. NaOH solution yields picric acid, m.p. 122°C.<br>- On boiling with KOH in methanol yields trinitroanisole, m.p. 68°C.<br>- With KOH in ethanol yields trinitrophenetole, m.p. 78°C.<br>- With $NH_3$ in alcohol yields picramide, m.p. 189°C.<br>- On warming with aniline yields trinitrodiphenylamine, m.p. 177°C. |
| 2-Nitro-p-dibromobenzene 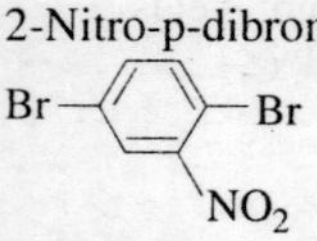 | 85 | | - On boiling with KOH in methanol yields 4-bromo-2-nitroanisole, m.p. 86°C. |
| m-Dinitrobenzene 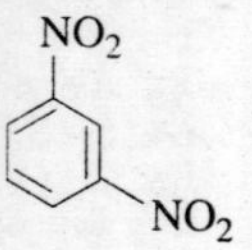 | 90 | Pale yellow, insoluble in $H_2O$, soluble in alcohol | - On reduction with $NH_4SH$ in alcohol yields m-nitroaniline, m.p. 114°C.<br>- On redn. with Sn +HCl yields m-phenylene diamine, m.p. 63°C.<br>- On boiling with alkaline $K_3Fe(CN)_6$ yields 2, 4-dinitrophenol, m.p. 114°C.<br>- Gives violet colour on adding a trace of glucose to boiling solution of compound in very dil. NaOH solution. |
| 6-Nitro-o-toluidine | 92 | See primary amines (Table 6.15b) | |
| 4, 6-Dinitro-m-xylene 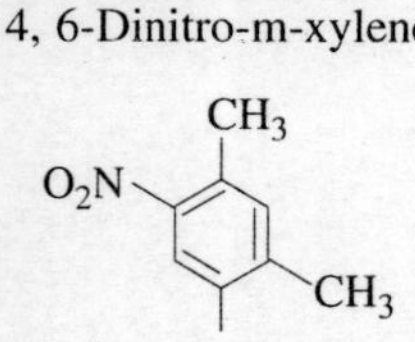 | 93 | Yellow, sparingly soluble in cold alcohol | - On redn. with Sn + HCl yields diamino-m-xylene, m.p. 104°C.<br>- On warming with conc. $HNO_3$ + $H_2SO_4$ yields trinitroxylene, m.p. 125°C. |

| *Name and structure* | *M.P. (°C)* | *Properties* | *Derivatives and tests* |
|---|---|---|---|
| p-Nitrobenzyl alcohol | 93 | See alcohols (Table 6.6b) | |
| 2, 4-Dinitroanisole 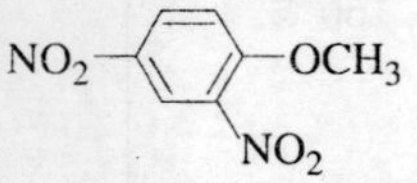 | 95 | Yellow, insoluble in $H_2O$, soluble in alcohol, ether | - With conc. $HNO_3$ + $H_2SO_4$ at 30–50° for 10–15 min, yields 2, 4, 6-trinitro anisole, m.p. 68°C.<br>- On heating with conc. NaOH solution yields 2, 4-dinitrophenol, m.p. 114°C. |
| m-Nitrophenol | 97 | See Phenols (Table 6.4b) | |
| 3-Nitro-o-toluidine | 97 | See Primary amines (Table 6.15b) | |
| p-Nitro benzyl bromide 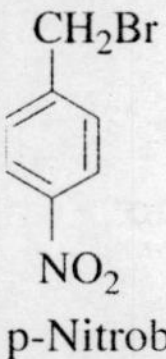 | 99 | Pale yellow, insoluble in $H_2O$, sparingly soluble in alcohol, sol. in ether | - On oxidation ($KMnO_4$) yields p-nitrobenzoic acid, m.p. 240°C.<br>- On boiling with $Na_2CO_3$ solution yields p-nitrobenzyl alcohol, m.p. 93°C. |
| p-Nitrobenzaldehyde | 106 | See carbonyl compounds (Table 6.3b) | |
| 4-Nitro-o-toluidine | 107 | See Primary amines (Table 6.15b) | |
| 2, 4-Dinitrophenol | 114 | See Phenols (Table 6.4b) | |
| m-Nitroaniline | 114 | See Primary amines (Table 6.15b) | |
| p-Nitrophenol | 114 | See Phenols (Table 6.4b) | |
| 4-Nitrodiphenyl 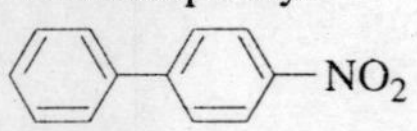 | 114 | | - On oxidation with $CrO_3$ in acetic acid gives p-nitrobenzoic acid, m.p. 240°C.<br>- On redn. gives p-aminobiphenyl, m.p. 53°C. |
| 3-Nitro-p-toluidine | 116 | See primary amines (Table 6.15b) | |
| o-Dinitrobenzene 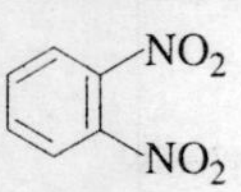 | 117 | Colourless, insoluble in $H_2O$, | - With $NH_4SH$ in alcohol yields o-nitro aniline, m.p. 71°C.<br>- On reduction (Sn + HCl) yields o-phenylene diamine, m.p. 102°C. |
| 5-Nitro-1-naphthylamine 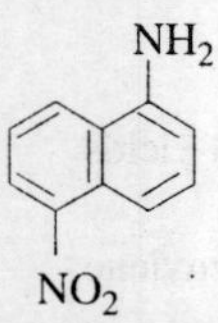 | 119 | Yellow, insoluble in $H_2O$ | - Acetyl derivative, m.p. 220°C. |
| Picric acid | 122 | See Phenols (Table 6.4b) | |
| 1, 3, 5-Trinitrobenzene 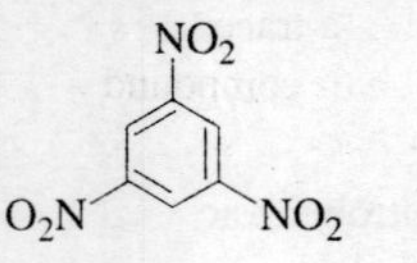 | 122 | Very pale yellow, insoluble in $H_2O$, sparingly soluble in alcohol, ether | - With dil. NaOH solution gives red colour, which disappears on acidification<br>- With $Na_2CO_3$ soln. + $K_3Fe(CN)_6$ yields picric acid, 122°C.<br>- On reduction (Sn + HCl) yields unstable triaminobenzene (triacetyl derivative, m.p. 208°C). |
| 3-Nitrosalicylic acid | 125 (hydrated)<br>148 (anhydrous) | See carboxylic acids (Table 6.2b) | |

| *Name and structure* | *M.P. (°C)* | *Properties* | *Derivatives and tests* |
|---|---|---|---|
| 1-Nitro-2-naphthyl amine 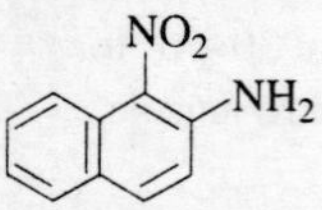 | 126 | Yellow, soluble in organic solvents | - Acetyl derivative, m.p. 125°C.<br>- Benzoyl derivative, m.p. 168°C.<br>- p-Tosyl derivative, m.p. 160°C. |
| p-Bromonitrobenzene 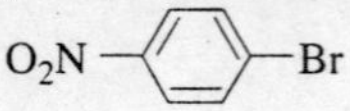 | 126 | Colourless or pale yellow, insoluble in $H_2O$, sparingly soluble in alcohol, ether | - With conc. $HNO_3 + H_2SO_4$ at 30–50° for 10-15 min gives 2, 4-dinitro bromobenzene, m.p. 74°C.<br>- On reduction (Sn + HCl) yields p-bromoaniline, m.p. 66°C. |
| 5-Nitro-o-toluidine | 130 | See Primary amines (Table 6.15b) | |
| 2, 6-Dinitroaniline | 138 | See Primary amines (Table 6.15b) | |
| 2, 4-Dinitro-1-naphthol | 138 | See Phenols (Table 6.4b) | |
| m-Nitrobenzoic acid | 141 | See carboxylic acids (Table 6.2b) | |
| m-Nitrobenzamide | 142 | See amides (Table 6.18b) | |
| 3-Nitro-4-acetamido-toluene (2-Nitro-4-methyl-acetanilide) 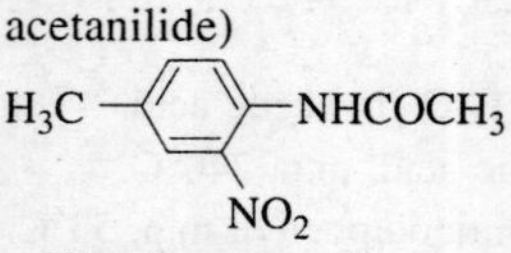 | 144 | Yellow, soluble in hot $H_2O$ | - On oxidation ($KMnO_4$) yields 3-nitro-4-acetamidobenzoic acid, m.p. 220°C.<br>- On hydrolysis yields 3-nitro-p-toluidine, m.p. 116°C. |
| o-Nitrobenzoic acid | 146 | See carboxylic acids (Table 6.2b) | |
| p-Nitroaniline | 147 | See primary amines (Table 6.15b) | |
| 3-Nitrosalicylic acid | 148 | See carboxylic acids (Table 6.2b) | |
| m-Nitrobenzanilide 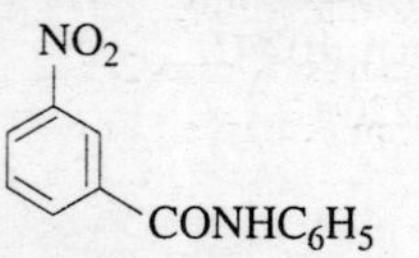 | 154 | Colourless, sparingly soluble in $H_2O$ | - On hydrolysis with conc. HCl or 50% $H_2SO_4$ yields m-nitrobenzoic acid m.p. 141°C. |
| p-Nitrophenylacetic acid 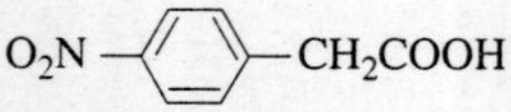 | 154 | Yellow, sparingly soluble in $H_2O$ soluble in alcohol, ether | - Anilide, m.p. 212°C.<br>- Amide, m.p. 198°C. |
| m-Nitroacetanilide 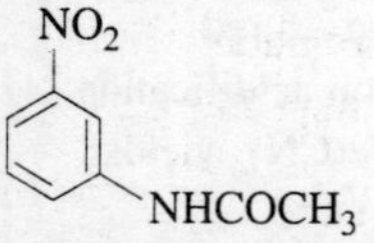 | 154 | See anilides (Table 6.18b) | |
| o-Nitrobenzanilide 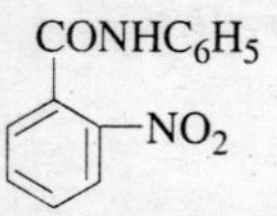 | 155 | Colourless, soluble in hot $H_2O$ | - On hydrolysis gives o-nitrobenzoic acid, m.p. 146°C. |
| p-Nitrophenylhydrazine 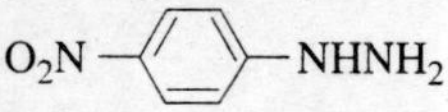 | 157 | Primary amines, (Table 6.15b) | |

| *Name and structure* | *M.P. (°C)* | *Properties* | *Derivatives and tests* |
|---|---|---|---|
| 4-Nitrophthalic acid 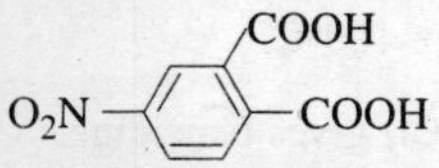 | 165 | See carboxylic acids (Table 6.2b) | |
| 1, 8-Dinitronaphthalene 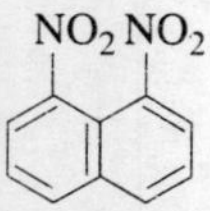 | 170 | Yellow, sparingly soluble in alcohol, ether | - On boiling with fuming $HNO_3$ + conc. $H_2SO_4$ (*caution*) yields 1, 3, 8-trinitro-naphthalane, m.p. 218°C.<br>- On redn. (Sn + HCl) in alcohol yields 1, 8-diaminonaphthalene, m.p. 66°C. |
| p-Dinitrobenzene 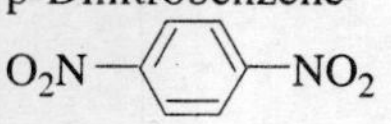 | 172 | Colourless, insoluble in $H_2O$, alcohol | - With alcoholic $NH_4SH$ yields p-nitroaniline, m.p. 147°C.<br>- On boiling with 5% NaOH solution gives p-Nitrophenol, m.p. 114°C.<br>- On redn. (Sn + HCl) yields p-phenylene-diamine, m.p. 140°C. |
| 3, 5-Dinitrosalicylic acid 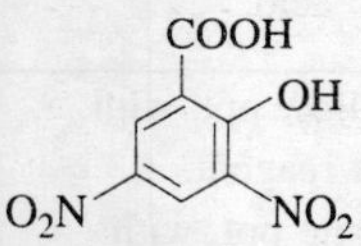 | 173 | See carboxylic acids (Table 6.2b) | |
| o-Nitrobenzamide | 176 | See amides (Table 6.18b) | |
| 2, 4-Dinitroaniline | 180 | See primary amines (Table 6.15b) | |
| 2, 4-Dinitrobenzoic acid | 183 | See carboxylic acids (Table 6.2b) | |
| Picramide | 189 | See primary amines (Table 6.15b) | |
| 4-Nitro-2-methylacetanilide 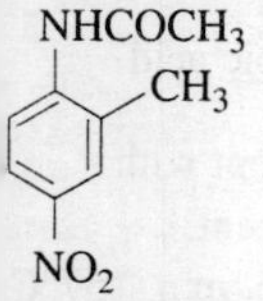 | 196 | Colourless soluble in hot $H_2O$ | - On hydrolysis with 50% $H_2SO_4$ (refluxing) gives 5-nitro-o-toluidine, m.p. 130°C.<br>- On oxidation ($KMnO_4$) yields 5-nitro-2 -acetamino-benzoic acid, m.p. 215°C. |
| 2, 4-Dinitrophenylhydrazine | 198 | See primary amines (Table 6.15b) | |
| m-Nitrocinnamic acid | 200 | See carboxylic acids (Table 6.2b) | |
| p-Nitrobenzamide | 201 | See amides (Table 6.18b) | |
| 3, 5-Dinitrobenzoic acid | 205 | See carboxylic acids (Table 6.2b) | |
| p-Nitrobenzanilide 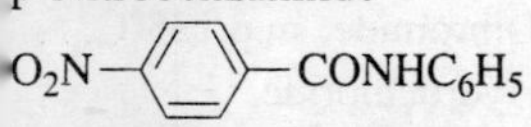 | 211 | Colourless, soluble in alcohol, ether | - On hydrolysis with NaOH solution yields p-nitrobenzoic acid, m.p. 240°C. |
| 1, 5-Dinitronaphthalene 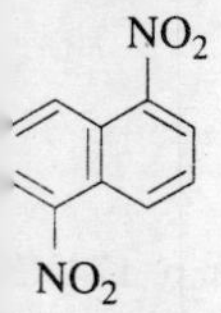 | 214 | Yellow, sparingly soluble in alcohol, ether | - On reduction (Sn + HCl) yields 1, 5-diamino-naphthalene, m.p. 190°C.<br>- On boiling for 5 min with fuming $HNO_3$ + conc. $H_2SO_4$ yields 1, 4, 5-trinitro-naphthalene, m.p. 154°C. |
| 3-Nitrophthalic acid | 218 | See carboxylic acids (Table 6.2b) | |

| *Name and structure* | *M.P. (°C)* | *Properties* | *Derivatives and tests* |
|---|---|---|---|
| 2, 4, 6-Trinitrobenzoic acid | 220 (d) | See carboxylic acids (Table 6.2b) | |
| 5-Nitrosalicylic acid | 230 | See carboxylic acids (Table 6.2b) | |
| 4, 4′-Dinitrodiphenyl 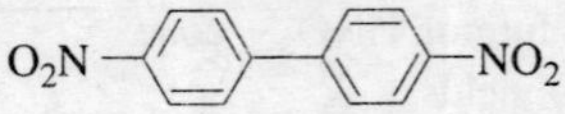 | 233 | Yellow, insoluble in $H_2O$ | - On reduction (Sn + HCl) gives benzidine, m.p. 127°C. |
| o-Nitrocinnamic acid | 240 | See carboxylic acids, (Table 6.2b) | |
| p-Nitrobenzoic acid | 240 | See carboxylic acids, (Table 6.2b) | |
| p-Nitrocinnamic acid | 285 | See carboxylic acids, (Table 6.2b) | |

## **Table 6.15(a)** Amines—Primary (Liquids)

| *Name and structure* | *B.P. (°C)* | *Properties* | *Derivatives (M.P.°C)* | | | | | *Specific tests and derivatives* |
|---|---|---|---|---|---|---|---|---|
| | | | *Acetyl* | *Benzoyl* | *p-Toluene sulphonyl* | *2, 4-dinitrophenyl* | *Picrate* | |
| Methylamine<br>$CH_3NH_2$ | Gas | Colourless, soluble in $H_2O$, alcohol | – | 80 | 75 | 178 | 215 | - Gives yellow ppt. with Nessler's reagent. |
| Ethylamine<br>$C_2H_5NH_2$ | 18 | Colourless, ammonical odour, soluble in $H_2O$, alcohol, ether | – | 71 | 63 | 114 | 166 | - Gives white ppt. with Nesslers reagent. |
| Isopropylamine<br>$(CH_3)_2CHNH_2$ | 32 | Colourless, soluble in $H_2O$, alcohol, ether | – | 100 | 51 | 94 | 150 | - Hydrochloride, m.p. 139°C (Hygroscopic). |
| tert. Butylamine<br>$(CH_3)_3CNH_2$ | 46 | | - | 134 | - | - | 198 | - Hydrochloride m.p. 310 |
| n-Propylamine<br>$CH_3CH_2CH_2NH_2$ | 49 | Colourless, soluble in $H_2O$, alcohol, ether | - | 85 | 52 | 96 | 135 | - Gives brown ppt with Nessler's reagent.<br>- Hydrochloride, m.p. 157°C. |
| Cyclopropylamine<br>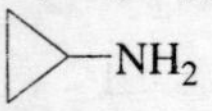 | 50 | | - | 99 | - | - | 149 | - Hydrochloride m.p. 100°C. |
| Allylamine<br>$CH_2{=}CHCH_2NH_2$ | 56 | Colourless, soluble in $H_2O$, alcohol, ether | - | - | 64 | 76 | 140 | - Decolourises $KMnO_4$ solution.<br>- Dibromide, m.p.135°C.<br>- Hydrochloride, m.p. 105-110°C. |
| sec. Butylamine<br>$CH_3CH_2CH—NH_2$<br>\|<br>$CH_3$ | 63 | Colourless, soluble in $H_2O$ | - | 76 | 55 | - | 140 | |
| Iso-butylamine<br>$(CH_3)_2CHCH_2NH_2$ | 68 | Colourless, soluble in $H_2O$, alcohol, ether | - | 57 | 78 | 80 | 151 | - Hydrochloride, m.p. 174°C (hygroscopic). |
| *n*-Butylamine<br>$CH_3CH_2CH_2CH_2NH_2$ | 77 | Colourless, soluble in $H_2O$, alcohol, ether | - | 42 | - | - | 151 | - Hydrochloride, m.p. 195°C. |

| *Name and structure* | *B.P. (°C)* | *Properties* | *Derivatives (M.P.°C)* Acetyl | Benzoyl | p-Toluene sulphonyl | 2, 4-dinitrophenyl | Picrate | *Specific tests and derivatives* |
|---|---|---|---|---|---|---|---|---|
| iso-Amylamine $(CH_3)_2CHCH_2CH_2NH_2$ | 95 | | - | - | 65 | 91 | 138 | - Hydrochloride, m.p. m.p. 215°C. |
| n-Amylamine $CH_3(CH_2)_4NH_2$ | 104 | Colourless, soluble in $H_2O$ | - | - | - | 81 | 139 | |
| Ethylenediamine $(CH_2NH_2)_2$ | 116 (m.p. 8°) | Colourless, soluble in $H_2O$, alcohol | 172 | 249 | 160 | 302 | 233 | |
| Cyclohexylamine 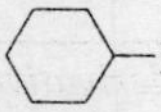 $-NH_2$ | 134 | | 104 | 148 | 87 | 156 | 158 | |
| Ethanolamine (2-hydroxy ethylamine) $HOCH_2CH_2NH_2$ | 171 | | | | | | 160 | |
| Aniline 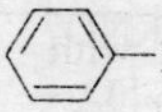 $-NH_2$ | 184 | Colourless or reddish, characteristic odour, soluble in alcohol, ether, sparingly soluble in $H_2O$ | 114 | 160 | 103 | 156 | – | - With bromine water gives tribromo derivative m.p. 119°C.<br>- Hydrochloride, m.p. 198°C.<br>- Azo β-naphthol derivative m.p. 134°C. |
| Benzylamine 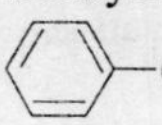 $-CH_2NH_2$ | 184 | Colourless, soluble in $H_2O$, alcohol, ether | 60 | 106 | 116 | 116 | 199 | - Hydrochloride, m.p. 253°C.<br>- Oxalate, m.p. 201°C (d). |
| 1-Phenylethylamine $CH_3CH—NH_2$ 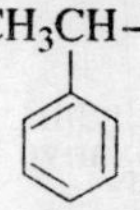 | 185 | Colourless, soluble in alcohol | 57 | 120 | – | 118 | 189 | |
| 2-Phenylethylamine $C_6H_5CH_2CH_2NH_2$ | 198 | | | 116 | - | 154 | 167 | - |
| o-Toluidine 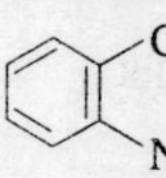 $CH_3$, $NH_2$ | 199 | Colourless or reddish, sparingly soluble in $H_2O$, soluble in alcohol, ether | 112 | 144 | 109 | 126 | 213 | - On shaking ethereal solution with dil. bleaching powder soln. gives brown colour.<br>- Hydrochloride, m.p. 215°C.<br>- Gives isocyanide test<br>- With bromine water gives dibromo deriv. m.p. 50°C.<br>- Azo β-naphthol deriv., m.p. 128°C. |
| m-Toluidine 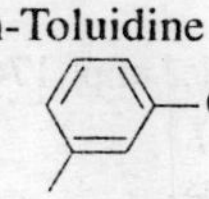 $-CH_3$, $H_2N$ | 202 | Colourless, insoluble in $H_2O$, soluble in alcohol, ether | 65 | 125 | 114 | 160 | 200 | - Hydrochloride 228°C.<br>- On shaking ethereal solution with dil bleaching powder solution water layer becomes |

| *Name and structure* | *B.P. (°C)* | *Properties* | *Derivatives (M.P.°C)* *Acetyl* | *Benzoyl* | *p-Toluene sulphonyl* | *2, 4-dinitrophenyl* | *Picrate* | *Specific tests and derivatives* |
|---|---|---|---|---|---|---|---|---|
| | | | | | | | | yellow brown & ether becomes red.<br>- Sol in 50% $H_2SO_4$ gives yellowish colour.<br>- Brown colour with $K_2Cr_2O_7$.<br>- Red colour with $HNO_3$<br>- Azo $\beta$-naphthol deriv., m.p. 140°C. |
| (–) Menthylamine 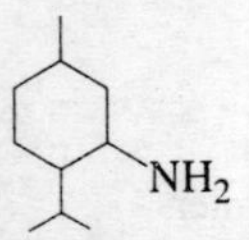 | 207 | | 145 | 156 | – | – | 215 | |
| o-Chloroaniline 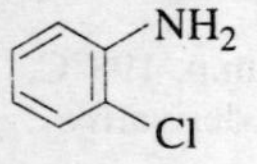 | 209 | Colourless, insoluble in $H_2O$, soluble in ether, alcohol | 87 | 99 | 105 | 150 | 134 | - On adding (1 mole) $KNO_3$ to solution in conc. $H_2SO_4$— yields 5-Nitro derivative., m.p. 117°C. |
| 4-m-Xylidine (2, 4-dimethylaniline) 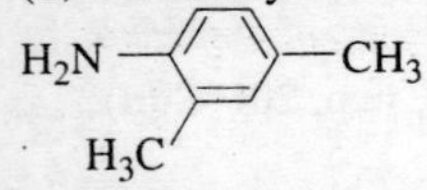 | 212 | Colourless, sparingly soluble in $H_2O$ | 133 | 192 | - | 156 | 209 | - Hydrochloride, m.p. 235°C.<br>- Azo-$\beta$-naphthol derivative m.p. 166°C. |
| p-Xylidine (2, 5-Dimethylaniline) 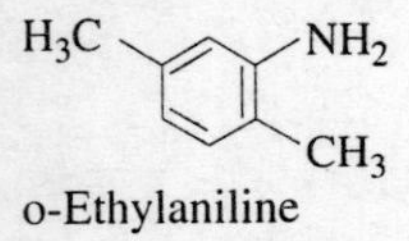 | 214 (m.p.15) | Colourless, sparingly-solution in $H_2O$ | 139 | 140 | 119 | 150 | 171 | - Hydrochloride, m.p. 228°C.<br>- Azo-$\beta$-naphthol derivative m.p. 150°C. |
| o-Ethylaniline 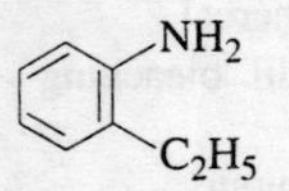 | 216 | | 111 | 147 | – | – | 95 | |
| 2-m-Xylidine (2, 6-Dimethylaniline) 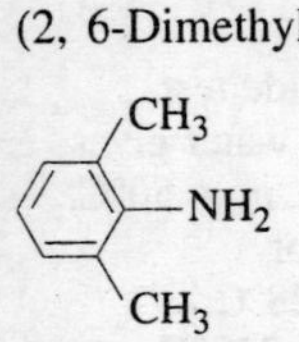 | 216 | | 177 | 168 | 212 | – | 180 | |
| 5-m-Xylidine (3, 5-Dimethylaniline) | 220 | Colourless, soluble in alcohol | 144 | 136 | – | – | 209 | |

| Name and structure | B.P. (°C) | Properties | Derivatives (M.P.°C) | | | | | Specific tests and derivatives |
|---|---|---|---|---|---|---|---|---|
| | | | Acetyl | Benzoyl | p-Toluene sulphonyl | 2, 4-dinitrophenyl | Picrate | |
| 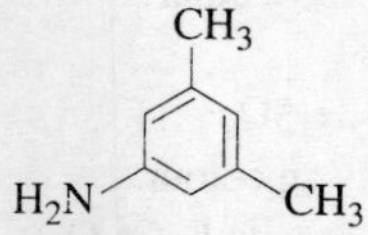 | | | | | | | | |
| o-Anisidine 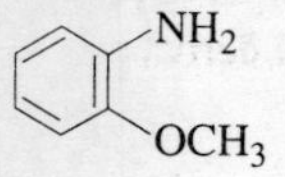 | 225 (m.p. 5°) | Colourless, sparingly soluble in $H_2O$, soluble in alcohol, ether | 85 | 60 | 127 | 151 | 200 | - On boiling with 48% HBr gives o-aminophenol, m.p. 174°C.<br>- Azo-β-naphthol deriv. m.p. 180°C. |
| N-Methyl-N-phenylhydrazine $C_6H_5N(CH_3)NH_2$ | 227 | Insoluble in $H_2O$ | 92 | 152 | - | - | - | - Reduces Fehling's solution on warming. |
| o-Bromoaniline 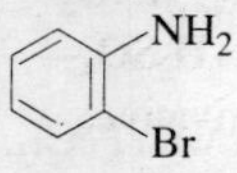 | 229 (m.p. 31°) | Colourless, insoluble in $H_2O$, soluble in alcohol, ether | 99 | 116 | 90 | 161 | 129 | |
| o-Phenetidine 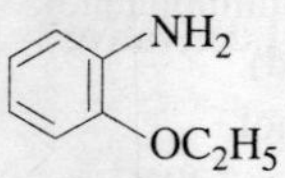 | 229 | Colourless, insoluble in $H_2O$, soluble in alcohol, ether | 79 | 104 | 164 | 164 | 119 | - Hydrochloride, m.p. 214°C.<br>- On boiling with 48% HBr gives o-aminophenol, m.p. 174°C. |
| m-Chloroaniline 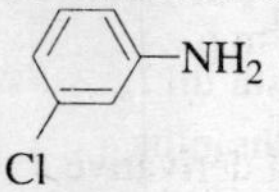 | 230 | Colourless, insoluble in $H_2O$, soluble in alcohol, ether | 72 | 121 | 138 | 184 | 177 | |
| Mesidine (2, 4, 6-Trimethylaniline) 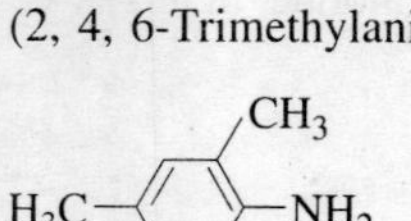 | 232 | | 216 | 204 | 167 | - | 193 | |
| 3-Bromo-4-aminotoluene 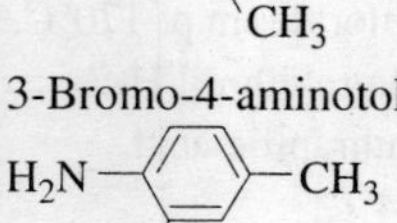 | 240 (m.p. 26) | | 117 | 149 | - | - | - | - Hydrochloride, m.p. 221°C. |
| Phenylhydrazine 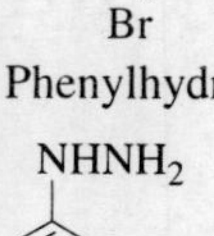 | 243 (m.p. 19) | Colourless or yellow-brown, sparingly soluble in $H_2O$, alcohol and ether | 128 | 168 | - | - | - | - Reduces Fehling's solution in cold.<br>- Hydrochloride, m.p. 240°C.<br>- With HCl & $CuSO_4$ (warming) yields chlorobenzene, B.P. 132°C. |

| *Name and structure* | *B.P. (°C)* | *Properties* | *Derivatives (M.P.°C)* | | | | | *Specific tests and derivatives* |
|---|---|---|---|---|---|---|---|---|
| | | | *Acetyl* | *Benzoyl* | *p-Toluene sulphonyl* | *2, 4-dinitrophenyl* | *Picrate* | |
| m-Phenetidine 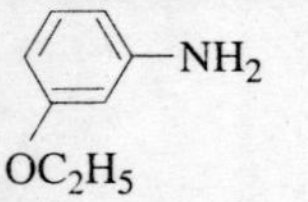 | 248 | Colourless, insoluble in $H_2O$, soluble in alcohol, ether | 97 | 103 | 157 | - | 158 | |
| m-Anisidine 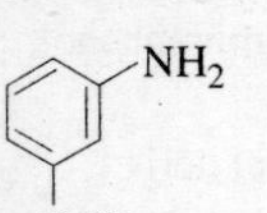 | 251 | Colourless, sparingly soluble in $H_2O$, soluble in alcohol, ether | 81 | – | 68 | 138 | 169 | |
| m-Bromoaniline 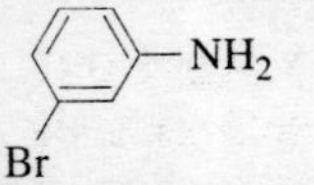 | 251 (m.p. 18) | Colourless, insoluble in $H_2O$, soluble in alcohol, ether | 87 | 136 | – | 178 | 180 | |
| p-Phenetidine 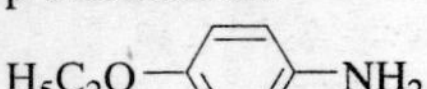 | 254 | Colourless, insoluble in $H_2O$, soluble in alcohol, ether | 134 | 173 | 106 | 118 | 69 | - Hydrochloride, m.p. 234°C.<br>- Gives red colour with $FeCl_3$.<br>- On boiling with 48% HBr gives p-aminophenol, m.p. 184°C (d).<br>- Azo-$\beta$-naphthol derivative m.p. 139°C. |
| Methyl anthranilate 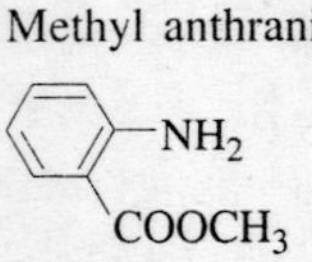 | 260 (d) (m.p. 25°) | Colourless, orange blossom odour, sparingly soluble in $H_2O$, soluble in alcohol, ether | 101 | 100 | – | – | 106 | - Gives dye test.<br>- On heating with dil. HCl gives anthranilic acid, m.p. 144°C.<br>- Hydrochloride, m.p. 178°C. |
| p-Aminodiethylaniline 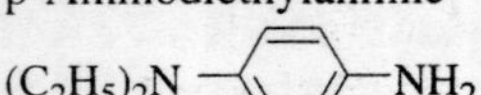 | 262 | Colourless, darkens rapidly in air | 104 | 172 | 112 | - | - | |
| Ethyl anthranilate 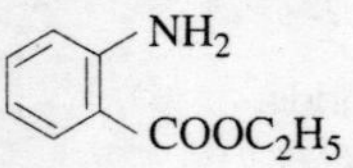 | 265 (d) (m.p. 13) | Colourless, insoluble in $H_2O$, soluble in alcohol, ether | 61 | 98 | 112 | - | - | - Gives dye test.<br>- Hydrochloride, m.p. 170°C.<br>- On boiling with dil HCl gives anthranilic acid, m.p. 144°C. |
| Ethyl m-aminobenzoate 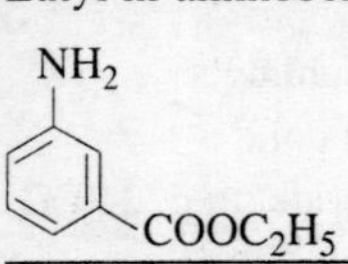 | 294 | Colourless, insoluble in $H_2O$, sparingly soluble in hot $H_2O$, soluble in alcohol | 110 | 114 | - | - | - | - Hydrochloride, m.p. 185°C.<br>- Gives dye test. |

**Table 6.15(b)** Amines—Primary (Solids)

| *Name and structure* | *M.P. (°C)* | *Properties* | *Derivatives (M.P.°C)* *Acetyl* | *Benzoyl* | *p-Toluene sulphonyl* | *2, 4-dinitrophenyl* | *Picrate* | *Specific tests and derivatives* |
|---|---|---|---|---|---|---|---|---|
| o-Aminoacetophenone 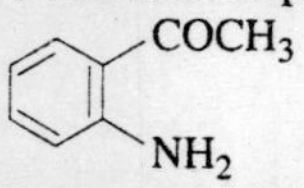 | 20 (b.p. 252) | | 76 | 98 | 148 | – | – | |
| Methyl anthranilate | 25 (b.p. 260) | See liquids | | | | | | |
| Tetramethylenediamine (Putriscine) $(CH_2CH_2NH_2)_2$ | 27 (b.p. 159) | | 137 (di) | 177 (di) | 224 | – | 249 | |
| p-Amino-N, N-dimethyl aniline 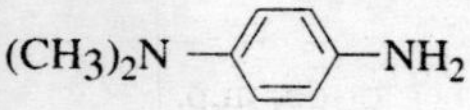 | 41 (b.p. 262) | Colourless or grey, characteristic odour, insoluble in $H_2O$, soluble in alcohol, ether | 132 | 228 | – | – | 186 | - Gives dye test.<br>- Hydrochloride, m.p. 220°C.<br>- Solution of dihydrochloride gives purple colour with excess $AgNO_3$ in cold and odour of quinone on warming. |
| Hexamethylenediamine $(CH_2CH_2CH_2NH_2)_2$ | 42 (b.p. 205) | | 126 | 155 | – | – | 220 | |
| p-Toluidine 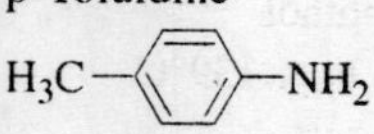 | 44 (b.p. 200) | Colourless or grey, characteristic odour, insoluble in $H_2O$, soluble in alcohol, ether | 153 | 158 | 117 | 136 | 181 | - Hydrochloride, m.p. 240°C.<br>- With $Br_2$ water gives dibromo derivative, m.p. 73°C.<br>- Azo-$\beta$-naphthol derivative m.p. 134°C.<br>- Gives yellow colour with $K_2Cr_2O_7$; blue with $HNO_3$, changing gradually through violet and red to brown. |
| o-Ethylaniline 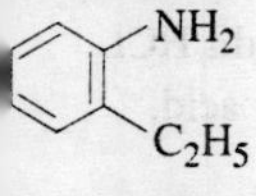 | 46 (b.p. 216) | | 112 | 146 | – | – | 194 | |
| 4-o-Xylidine (3, 4-Dimethylaniline) 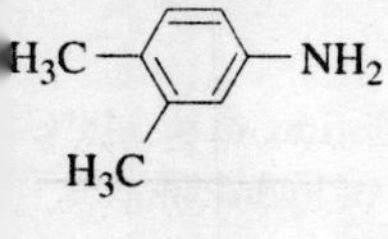 | 48 | | 99 | 118 | – | 141 | – | |

| *Name and structure* | *M.P. (°C)* | *Properties* | *Derivatives (M.P.°C)* Acetyl | Benzoyl | p-Toluene sulphonyl | 2, 4-dinitrophenyl | Picrate | *Specific tests and derivatives* |
|---|---|---|---|---|---|---|---|---|
| o-Phenylaniline (2-aminobiphenyl) 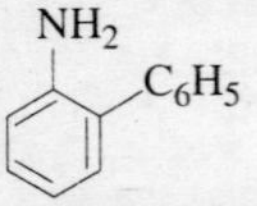  | 49 | Insoluble in $H_2O$ | 119 | 86 | – | – | – | |
| 1-Naphthylamine 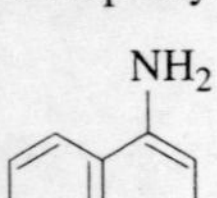  | 50 (b.p. 300) | Colourless, darkens on standing, obnoxious odour, insoluble in $H_2O$, soluble in alcohol, ether | 160 | 161 | 157 | 190 | 161 | - Hydrochloride, m.p. 286°C.<br>- Solution of hydrochloride gives a blue ppt. with $FeCl_3$.<br>- Azo-$\beta$-naphthol deriv. m.p. 274°C. |
| 2, 5-Dichloroaniline 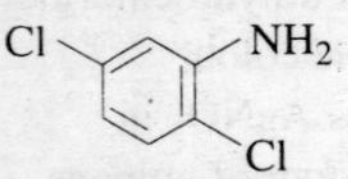  | 50 | | 133 | 120 | – | – | – | |
| p-Phenylaniline (4-Aminobiphenyl) 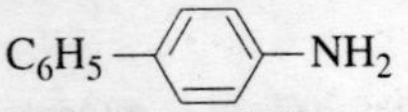  | 53 | | 172 | 230 | 255 | – | – | |
| 2-Aminopyridine 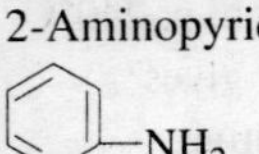  | 57 | | 71 | 165 | – | – | 217 | |
| p-Anisidine 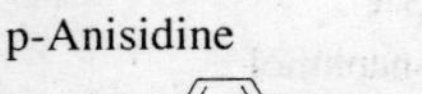  | 57 (b.p. 246) | Colourless, soluble in hot $H_2O$, alcohol, ether | 127 | 154 | 114 | 141 | 117 | - Hydrochloride, m.p. 216°C.<br>- Solution of hydrochloride gives violet colour with $FeCl_3$.<br>- On boiling with 48% HBr yields p-aminophenol, m.p. 184°C.<br>- Azo-$\beta$-naphthol derivative m.p. 139°C. |
| 4-Bromo-2-methylaniline   | 59 | | 156 | 115 | – | – | – | |
| p-Iodoaniline 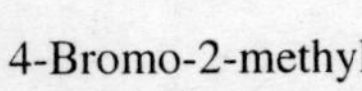  | 62 | | 184 | 222 | – | – | – | |
| m-Phenylenediamine 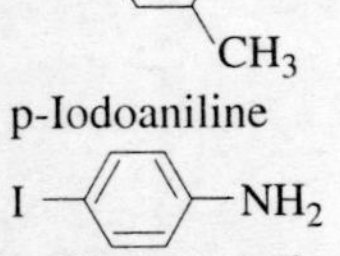  | 63 | Colourless, darkens rapidly in air, soluble in hot $H_2O$, alcohol, ether | 191 | 240 | 172 | – | 184 | - Hydrochloride, m.p. 348°C.<br>- Solution of hydrochloride with $NaNO_2$ gives brown colour. |

| *Name and structure* | *M.P. (°C)* | *Properties* | *Derivatives (M.P.°C)* Acetyl | Benzoyl | p-Toluene sulphonyl | 2, 4-dinitrophenyl | Picrate | *Specific tests and derivatives* |
|---|---|---|---|---|---|---|---|---|
| | | | | | | | | - With $Br_2$ in dil HCl gives tribromo derivative m.p. 158°C. |
| 2, 4-Dichloroaniline<br>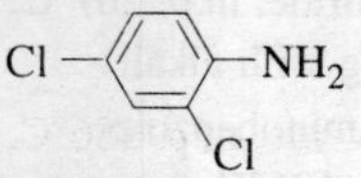 | 63 (b.p. 245) | Colourless, sparingly soluble in $H_2O$, soluble in alcohol, ether | 146 | 117 | 126 | 116 | 106 | - Azo-$\beta$-naphthol deriv. m.p. 183°C. |
| p-Bromoaniline<br>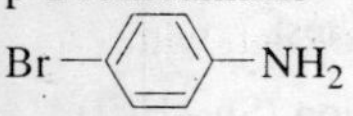 | 66 (b.p. 245) | Colourless, insoluble in $H_2O$, soluble in alcohol, ether | 167 | 204 | 101 | 158 | 180 | |
| p-Chloroaniline<br>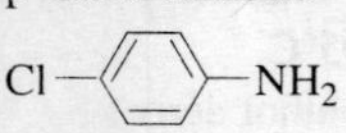 | 70 (b.p. 230) | Colourless, soluble in hot $H_2O$, alcohol, ether | 178 | 192 | 95 | 167 | 178 | - Azo-$\beta$-naphthol derivative m.p. 160°C. |
| o-Nitroaniline<br>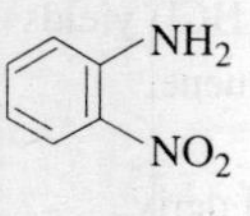 | 71 | Orange yellow, sparingly soluble in $H_2O$, soluble in hot alcohol, ether | 93 | 98 | – | – | 73 | - On reduction (Zn dust + NaOH) gives o-phenylene-diamine, m.p. 102°C.<br>- Azo-$\beta$-naphthol derivative, m.p. 212°C. |
| 2-Nitro-p-toluidine (4-Amino-2-nitrotoluene)<br>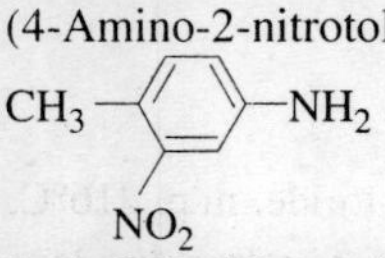 | 77 | Yellow, sparingly soluble in $H_2O$, soluble in hot alcohol, ether | 148 | 172 | 164 | | | - Gives dye test. |
| 2, 4, 6-Trichloroaniline<br>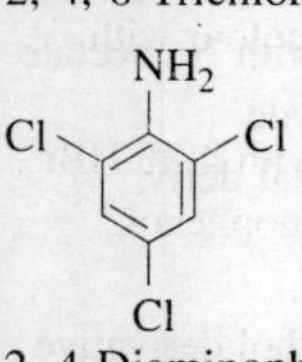 | 77 | Colourless, insoluble in HCl, soluble in alcohol, ether | 204 | 174 | – | – | 83 | - On diazotizing in alcohol + $H_2SO_4$ and then warming yields 1, 3, 5-tricholo-robenzene, m.p. 63°C.<br>- Gives dye test. |
| 2, 4-Diaminophenol<br>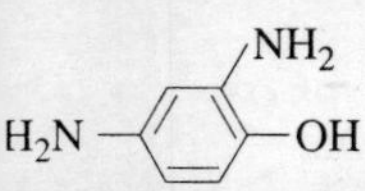 | 79 | See phenols (Table 6.4b) | | | | | | |
| 2, 4-Dibromoaniline<br>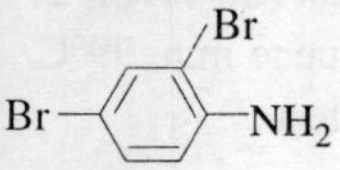 | 79 | Colourless, soluble in alcohol | 146 | 134 | – | – | 124 | Gives dye test. |
| 2, 4-Diaminochloro-benzene<br>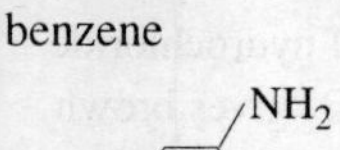 | 88 | | 243 | 178 | 215 | – | – | |
| m-Aminoacetanilide | 89 | | 191 | – | 241 | – | – | - Hydrochloride, m.p. 280°C. |

| *Name and structure* | *M.P. (°C)* | *Properties* | *Derivatives (M.P.°C)* | | | | | *Specific tests and derivatives* |
|---|---|---|---|---|---|---|---|---|
| | | | *Acetyl* | *Benzoyl* | *p-Toluene sulphonyl* | *2, 4-dinitrophenyl* | *Picrate* | |
| 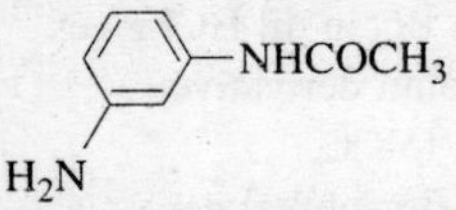 | | | | | | | | |
| Ethyl p-aminobenzoate (Benzocaine) 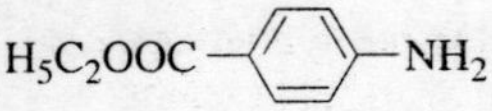 | 92 | Colourless, insoluble in $H_2O$, soluble in alcohol, ether | 110 | 149 | – | – | 131 | - Hydrochloride, m.p. 207°C.<br>- On boiling with alkali yields p-aminobenzoic acid, m.p. 186°C.<br>- Gives dye test. |
| 6-Nitro-o-toluidine 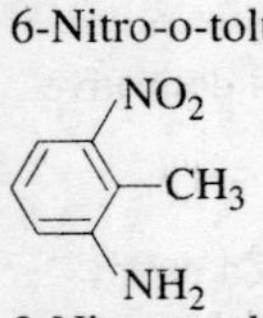 | 92 | Light yellow, insoluble in $H_2O$, sparingly soluble in hot $H_2O$, soluble in alcohol, ether | 157 | 167 | – | – | – | - On reduction (Sn +HCl) yields 2, 6-diamino toluene 105°C.<br>- Azo-$\beta$-naphthol deriv., m.p. 215°C. |
| 3-Nitro-o-toluidine 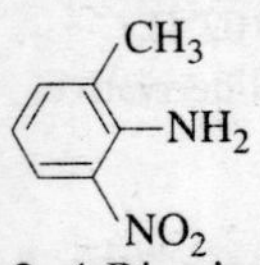 | 95 | Orange, slightly soluble in $H_2O$, soluble in alcohol, ether | 158 | – | – | – | – | - On redn. (Sn + HCl) yields 2, 3-diaminotoluene, m.p. 61°C.<br>- Azo-$\beta$-naphthol deriv., m.p. 168°C. |
| 2, 4-Diaminotoluene (m-Toluylenediamine) 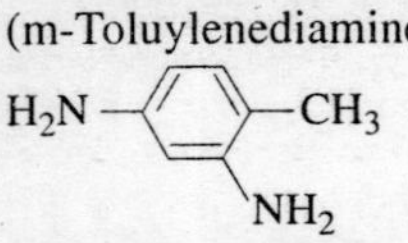 | 99 | Darkens rapidly in air, soluble in hot $H_2O$ | 224 | 224 | 192 | 184 | 170 | Hydrochloride m. p. 305°C. |
| o-Phenylenediamine 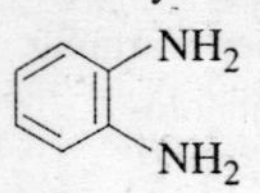 | 102 (b.p. 257) | Colourless, sparingly soluble in $H_2O$, soluble in hot $H_2O$, alcohol, ether | 185 | 301 | 201 | – | 208 | - Hydrochloride, m.p. 260°C.<br>- On heating with Na acetate + benzaldehyde yields dibenzal deriv. m.p. 106°C. |
| p-Aminoacetophenone 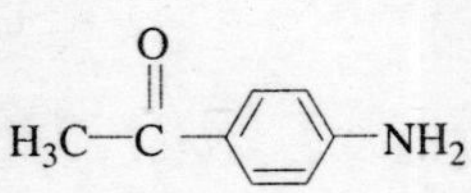 | 106 | | 167 | 205 | 203 | – | – | |
| 4-Nitro-o-toluidine (2-Amino-4-nitrotoluene) 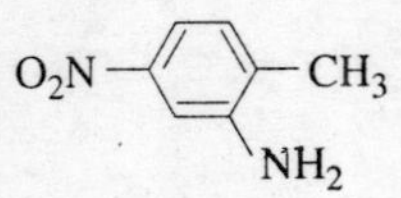 | 107 | | 151 | 183 | – | – | – | - On reduction (Sn +HCl) 2, 4-diamotoluene m.p. 99°C.<br>- Gives dye test. |

| *Name and structure* | *M.P. (°C)* | *Properties* | *Derivatives (M.P.°C)* | | | | | *Specific tests and derivatives* |
|---|---|---|---|---|---|---|---|---|
| | | | *Ace-tyl* | *Benz-oyl* | *p-Tol-uene sulph-onyl* | *2, 4-dini-troph-enyl* | *Pic-rate* | |
| β-Glucosamine 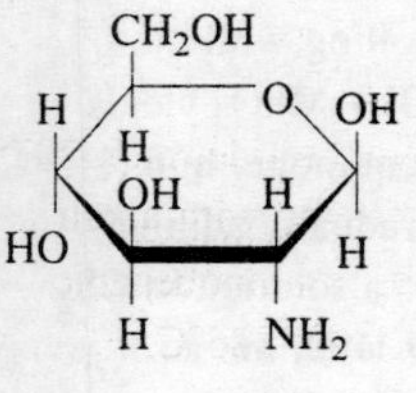 | 110 | Soluble in $H_2O$, sparingly soluble in alcohol, insoluble in ether | 205 (N-acetyl) | 203 | – | – | – | - Reduces Fehling's solution<br>- Hydrochloride, m.p. 200°C.<br>- Oxime, m.p. 127°C(d).<br>- Semicarbazone, m.p., 165°C.<br>- With phenylhydrazine in dil acetic acid yields D-glucosazone, m.p. 205°C. |
| β-Naphthylamine 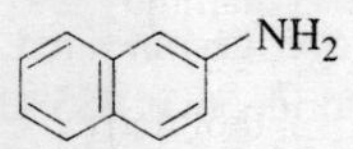 | 112 (b.p. 294) | Yellow, insoluble in $H_2O$, sparingly soluble in alcohol, ether | 132 | 162 | 133 | 179 | 195 | Hydrochloride, m.p. 260°C.<br>- Azo-β-naphthol deriv. m.p. 174°C. |
| m-Nitroaniline 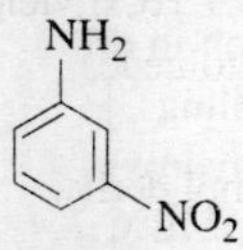 | 114 (b.p. 285) | Yellow, insoluble in $H_2O$, sparingly sol. in alcohol, ether, soluble in hot $H_2O$ | 154 | 155 | 138 | - | 143 | On reduction (Sn + HCl) yields m-phenylene-diamine, m.p. 63°C.<br>- Azo-β-naphthol deriv., m.p. 194°C. |
| 3-Nitro-p-toluidine (4-Amino-3-nitrotoluene) 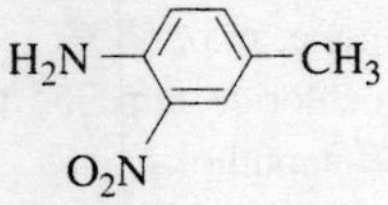 | 116 | Red, sparingly soluble in $H_2O$, soluble in alcohol | 96 | 148 | 145 | | | - On reduction (Zn dust+ NaOH in dil. alcohol) yields 3, 4-diaminotoluene, m.p. 89°C. |
| 2, 4, 6-Tribromoaniline 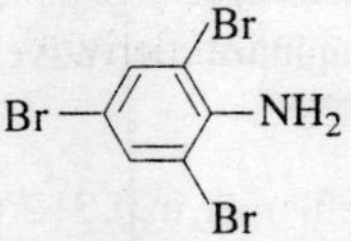 | 119 | Colourless, insoluble in $H_2O$, HCl, soluble in hot alcohol, ether | 232 | 198 | – | – | – | - On diazotising in alcohol + $H_2SO_4$ and then warming yields 1, 3, 5-tribromo-benzene, m.p. 120°C. |
| 2, 6-Diaminopyridine 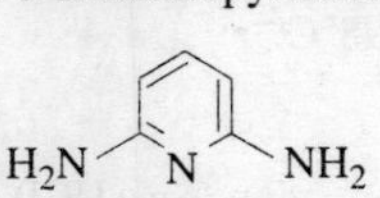 | 121 | | 203 (di) | 176 (di) | – | – | 240 | |
| m-Aminophenol | 122 | See phenols (Table 6.4b) | | | | | | |
| p-Aminobenzophenone 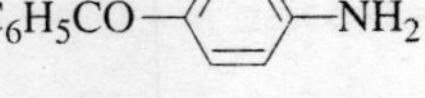 | 124 | | 153 | 152 | – | – | – | |
| p-Aminoazobenzene  | 125 | Yellow, slightly soluble in hot $H_2O$, soluble in hot alcohol, | 144 | 211 | – | – | – | - On redn. (Sn + HCl) gives p-phenylenediamine, m.p. 140°C. |

| *Name and structure* | *M.P. (°C)* | *Properties* | *Derivatives (M.P.°C)* Acetyl | Benzoyl | p-Toluene sulphonyl | 2, 4-dinitrophenyl | Picrate | *Specific tests and derivatives* |
|---|---|---|---|---|---|---|---|---|
| | | ether | | | | | | - Azo-$\beta$-naphthol derivative, m.p. 202°C. |
| Benzidine<br>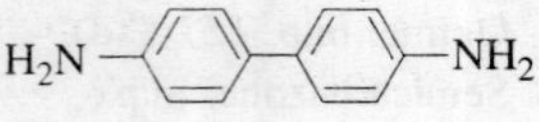 | 127 | Colourless, insoluble in $H_2O$, sparingly soluble in hot $H_2O$, hot alcohol | 317 | 352 | 243 | – | – | - On boiling with dil. $H_2SO_4 + MnO_2$ yields benzoquinone, m.p. 115°C.<br>- On gradually adding $Br_2$ to a solution in $CS_2$ under $H_2O$, the aq. layer becomes first blue, then green and finally colourless; the $CS_2$ layer becomes red.<br>- Hydrochloride, m.p. 385°C.<br>- Diazo-$\beta$-naphthol derivative m.p. 302°C.<br>- On diazotisation in dil $H_2SO_4$ and boiling yields p, p′-dihydroxy diphenyl, m.p. 272°C. |
| o-Tolidine<br>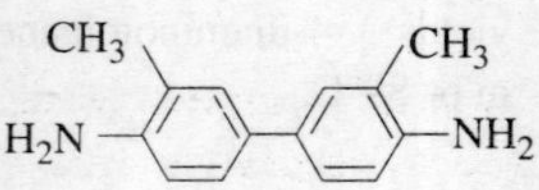 | 129 | Colourless sparingly soluble in $H_2O$, soluble in alcohol, ether | 314 | 265 | – | – | 185 | On boiling with dil. $H_2SO_4 + MnO_2$ yields toluquinone, m.p. 68°C.<br>- Dihydrochloride, m.p. 355°C<br>- Diazo-$\beta$-naphthol derivative m.p. 297°C. |
| 5-Nitro-o-toluidine<br>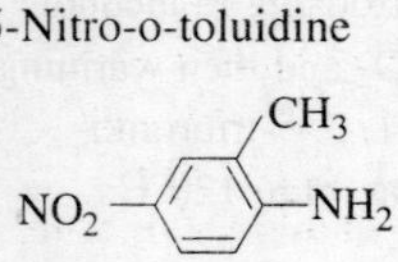 | 130 | Yellow, almost insoluble in $H_2O$, soluble in alcohol | 200 | 174 | – | – | – | - With Sn + HCl gives 2, 5-diaminotoluene m.p. 64°C.<br>- Azo-$\beta$-naphthol derivative, m.p. 248°C. |
| o-Dianisidine<br>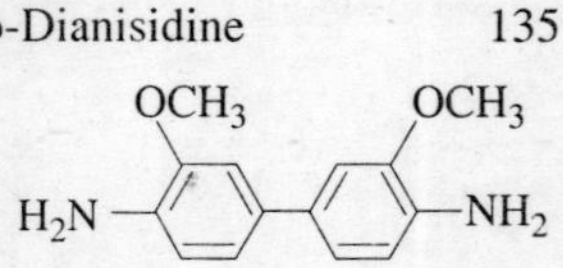 | 135 | | 243 | 236 | – | – | 225 | - Dihydrochloride, m.p. 312°C<br>- Diazo-$\beta$-naphthol derivative m.p. 298°C. |
| 2, 6-Dinitroaniline<br>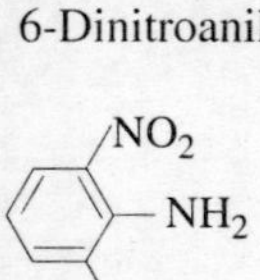 | 138 | Yellow, slightly soluble in cold alcohol | 197 | – | – | – | – | - With Sn + HCl yields 1, 2, 3-triaminobenzene, m.p. 103°C.<br>- Gives dye test. |

| *Name and structure* | *M.P. (°C)* | *Properties* | *Derivatives (M.P.°C)* | | | | | *Specific tests and derivatives* |
|---|---|---|---|---|---|---|---|---|
| | | | *Ace-tyl* | *Benz-oyl* | *p-Toluene sulphonyl* | *2, 4-dinitrophenyl* | *Picrate* | |
| p-Phenylenediamine  | 140 | Colourless, darkens on exposure, sparingly soluble in $H_2O$, soluble in alcohol, ether | 303 | > 300 | 266 | 177 | 202 | - Dihydrochloride, m.p. > 390°C. - Solution of dihydrochloride with excess $AgNO_3$ solution in cold gives green colour rapidly changing to brown through red. - On boiling with dil. $FeCl_3$ or $H_2SO_4 + MnO_2$ gives benzoquinone, m.p. 115°C. |
| p-Nitroaniline 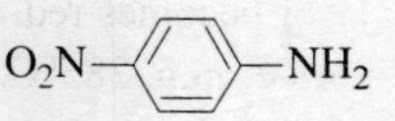 | 147 | Yellow, insoluble in $H_2O$, sparingly soluble in alcohol, ether | 215 | 199 | 191 | – | 100 | - With Sn +HCl yields p-phenylenediamine, m.p. 140°C. - On boiling with conc. NaOH solution yields p-nitrophenol, m.p. 114°C. - Azo-$\beta$-naphthyl deriv., m.p. 250°C. |
| p-Nitrophenylhydrazine 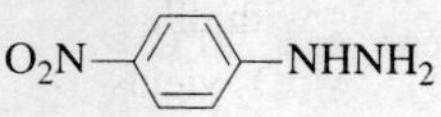 | 157(d) | Oragne red, insoluble in $H_2O$, soluble in hot $H_2O$, alcohol, ether | 205 | 193 | – | – | 119 | - With Sn +HCl yields p-Phenylenediamine, m.p. 140°C. |
| p-Aminoacetanilide 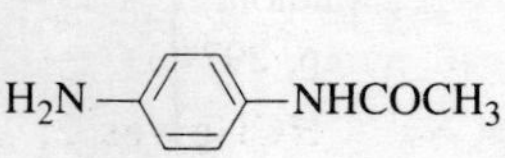 | 163 | Colourless, darkens in air, soluble in hot $H_2O$, alcohol, ether | 303 | – | – | – | – | - On boiling with HCl, yields p-phenylenediamine HCl, - Azo-$\beta$-naphthol derivative, m.p. 261°C. |
| Picramic acid 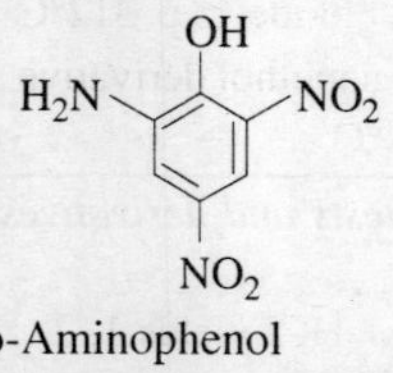 | 168 | Brownish-red, insoluble in $H_2O$, soluble in alcohol, sparingly soluble in ether | 201 (N-acety) | 220 (N. benzoyl) | – | – | – | - Soluble in dil. NaOH with orange colour. |
| o-Aminophenol | 174 | See phenols (Table 6.4b) | | | | | | |
| 4-Amino-2-methylphenol 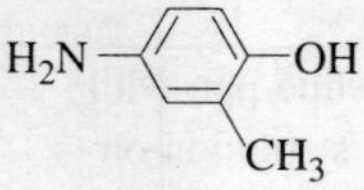 | 175 | Soluble in $H_2O$, ether | 103 (di) | 194 (di) | 109 | – | – | - Also see Phenols (Table 6.4b) |

| *Name and structure* | *M.P. (°C)* | *Properties* | *Derivatives (M.P.°C)* | | | | | *Specific tests and derivatives* |
|---|---|---|---|---|---|---|---|---|
| | | | *Ace-tyl* | *Benz-oyl* | *p-Tol-uene sulph-onyl* | *2, 4-dini-troph-enyl* | *pic-rate* | |
| 2, 4-Dinitroaniline 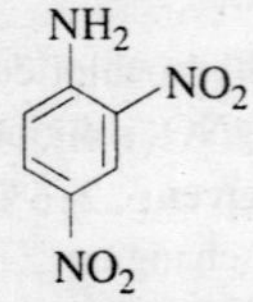 | 180 | Yellow with blue reflex, slightly soluble in hot $H_2O$ | 120 | 200 (220) | 219 | – | – | - On boiling with conc. NaOH solution yields 2, 4-dinitrophenol, m.p. 114°C. - Azo-β-naphthol derivative, m.p. 302°C. |
| p-Aminophenol | 184 | See phenols (Table 6.4b) | | | | | | |
| 2, 4, 6-Trinitroaniline (Picramide) 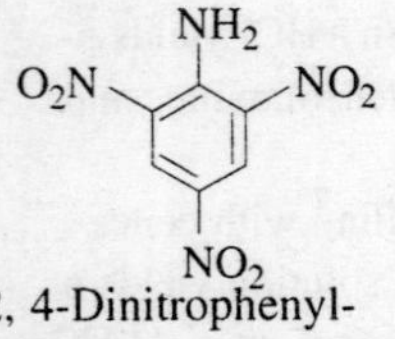 | 189 | Yellow with blue reflex | 230 | 196 | – | – | – | - On boiling with dil. NaOH soln yields picric acid. m.p. 122°C. |
| 2, 4-Dinitrophenyl-hydrazine 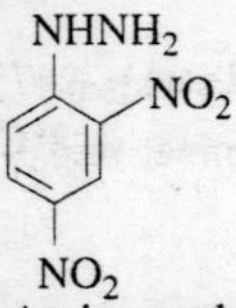 | 198 (d) | Red, insoluble in $H_2O$, ether, sparingly soluble in alcohol, soluble in dil. acids | 197 | 206 | – | – | – | - On boiling with conc. NaOH solution gives 2, 4-dinitrophenol, m.p. 114°C. - On heating with $CuSO_4$ solution gives m-dinitrobenzene, m.p. 90°C. |
| 2-Amino anthraquinone 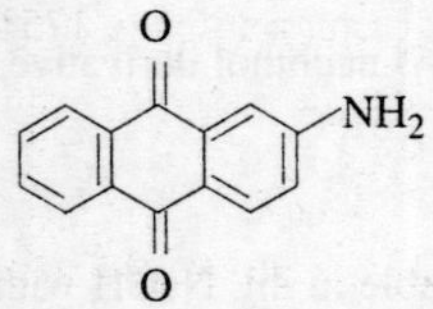 | 302 | | 257 | 228 | 304 | – | – | |

**Table 6.16(a)** Amines—Secondary (Liquids)

| *Name and structure* | *B.P. (°C)* | *Properties* | *Derivatives (M.P.°C)* | | | | | *Specific tests and derivatives* |
|---|---|---|---|---|---|---|---|---|
| | | | *Ace-tyl* | *Benz-oyl* | *p-Tol-uene sulph-onyl* | *2, 4-dini-troph-enyl* | *Pic-rate* | |
| Dimethylamine $(CH_3)_2NH$ | 7 gas | Soluble in alcohol and $CHCl_3$ | – | 42 | 79 | 87 | 158 | - Gives white ppt. with Nessler's solution on dilution with $H_2O$. - Hydrochloride, m.p. 170°C (Hygroscopic). |

| Name and structure | B.P. (°C) | Properties | Derivatives (M.P.°C) Acetyl | Benzoyl | p-Toluene sulphonyl | 2, 4-dinitrophenyl | Picrate | Specific tests and derivatives |
|---|---|---|---|---|---|---|---|---|
| Diethylamine $(C_2H_5)_2NH$ | 55 | Colourless, ammonical fishy odour, soluble in $H_2O$, alcohol, ether | – | 42 | 60 | 80 | 155 | - Gives white ppt. with Nessler's solution.<br>- Gives yellow ppt. with a solution of $HgI_2$ in KI.<br>- Hydrochloride, m.p. 216°C (hygroscopic) |
| Ethyleneimine (Aziridine)  | 56 | | – | – | 52 | – | 142 | |
| Di-isopropylamine $[(CH_3)_2CH]_2NH$ | 84 | Colourless, ammonical odour | – | – | – | – | 140 | |
| Pyrrolidine 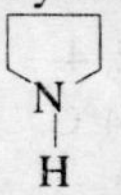 | 89 | Colourless, soluble in $H_2O$ | – | – | 123 | – | 112 | |
| Piperidine  | 107 | Colourless, unpleasant odour, soluble in $H_2O$ and alcohol | – | 48 | 103 (96) | 93 | 152 | - Hydrochloride, m.p. 247°C.<br>- Nitroso deriv. b.p. 218°C. |
| Di-n-propylamine $(CH_3CH_2CH_2)_2NH$ | 110 | Colourless, soluble in $H_2O$, alcohol, ether | – | – | – | 40 | 75 | |
| Morpholine 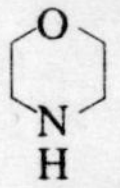 | 130 | Soluble in all common solvents | – | 75 | 147 | – | 146 | - Hydrochloride, m.p. 175°C. |
| Pyrrole  | 131 | Colourless, darkens on exposure to light soluble in $H_2O$, alcohol, ether | – | – | – | – | 69 (d) | |
| Di-isobutylamine $[(CH_3)_2CHCH_2]_2NH$ | 137 | Colourless, sparingly soluble in $H_2O$, soluble in alcohol, ether | 86 | – | – | 112 | 119 | - Hydrochloride, m.p. 262°C. |
| Di-n-butylamine $(CH_3CH_2CH_2CH_2)_2NH$ | 159 | Colourless, soluble in $H_2O$, alcohol, ether | – | – | – | – | 60 | - Nitroso deriv., m.p. 61°C. |
| Coniine | 166 | See Alkaloids (Table 6.25a) | | | | | | |
| N-Methylbenzylamine $C_6H_5CH_2NHCH_3$ | 181 | | – | – | 95 | – | – | |

| *Name and structure* | *B.P. (°C)* | *Properties* | *Derivatives (M.P.°C)* Acetyl | Benzoyl | p-Toluene sulphonyl | 2, 4-dinitrophenyl | Picrate | *Specific tests and derivatives* |
|---|---|---|---|---|---|---|---|---|
| Di-isoamylamine $[(CH_3)_2CH{\cdot}CH_2CH_2]_2NH$ | 186 | Colourless, insoluble in $H_2O$ | – | – | – | – | – | - Hydrochloride, m.p. 276°C. |
| N-Methylaniline $C_6H_5NHCH_3$ | 194 | Colourless, insoluble in $H_2O$, soluble in alcohol, ether | 102 | 63 | 95 | 167 | 147 | - Hydrochloride, m.p. 121°C. |
| N-Ethylaniline $C_6H_5NHC_2H_5$ | 205 | Colourless, insoluble in $H_2O$, soluble in alcohol, ether | 54 | 60 | 88 | 95 | 138 | - Hydrochloride, m.p. 176°C. |
| N-Methyl-m-toluidine 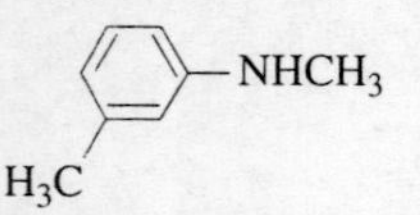 | 206 | | 66 | – | – | – | – | |
| N-Methyl-o-toluidine 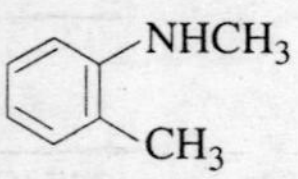 | 207 | | 55 | 66 | 120 | 155 | 90 | |
| N-Methyl-p-toluidine 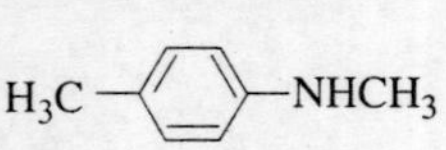 | 210 | | 83 | 53 | 60 | – | 131 | |
| N-Ethyl-p-toluidine 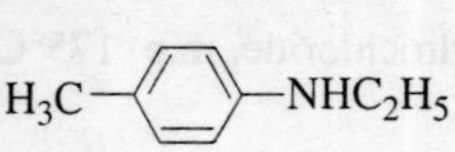 | 217 | | – | 40 | 71 | 120 | – | |
| N-Ethyl-o-toluidine 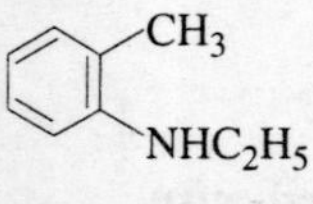 | 218 | | – | 72 | 75 | 114 | – | |
| N-Ethyl-m-toluidine 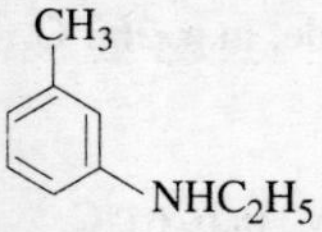 | 221 | | – | 72 | – | – | 132 | - Hydrochloride, m.p. 159°C. |
| N-Propylaniline  | 222 | Colourless, insoluble in $H_2O$ | 48 | – | 56 | – | – | |

| *Name and structure* | *B.P. (°C)* | *Properties* | *Derivatives (M.P.°C)* Acetyl | Benzoyl | p-Toluene sulphonyl | 2, 4-dinitrophenyl | Picrate | *Specific tests and derivatives* |
|---|---|---|---|---|---|---|---|---|
| 1, 2, 3, 4-Tetrahydro isoquinoline 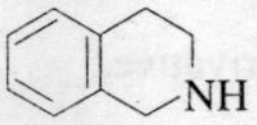 | 233 | | 46 | 129 | – | – | 200 | - Reduces Tollen's reagent. |
| N-Butylaniline $C_6H_5NHCH_2CH_2CH_2CH_3$ | 236 | Colourless, insoluble in $H_2O$ | – | 56 | 54 | – | – | |
| 1, 2, 3, 4-Tetrahydro-quinoline 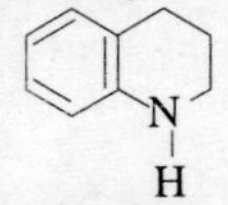 | 250 (m.p. 20) | | – | 75 | – | – | 141 | |
| Dicyclohexylamine $(C_6H_{11})_2NH$ | 254 (m.p. 20) | | 103 | 153 | 119 | – | 173 | |
| Dibenzylamine $(C_6H_5CH_2)_2NH$ | 300 (d) | Colourless, insoluble in $H_2O$ | – | 112 | 81 | 105 | – | - Hydrochloride, m.p. 256°C. |

**Table 6.16(b)** Amines—Secondary (Solids)

| *Name and structure* | *M.P. (°C)* | *Properties* | *Derivatives (M.P.°C)* Acetyl | Benzoyl | p-Toluene sulphonyl | 2, 4-dinitrophenyl | Picrate | *Specific tests and derivatives* |
|---|---|---|---|---|---|---|---|---|
| N-Benzylaniline  | 37 (b.p. 306) | Colourless, insoluble in $H_2O$, soluble in alcohol | 58 | 107 | 140 | 168 | – | |
| Indole 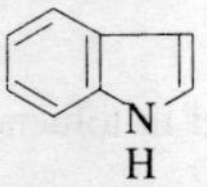 | 52 | Colourless, soluble in hot $H_2O$, hot alcohol | – | 68 | – | – | 187 | - N-Nitroso derivative, m.p. 171°C. |
| Diphenylamine 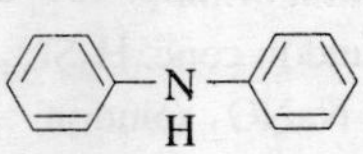 | 54 (b.P. 310) | Colourless, insoluble in $H_2O$, soluble in alcohol, ether | 101 | 180 | 142 | – | 182 | - N-Nitroso derivative, m.p. 66°C. - Sulphate salt 123-25°C. |
| N-Phenyl-1-naphthyl-amine 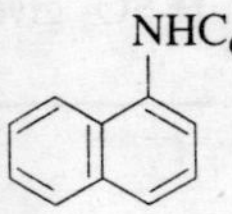 | 62 | | 115 | 152 | – | – | – | |

| *Name and structure* | *M.P. (°C)* | *Properties* | *Derivatives (M.P.°C)* Acetyl | Benzoyl | p-Toluene sulphonyl | 2, 4-dinitrophenyl | Picrate | *Specific tests and derivatives* |
|---|---|---|---|---|---|---|---|---|
| Di-p-tolylamine 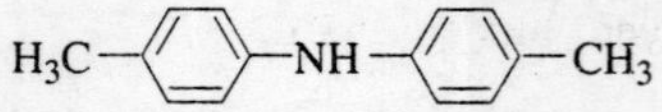  | 79 (b.p. 330°) | Colourless, sparingly $H_2O$, soluble in alcohol | 85 | 125 | – | – | – | - Gives yellow colour in conc. $H_2SO_4$ with trace of $HNO_3$. - N-nitroso derivative, m.p. 101°C. |
| p-Methylaminophenol | 85 | See phenols (Table 6.4b) | | | | | | |
| 3-Methyl indole 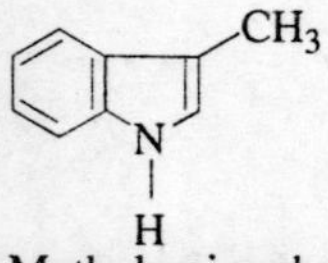  | 95 | Foul odour | 68 | – | – | – | 170 | - Hydrochloride, m.p. 167°C. |
| o-Methylaminophenol | 96 | See phenols (Table 6.4b) | | | | | | |
| Piperazine 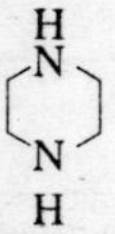  | 104 | Colourless, soluble in $H_2O$, alcohol, insoluble in ether | 134 (di) | 196 (di) | – | – | 280 | |
| N-Phenyl-2-naphthylamine 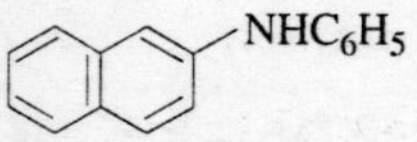  | 108 | | 93 | 136 | – | – | – | |
| 1, 1′-Dinaphthylamine 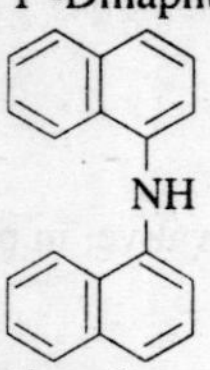  | 115 | | 217 | – | – | – | 169 | |
| Carbazole 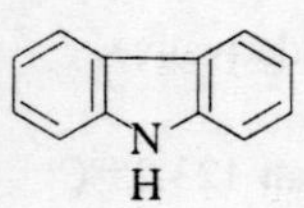  | 245 | | 69 | 98 | - | – | 186 | - Picrate obtained in toluene solution. - To solution of the compound in conc. $H_2SO_4$ add aq. $NaNO_2$ solution (2-3%)—deep green colour. - Compound + isatin solution in conc. $H_2SO_4$ gives blue colour. |

**Table 6.17(a)** Amines—Tertiary (Liquids)

| *Name and structure* | *B.P. (°C)* | *Properties* | *Derivatives* | | *Specific tests and derivatives* |
|---|---|---|---|---|---|
| | | | *Meth iodide* | *Picrate* | |
| Trimethylamine $(CH_3)_3N$ | gas | Colourless, soluble in $H_2O$ | 230(d) | 216 | - Hydrochloride, m.p. 278°C.<br>- Gives reddish yellow ppt with Nessler's solution. |
| Triethylamine $(C_2H_5)_3N$ | 89 | Colourless, soluble in $H_2O$ alcohol, ether | 280 | 173 | - Hydrochloride, 253°C. |
| Pyridine 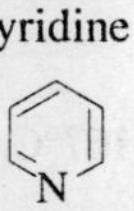 | 116 | Colourless, soluble in $H_2O$, alcohol, ether, characteristic odour | 117 | 167 | |
| Pyrimidine  | 124 | soluble in water | – | 156 | |
| α-Picoline (2-Methylpyridine) 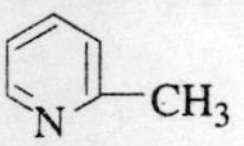 | 129 | Colourless, soluble in $H_2O$, alcohol, ether, characteristic odour | 230 | 169 | - On boiling with dil. $KMnO_4$ yields picolinic acid, m.p. 137°C. |
| β-Picoline (3-Methylpyridine) 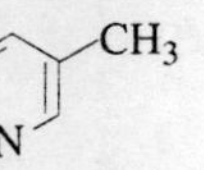 | 143 | Colourless, soluble in $H_2O$, alcohol, ether, characteristic odour | 92 | 150 | - On oxidation ($KMnO_4$) gives nictotinic acid, m.p. 228°C. |
| γ-Picoline (4-Methylpyridine) 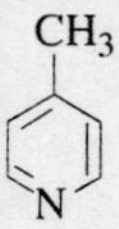 | 143 | Colourless, soluble in $H_2O$, alcohol, ether, characteristic odour | 152 | 167 | - On oxidation ($KMnO_4$) gives isonicotinic acid acid, m.p. 308°C. |
| 2, 6-Lutidine (2, 6-Dimethylpyridine) 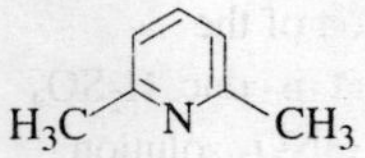 | 143 | | 238 | 163 (168) | |
| Tri-n-propylamine $(n\text{-}C_3H_7)_3N$ | 156 | | 208 | 116 | |
| Collidine 2, 4, 6-Trimethyl pyridine 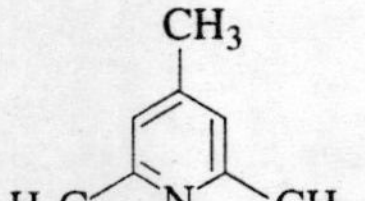 | 171 | | – | 156 | |

| *Name and structure* | *B.P. (°C)* | *Properties* | *Derivatives* | | *Specific tests and derivatives* |
|---|---|---|---|---|---|
| | | | *Meth iodide* | *Picrate* | |
| N, N-Dimethyl-o-toluidine 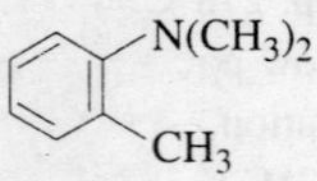 | 185 | Colourless, insoluble in $H_2O$, soluble in alcohol | 210 | 122 | - Hydrochloride, m.p. 156°C. |
| N, N-Dimethylaniline 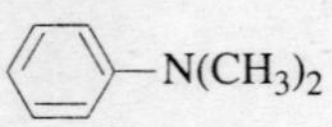 | 193 | Yellow, insoluble in $H_2O$, soluble in alcohol, ether | 228 (d) | 162 | - Gives green colour on treating the solution of the compd. in dil. HCl with a crystal of $NaNO_2$.<br>- Gives deep dichromate colour on treating the solution of the compd. in conc. $H_2SO_4$ with very dil. $HNO_3$.<br>- p-Nitroso compound, m.p. 87°C. |
| N-Ethyl-N-methyl aniline $CH_3NC_2H_5$ 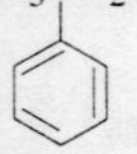 | 201 | Colourless, insoluble in $H_2O$, soluble in alcohol | 125 | 134 | - p-Nitroso compound, m.p. 66°C.<br>- Hydrochloride, m.p. 114°C. |
| N-Methyl-2-pyrrolidone 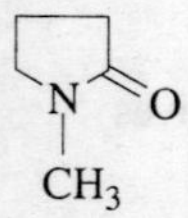 | 202 | | – | – | - Hydrochloride, m.p. 80°C. |
| N, N-Diethyl-o-toluidine 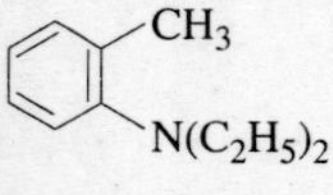 | 210 | | 224 | 180 | |
| N, N-Dimethyl-p-toluidine 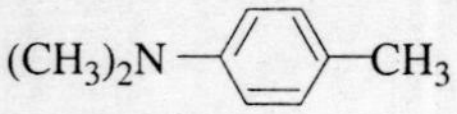 | 210 | Insoluble in $H_2O$, soluble in alcohol | 220 (d) | 129 | |
| Tri-n-butylamine $(n\text{-}C_4H_9)_3N$ | 211 | | 186 | 106 | |
| N, N-Dimethyl-m-toluidine 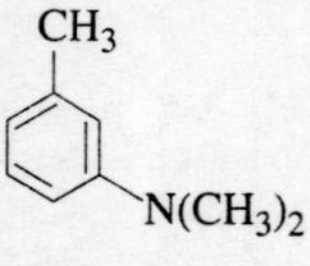 | 212 | | 177 | 131 | |
| N, N-Diethylaniline $C_6H_5N(C_2H_5)_2$ | 218 | Insoluble in $H_2O$, soluble in alcohol, ether | 102 | 142 | |

| Name and structure | B.P. (°C) | Properties | Derivatives | | Specific tests and derivatives |
|---|---|---|---|---|---|
| | | | Meth iodide | Picrate | |
| N, N-Diethyl-p-toluidine 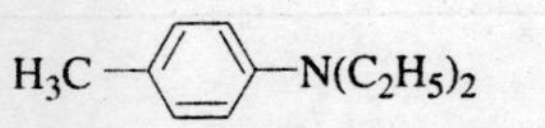 | 229 | | 184 | 110 | - Hydrochloride, m.p. 157°C. |
| N, N-Diethyl-m-toluidine 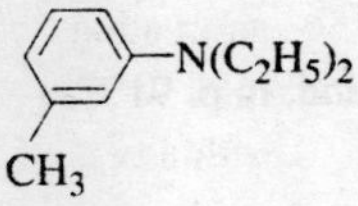 | 231 | | – | 97 | |
| Quinoline 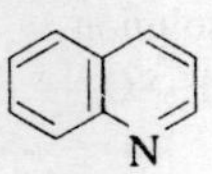 | 238 | | 72 (hydrated form) 133 (anhydrous form) | 203 | - Hydrochloride, m.p. 143°, (anhydrous) m.p. 93°C. (hydrated)<br>- On adding bleaching powder soln. to the soln. in boric acid yields chloroquinoline, m.p. 112°C, which with NaOH solution yields carbostyril, m.p. 199°C. |
| Isoquinoline 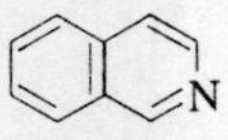 | 240 (m.p. 24) | | 159 | 222 | - Acid sulphate, m.p. 205°C. |
| Di-n-Propylaniline 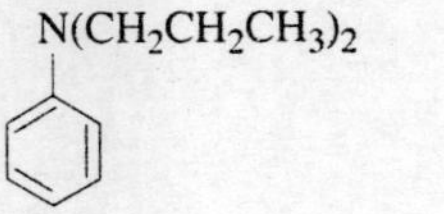 | 245 | | 156 | | - p-Nitroso deriv., m.p. 42°C. |
| Nicotine | 247 | See Alkaloids (Table 6.25a) | | | |
| 2-Methylquinoline (Quinaldine) 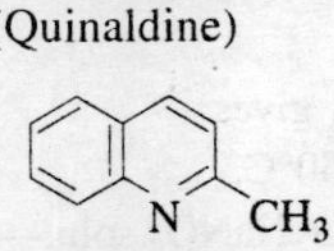 | 247 | | 195 | 191 | - On boiling with conc. $HNO_3$ gives nitroquinaldinic acid, m.p. 219°C. |
| 4-Methylquinoline (Lepidine) 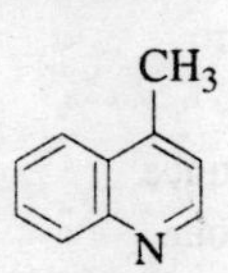 | 263 | | 174 | 212 | |
| Di-n-butylaniline 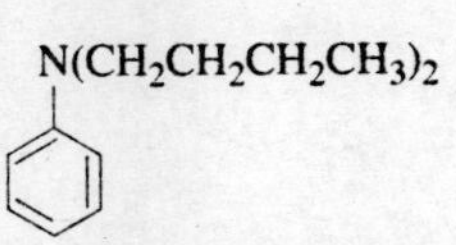 | 271 | | | 125 | |
| N-Methyldiphenylamine $C_6H_5N(CH_3)C_6H_5$ | 296 | | | | - p-Nitroso derivative, m.p. 44°C.<br>- with $HNO_3$ gives violet colour. |

**Table 6.17(b)** Amines—Tertiary (Solids)

| *Name and structure* | *M.P. (°C)* | *Properties* | *Derivatives* | | *Specific tests and derivatives* |
|---|---|---|---|---|---|
| | | | *Meth iodide* | *Picrate* | |
| 2, 2′-Bipyridyl 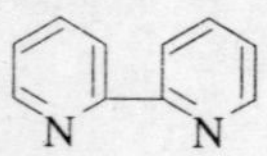  | 69 | Sparingly soluble in $H_2O$ | – | 158 | |
| Dibenzylaniline 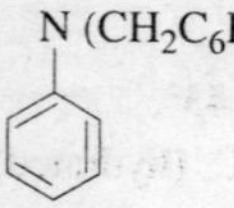  | 70 | | 135 | 131(d) | p-Nitroso compound, m.p. 91°C. |
| p-Dimethylamino benzaldehyde | 74 | See carbonyl compounds (Table 6.3b) | | | |
| p-Dimethylaminophenol | 75 | See phenols (Table 6.4b) | | | |
| 8-Hydroxyquinoline | 76 | See phenols (Table 6.4b) | | | |
| p-Nitroso-N, N-diethyl aniline | 84 | See nitroso compound (Table 6.21b) | | | |
| p-Nitroso-N, N-dimethyl aniline | 87 | See nitroso compounds (Table 6.21b) | | | |
| Tribenzylamine $(C_6H_5CH_2)_3N$ | 91 | | 184 | 190 | - Sulphate, m.p. 106°C. |
| Acridine 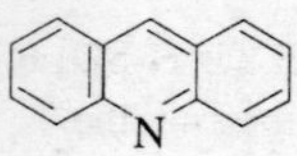  | 106 and 108 (two stable forms) | Colourless, soluble in $H_2O$, alcohol, ether | 224 | 208 | |
| Antipyrene 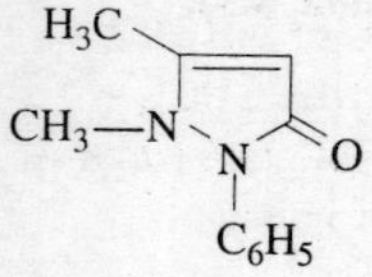  | 113 | Colourless, soluble in $H_2O$, alcohol, sparingly soluble in ether | | 188 | - Gives orange colour with $FeCl_3$.<br>- With $Br_2$ in $CHCl_3$ gives dibromide, m.p. 150°C.<br>- Compd + dil. HCl + $NaNO_2$ soln.—green crystals of nitroso compd m.p. 200°C(d).<br>- On nitration—gives nitro derivative, m.p. 273°C. |
| Triphenylamine $(C_6H_5)_3N$ | 127 | Colourless, insoluble in $H_2O$, sparingly soluble in alcohol, ether | – | – | - With conc. $H_2SO_4$ and trace of $HNO_3$ gives blue colour.<br>- Nitration (Fuming $HNO_3$)—Trinitro derivative, m.p. 280°C. |

| *Name and structure* | *B.P. (°C)* | *Properties* | *Derivatives* | | *Specific tests and derivatives* |
|---|---|---|---|---|---|
| | | | *Meth iodide* | *Picrate* | |
| Hexamethylene tetramine (Hexamine) 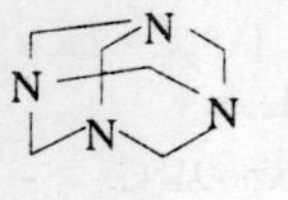 | 280 | | 190 | 157 (d) | - Dibromide, m.p. 196-200°C.<br>- Adds $Br_2$ to form a red tetrabormide which on keeping in air loses bromine to form yellow dibromide.<br>- Solution in HCl gives formaldehyde.<br>- Aq solution with conc. $HNO_3$ at 0° gives nitrate, m.p. 165°C. |

**Table 6.18** Amides, Imides, Anilides, Guanidines

| *Name and structure* | *B.P./M.P. (°C)* | *Properties* | *Derivatives* | *Specific test and derivatives* |
|---|---|---|---|---|
| Formamide $HCONH_2$ | B.P. 195(d) b.p. 110/11 mm | Colourless, soluble in $H_2O$, alcohol, sparingly soluble in ether | Anilide, m.p. 46°C. 1-naphthylamine derivative, m.p. 138°C. | - Gives white ppt. with $HgCl_2$ solution.<br>- With aq. $FeCl_3$ gives wine red colour changing to brown ppt. on boiling. |
| Guanidine $(NH_2)_2C{=}NH$ | M.P. (Available as nitrate salt) | Alkaline, deliquescent solid, almost insoluble in cold $H_2O$ | Carbonate, 197°C. Nitrate, 214°C. Acetate, 229°C Hydrochloride, 172°C | - On boiling with $Ba(OH)_2$ yields urea + $NH_3$.<br>- With NaOCl gives yellow colour changing to orange red.<br>- With conc. $HNO_3$ + $H_2SO_4$ in cold yields nitroguanidine m.p. 230°(d). |
| Formanilide $C_6H_5NHCHO$ | M.P. 46 (b.p. 216/120 mm) | Colourless, Faint odour, sparingly soluble in $H_2O$, soluble in alcohol, ether | Benzyl formanilide, m.p. 48°C obtained on warming with benzyl chloride | - On boiling with dil. HCl yields formic acid+$C_6H_5NH_2HCl$.<br>- On warming with $CHCl_3$ + KOH in alcohol gives odour of phenyl isocyanide. |
| Methyl carbanilate $NHCOOCH_3$ 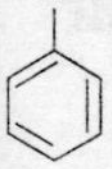 | M.P. 47 | | On boiling with aniline, gives carbanilide, m.p. 238°C. | |
| Ethyl carbanilate $NHCOOC_2H_5$ 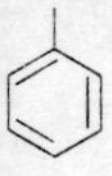 | M.P. 52 | | | - On boiling with aniline gives carbanilide, m.p. 238°C. |

| *Compound* | *M.P. (°C)* | *Properties* | *Derivatives* | *Specific test and derivs* |
|---|---|---|---|---|
| N-Ethyl acetanilide 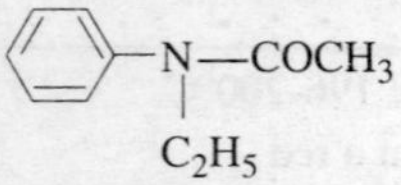 | M.P. 54 (b.p. 249) | | | - On boiling with HCl yields acetic acid, b.p. 118°C. and ethyl aniline hydrochloride. - With conc. $HNO_3$ +$H_2SO_4$ at 40° yields p-nitro derivative, m.p. 118°C which on hydrolysis yields p-nitroethylaniline, m.p. 95°C. |
| Allyl thiourea $CH_2$=$CHCH_2NHCSNH_2$ | M.P. 74 | Colourless, soluble in hot $H_2O$, alcohol, sparingly soluble in ether | | - On boiling with conc. NaOH yields $Na_2S$, $NH_3$ and allylamine, b.p. 53°C. - Yields $Ag_2S$ with ammonical $AgNO_3$ solution. - With ethyl iodide in alcohol forms addition product, m.p. 72°C. - With $CH_3COCl$ in acetone yields acetyl derivative m.p. 103°(d) which with NaOH yields N-acyl derivative, m.p. 95°C. - With $Br_2$ in alcohol yields bromopropylene-isothiourea HBr, m.p. 140°C. |
| Propionamide $CH_3CH_2CONH_2$ | M.P. 79 (b.p. 213°) | Colourless, soluble in $H_2O$, alcohol, ether | Propionanilide, 104°C Propionyl-$\alpha$-naphthylamine, 116°C | - With 75% $H_2SO_4$ at 120° yields propionic acid, b.p. 140°C. |
| Acetamide $CH_3CONH_2$ | M.P. 82 (b.p. 222°) | Colourless deliquiscent, solid, readily soluble in $H_2O$, alcohol, insoluble in ether | Picrate, 107°C Anilide, 114 $\alpha$-Naphthylamine derivative 159°C | - On distilling with conc. $H_2SO_4$, yields acetic acid, b.p. 118°C (smell of vinegar) |
| Acetoacetanilide $CH_3COCH_2.CO.NHC_6H_5$ | M.P. 85 | Colourless, soluble in hot $H_2O$ | Carbanilide, 238°C | - Gives violet colour with $FeCl_3$ solution. - On boiling with HCl yields acetic acid, b.p. 118°C (smell of vinegar) and aniline hydrochloride. - On heating with aniline yields acetone and carbanilide, m.p. 238°C. |

| *Name and structure* | *M.P. (°C)* | *Properties* | *Derivatives* | *Specific test and derivatives* |
|---|---|---|---|---|
| | | | | - On warming with conc. $H_2SO_4$ yields lepidone (4-methyl carbostyril), m.p. 224°C. |
| Acetoacet-o-anisidine 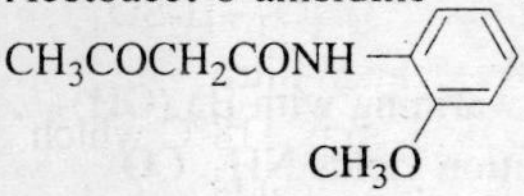 | 87 | | | - Gives violet colour with $FeCl_3$ solution. |
| o-Nitroacetanilide 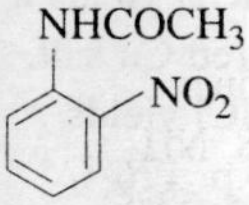 | 94 | Yellow, soluble in hot $H_2O$, alcohol, ether | | - On heating with NaOH solution yields o-nitroaniline, m.p. 71°C.<br>- On heating with conc. HCl gives smell of vinegar. |
| n-Butyranilide<br>$CH_3CH_2CH_2CONHC_6H_5$ | 95 | | | - On boiling with HCl yields n-butyric acid, b.p. 163°C. and aniline hydrochloride. |
| Acetoacet-p-toluidide 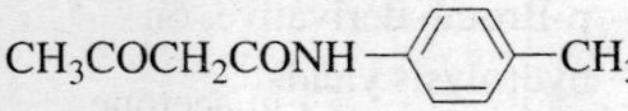 | 95 | | | - Gives violet colour with $FeCl_3$. |
| m-Toluamide 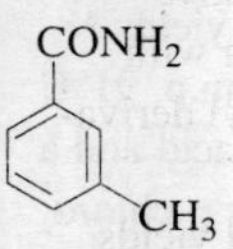 | 95 | White, sparingly soluble in ether, benzene | Anilide, m.p. 126°C | - On hydrolysis gives m-toluic acid, m.p. 110°C. |
| Semicarbazide<br>$NH_2CONHNH_2$ | 96 | Colourless, soluble in $H_2O$, alcohol, insoluble in ether | Hydrochoride, 173°(d) | - Reduces Fehling and Tollen's reagents.<br>- On heating with benzaldehyde in alcohol gives benzal derivative, m.p. 227°C (d) |
| Methyl urea<br>$CH_3NHCONH_2$ | 101 | | Xanthyl derivative, 230 | |
| Propionanilide<br>$CH_3CH_2CONHC_6H_5$ | 104 | | p-Nitro derivative, 182 | - On boiling with HCl yields propionic acid, b.p. 140°C and aniline hydrochloride.<br>- With conc. $HNO_3$ +$H_2SO_4$ at 0° yields p-nitro deriv., m.p. 182°C which on hydrolysis yields p-nitroaniline, m.p. 147°C. |
| Acetoacet-o-toluidine 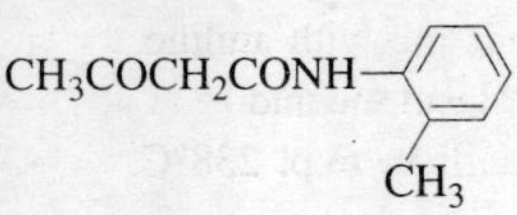 | 104 | | | - Gives violet colour with $FeCl_3$ solution.<br>- On warming with conc. $H_2SO_4$ yields 4, 8-dimethylcarbostyril, m.p. 217°C. |

| *Name and structure* | *M.P. (°C)* | *Properties* | *Derivatives* | *Specific test and derivatives* |
|---|---|---|---|---|
| Acetoacet-o-chloroaniline<br>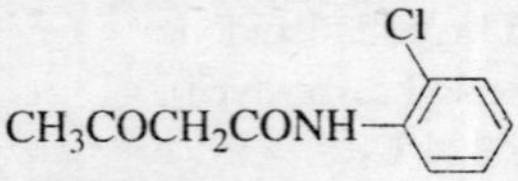 | 105 | | | - Gives violet colour with $FeCl_3$ solution. |
| Dicyanidamidine<br>$NH_2C(:NH).NHCONH_2$ | 105 | Colourless, soluble in $H_2O$, sparingly soluble in alcohol, insol. in ether | Hydrochloride, 173<br>Sulphate, 194<br>Picrate, 265 | - On heating to 160° evolves $NH_3$.<br>- On warming with Ba $(OH)_2$ solution yields $NH_3$, $CO_2$ and urea.<br>- With $CuSO_4$ + NaOH yields a sparingly soluble red Cu salt. |
| Iso-Valeranilide<br>$(CH_3)_2CHCH_2CONHC_6H_5$ | 110 | | | - On boiling with HCl yields iso-valeric acid, b.p. 176°C and aniline hydrochloride. |
| Acetanilide<br>$NHCOCH_3$<br>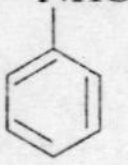 | 114 | Colourless, sparingly soluble in cold $H_2O$, soluble in hot $H_2O$, alcohol | p-Bromo, 167<br>p-Nitro, 210 | - On boiling with dil. HCl yields $CH_3COOH$ and aniline hydrochloride.<br>- p-Bromo derivative, on hydrolysis yields p-bromoaniline, m.p. 66°C.<br>- With NaOCl soln. yields N-chloro derivative, m.p. 91°C which with acetic acid and a trace of HCl yields p-chloro derivative, m.p. 179°C.<br>- p-Nitro derivative, on hydrolysis yields p-nitroaniline, m.p. 147°C. |
| n-Butyramide<br>$CH_3CH_2CH_2CONH_2$ | 114 | Colourless, sparingly soluble in $H_2O$, ether, soluble in alcohol | Xanthyalmide, 187 | - On heating with aniline gives n-butyranilide, m.p. 95°C. |
| Phenylacetanilide<br>$C_6H_5CH_2CONHC_6H_5$ | 118 | | 2, 4-Dinitro derivative, 189 | - On boiling with HCl yields phenylacetic acid, m.p. 76°C and aniline hydrochloride.<br>- With conc. $HNO_3$ in cold yields 2, 4-dinitrophenylacetic acid, m.p. 189°C. |
| Acetoacet-p-nitroanilide<br>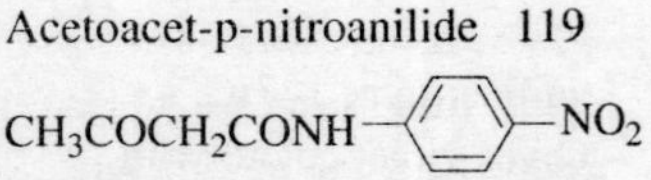 | 119 | | | - Violet colour with $FeCl_3$. |
| Chloracetamide<br>$ClCH_2CONH_2$ | 120 | | Xanthyl, m.p. 208 | |
| Cyanoacetamide<br>$CNCH_2CONH_2$ | 120 | | Xanthyl, m.p. 223 | |
| Dimethylcarbanilide<br>$C_6H_5N(CH_3)$.<br>$CO.N(CH_3)C_6H_5$ | 120<br>(b.p. 350) | | | - On boiling with HCl yields methylaniline hydrochloride. |

| *Name and structure* | *M.P. (°C)* | *Properties* | *Derivatives* | *Specific test and derives* |
|---|---|---|---|---|
| Nicotinamide<br>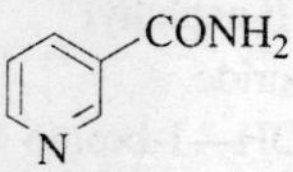 | 122 | Colourless, insoluble, in $H_2O$ soluble in benzene | Nicotinic acid, 236 | On hydrolysis gives nicotinic acid, m.p. 236°C. |
| Succinimide<br>$CH_2$—CO<br>NH<br>$CH_2$—CO | 126 | Colourless, soluble in $H_2O$, alcohol, sparingly soluble in ether, chloroform | Succinic acid, 185<br>Xanthyl, 246<br>Succinanil, 156 | - Gives fluorescein reaction (see succinic acid)<br>- On boiling with aniline gives succinanil, m.p. 156°C. |
| Benzamide<br>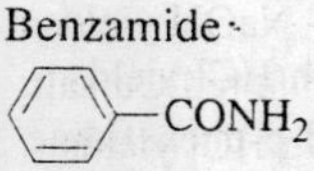 | 128 | Colourless, sparingly soluble in $H_2O$, soluble in absolute alcohol, hot $H_2O$ | Xanthyl, 224<br>Benzanilide, 164 | - On hydrolysis gives benzoic acid, m.p. 122°C. |
| Iso-butyramide<br>(2-Methylpropionamide)<br>$(CH_3)_2$ CHCONH$_2$ | 128 | Sparingly soluble in $H_2O$, Soluble in alcohol, ether | Anilide, 105<br>Xanthyl, 211 | - On hydrolysis gives iso-butyric acid, b.p. 155°C. |
| Acetylphenylhydrazine<br>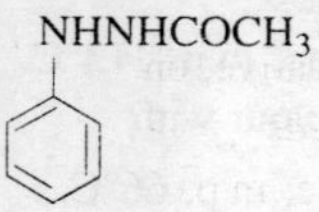 | 128 | Soluble in hot $H_2O$ | Dibromo, 146 | - On boiling with HCl gives phenylhydrazine hydro-chloride.<br>- Reduces Fehlings solution.<br>- With 2 moles of $Br_2$ in conc. HCl yields dibromo derivative, m.p. 146°C. |
| Urea<br>(carbamide)<br>$NH_2CONH_2$ | 132 | Colourless, odourless, soluble in $H_2O$, alcohol, sparingly soluble in ether | Nitrate, 163<br>Oxalate, 171 | - On heating to 150° yields $NH_3$ and biuret, m.p. 192°C.<br>- Evolves $N_2$ with NaOBr solution.<br>- Evolves $NH_3$ on boiling with dil. NaOH solution<br>- On warming with aniline hydrochloride solution yields phenyl urea, m.p.147°C and carbanilide, m.p. 238°C.<br>- Biuret (obtd. above) on dissolving in dil NaOH solution and treatment with a drop of dil. $CuSO_4$ solution gives purple colour. |
| 4-Aceto-m-xylidine<br> | 133 | Colourless, insoluble in $H_2O$, sparingly soluble in alcohol | | - On boiling with HCl yields acetic acid and m-xylidine hydrochloride.<br>- With fuming nitric acid yields 5-nitro derivative, m.p. 172°C.<br>- On adding slowly $KNO_3$ to solution in conc. $H_2SO_4$ yields 6-nitro derivative, m.p. 159°C. |

| *Name and structure* | *M.P. (°C)* | *Properties* | *Derivatives* | *Specific test and derivatives* |
|---|---|---|---|---|
| β-Acetnaphthyamine<br>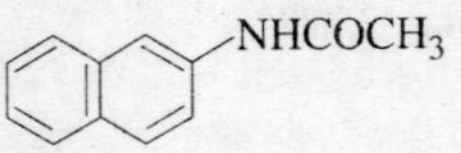 | 134 | Colourless, insoluble in $H_2O$, sparingly soluble in alcohol | | - On boiling with HCl yields acetic acid and β-naphthyl amine hydrochloride.<br>- With $Br_2$ in AcOH—1-bromo derivative, m.p. 140°C.<br>- With fuming $HNO_3$ yields two dinitro derives. One soluble in alcohol, m.p. 185°C. and the other insoluble in alcohol, m.p. 235°C. |
| Acet-p-phenetidide (Phenacetin)<br>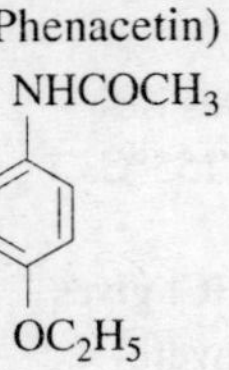 | 134 | Colourless, insoluble in $H_2O$, sparingly soluble in alcohol | | - On boiling with HCl yields acetic acid and p-phenetidine hydrochloride.<br>- On warming with 10% $HNO_3$ yields 3-nitro derivative, m.p. 103°C which on hydrolysis yields nitrophenetidine, m.p. 113°C. |
| Salicylamide<br>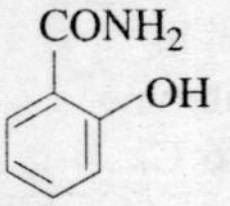 | 139 | Colourless, sparingly soluble in cold $H_2O$ soluble in alcohol, ether | Acetyl, 143<br>Benzoyl, 200 | - Gives violet colour with $FeCl_3$.<br>- On boiling with NaOH soln. yields salicylic acid, m.p. 159°C. |
| m-Tolylurea<br>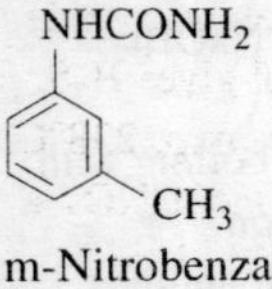 | 142 | | | - On alkaline hydrolyine yields m-toluidine b.p. 202°C. |
| m-Nitrobenzamide<br>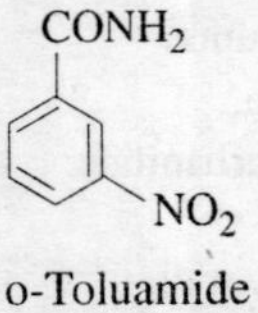 | 142 | Colourless, sparingly soluble in $H_2O$, soluble in alcohol, ether | | - On boiling with NaOH soln. m-nitrobenzoic acid, m.p. 141°C. |
| o-Toluamide<br>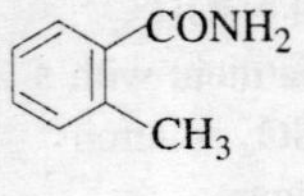 | 142 | Colourless, soluble in hot $H_2O$, alcohol | Benzoyl, 158<br>Anilide, 125 | - On hydrolysis yields o-toluic acid, m.p. 104°C. |
| o-Benzotoluidide<br>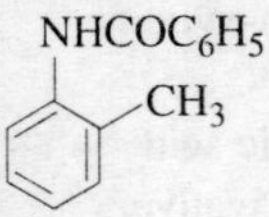 | 144 | | | - On boiling with HCl yields benzoic acid m.p. 122°C and o-toluidine hydrochloride.<br>- With hot $KMnO_4$ soln. yields benzoyl anthranilic acid, m.p. 185°C. |

| *Compound* | *M.P. (°C)* | *Properties* | *Derivatives* | *Specific test and derivalives* |
|---|---|---|---|---|
| Triphenylguanidine<br>$C_6H_5N{=}C(NHC_6H_5)_2$ | 145 | Colourless, insoluble in $H_2O$ | Hydrochloride, 241<br>Picrate, 180 | - On heating yields aniline, b.p. 183°C and carbodi-phenylimide, b.p. 330°C.<br>- On boiling with conc. KOH solutions yields aniline |
| Diphenylguanidine<br>$(C_6H_5NH)_2C{=}NH$ | 147 | Colourless, sparingly soluble in $H_2O$ | Perchlorate, 162 | - On boiling with HCl gives aniline hydrochloride, m.p. 198°C.<br>- With $Ac_2O$ at 100° yields acetanilide, m.p. 114°C. |
| Phenyl urea<br>$NHCONH_2$<br>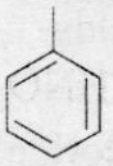 | 147 | Colourless, soluble in hot $H_2O$, alcohol, sparingly soluble in ether | Xanthyl deriv. 225 | - On heating above m.p. yields $NH_3$, $CO_2$ and carbanilide m.p. 238°C.<br>- With NaOH yields aniline and $NH_3$. |
| N-Benzylurea<br>$C_6H_5CH_2NHCONH_2$ | 149 | Colourless, soluble in hot $H_2O$ | | - On heating above 200° gives $CO_2$, $NH_3$ and dibenzylurea, m.p. 167°C. |
| Cinnamanilide<br>$CH{=}CHCONHC_6H_5$<br>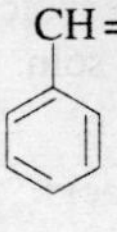 | 151 | Colourless, insoluble in $H_2O$, sparingly soluble in alcohol | | - On boiling with HCl, yields cinnamic acid, m.p. 133°C and aniline hydrochloride.<br>- On heating with acidic $KMnO_4$ yields benzaldehyde and benzoic acid, m.p. 122°C. |
| Sym. Diphenyl thiourea (Thiocarbanilide)<br>$C_6H_5NHCSNHC_6H_5$ | 153 | | | - With yellow mercuric oxide in warm alcohol gives $H_2S$ and carbanilide, m.p. 238°C. |
| p-Acetotoluidide<br>$NHCOCH_3$<br>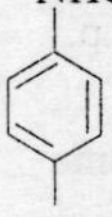<br>$CH_3$ | 153 | Colourless, insoluble in $H_2O$, soluble in absolute alcohol and ether | | - On nitration (conc. $HNO_3$ + $H_2SO_4$) at 30-40° gives 3-nitro derivative, m.p. 144°C. |
| Phenylthiourea<br>$C_6H_5NHCSNH_2$ | 154 | Colourless, insoluble in $H_2O$, soluble in hot $H_2O$, alcohol, sparingly soluble in ether | | - On boiling with aniline gives thiocarbanilide, m.p. 153°C. |
| m-Nitroacetanilide<br>$NHCOCH_3$<br>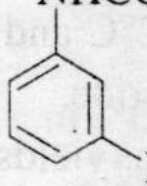<br>$NO_2$ | 154 | Colourless or yellowish brown, insoluble in $H_2O$, soluble in alcohol, chloroform | | - On reduction with Sn +HCl gives m-phenylenediamine, m.p. 63°C.<br>- On heating with NaOH solution gives m-nitroaniline, m.p. 114°C.<br>- On heating with dil. HCl—smell of vinegar. |

| *Name and structure* | *M.P. (°C)* | *Properties* | *Derivatives* | *Specific test and derivatives* |
|---|---|---|---|---|
| Phenylacetamide<br>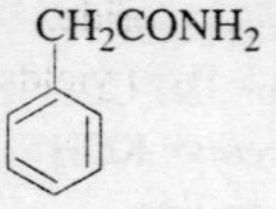 | 154 | Colourless, soluble in $H_2O$, ether, alcohol | Xanthylamide, 196<br>N-acetate, 129<br>N-benzoate, 130 | - On boiling with aniline yields phenylacetanilide, m.p. 118°C.<br>- On hydrolysis (NaOH solution) yields phenylacetic acid, m.p. 76°C. |
| Succianil<br>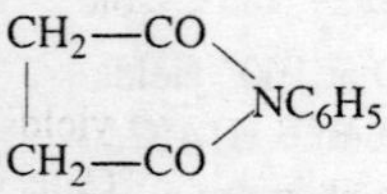 | 156 | | | - On boiling with HCl gives succinic acid, m.p. 185°C and aniline hydrochloride.<br>- On boiling with $Ba(OH)_2$ soln. yields succinanilic acid, m.p. 148°C. |
| N-Benzoyl-p-toluidine<br>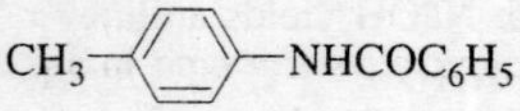 | 158 | | Dinitro, 186 | - On boiling with HCl yields benzoic acid, m.p. 122°C and p-toluidine hydrochloride.<br>- On nitration (fuming $HNO_3$ at 0°) yields dinitro derivative, m.p. 186°C.<br>- With $CrO_3$ in acetic acid yields p-benzoylaminobenzoic acid, m.p. 278°C. |
| p-Toluamide<br>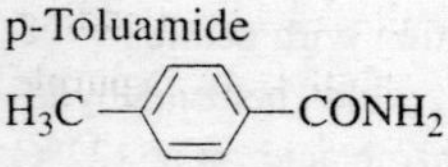 | 159 | Colourless, soluble in $H_2O$, alcohol, ether | N-acetyl, 147<br>Xanthyl, 224 | - On hydrolysis (NaOH solution) gives p-toluic acid m.p. 178°C. |
| $\alpha$-Acetylnaphthyl amide<br>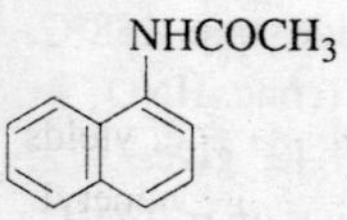 | 160 | | Dinitro, 250<br>Bromo, 193 | - On boiling with HCl yields acetic acid and $\alpha$-naphthylamine hydrochloride.<br>- On nitration (fuming $HNO_3$ in acetic acid) yields dinitro deriv., m.p. 250°C.<br>- With $Br_2$ in acetic acid (1 mol $Br_2$) yields 4- bromo derivative, m.p. 193°C. |
| p-Aminoacetanilide<br>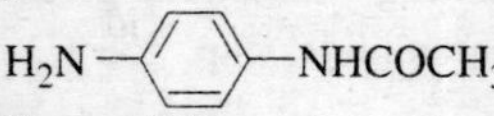 | 163 | See Primary Amines (Table 6.15b) | | |
| Benzanilide<br>$C_6H_5NHCOC_6H_5$ | 164 | Colourless or reddish, sparingly soluble in alcohol, ether, insoluble in $H_2O$ | p-Bromo, 204 | - On bromination ($Br_2$ in acetic acid) gives p-bromo derive, m.p. 204.<br>- On boiling with HCl yields benzoic acid, m.p. 122. |
| p-Bromoacetanilide<br>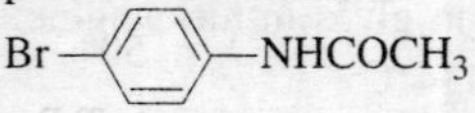 | 167 | Colourless, sparingly soluble in hot $H_2O$, soluble in alcohol, chloroform | Nitro, 104 | - With fuming $HNO_3$ (50°) in acetic acid gives 2-nitro derivative, m.p. 104°C<br>- On acid hydrolysis gives-p-bromo aniline m.p. 67°C. |

| *Name and structure* | *M.P. (°C)* | *Properties* | *Derivatives* | *Specific test and derivatives* |
|---|---|---|---|---|
| sym. Dibenzylurea $CO(NHCH_2C_6H_5)_2$ | 167 | | | - On heating with HCl yields benzylamine hydrochloride. |
| Benzoyl phenylhydrazine 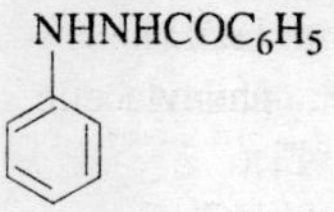 | 168 | Colourless, sparingly soluble in hot $H_2O$, ether, soluble in hot KOH solution | | - On oxidation with HgO yields red benzoylazobenzene, m.p. 80°C.<br>- On boiling with conc. HCl yields phenylhydrazine hydrochloride and benzoic acid, m.p. 122°C. |
| Malonamide $CH_2(CONH_2)_2$ | 170 | Colourless, soluble in $H_2O$, insoluble in absolute alcohol, ether | Anilide, 225<br>N, N-dibenzoate, 142 | - On heating with aniline yields malonanilide, m.p. 225°C.<br>- On heating with NaOH solution gives malonic acid, m.p. 133°C.<br>- On shaking the compd in aq. NaOH with 2 drops of aq. $CuSO_4$ gives red or violet-red colour. |
| Alloxan 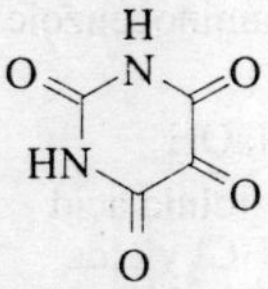 | 170 (d) | Crystalline solid, soluble in $H_2O$ | | - Gives blue colour with ferrous salt solution.<br>- On warming with $NH_2OH\cdot HCl$ solution and then adding $Na_2CO_3$ solution gives purple colour . |
| N-Bromosuccinimide 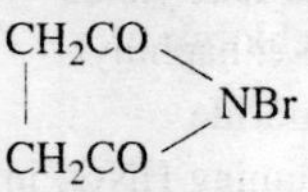 | 173 (180) | | | - Fuming $HNO_3$ in acetic acid at 50° gives nitro derivative m.p. 104°C. |
| p-Phenetyl urea 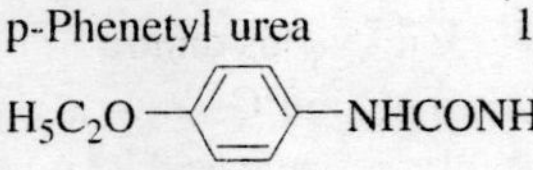 | 174 | Colourless, sparingly soluble in cold $H_2O$ | | - On heating above m.p. yields $NH_3$, $CO_2$ and di-p-phenetyl urea, m.p. 235°C.<br>- On boiling with 48% HBr yields p-aminophenol HBr salt. |
| o-Nitrobenzamide 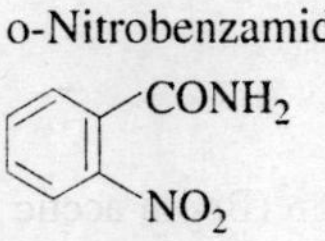 | 174 | Colourless, soluble in hot $H_2O$, hot alcohol, ether | | - On heating with NaOH soln. gives o-nitrobenzoic acid m.p. 146°C.<br>- On reduction (Sn + HCl, boil) gives anthranilic acid, m.p. 144°C.<br>- On heating with $P_2O_5$ gives o-nitrobenzonitrile, m.p. 110°C. |
| p-Chloroacetanilide 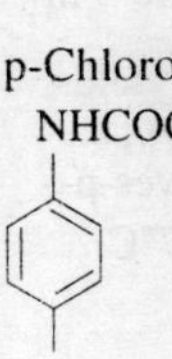 | 179 | Colourless, sparingly soluble in $H_2O$, soluble in alcohol, ether | Nitro, 104 | - With fuming $HNO_3$ at 50° gives 2-nitro derivative, m.p. 104°C.<br>- On heating with NaOH solution yields p-chloroaniline, m.p. 70°C. |

| *Name and structure* | *M.P. (°C)* | *Properties* | *Derivatives* | *Specific test and derivatives* |
|---|---|---|---|---|
| Thiourea<br>$NH_2CSNH_2$ | 180 | Colourless, soluble in $H_2O$, sparingly soluble in ether | | - With $K_4Fe(CN)_6$ and dil. acetic acid gives green colour changing to blue.<br>- On heating with benzyl chloride in dil. alcohol yields S-benzyl iso thiouronium chloride, m.p. 175°C.<br>- Hydrochloride m.p. 136<br>- On warming with ethyl sodio malonate in absolute alcohol yields thiobarbituric acid, m.p. 235°C. |
| p-Tolylurea<br>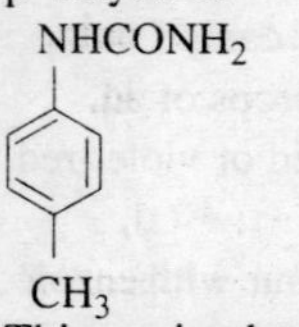 | 181 | Colourless, soluble in hot $H_2O$, alcohol | N-acetyl, 200<br>N-benzoyl, 223 | - On heating above m.p. yields $NH_3$, $CO_2$ and di-p-tolylurea, m.p. 268°C.<br>- On boiling with HCl yields p-toluidine hydrochloride. |
| Thiosemicarbazide<br>$NH_2CSNHNH_2$ | 182 | Colourless, soluble in $H_2O$, alcohol | Acetate, 165<br>Isopropylidine, 179<br>Hydrochloride, 188<br>Benzal, 159 | |
| Diacetyl-o-phenylene-diamine<br>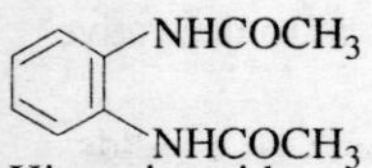 | 185 | Colourless, soluble in $H_2O$ | | - On boiling with HCl yields acetic acid and o-phenylene-diamine hydrochloride. |
| Hippuric acid<br>(Benzoyl glycine)<br>$C_6H_5CONHCH_2COOH$ | 187 | | p-Nitrobenzyl ester, 136<br>p-Bromophen-acyl ester, 151<br>Anilide, 208 | |
| 5, 5-Diethyl barbituric acid<br>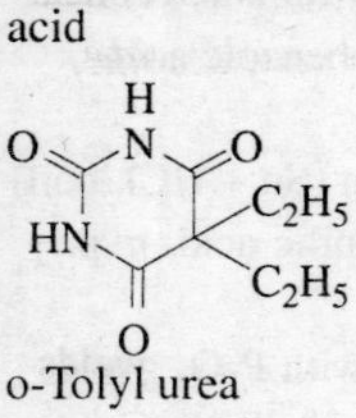 | 188<br>(d) | | Xanthyl deriv. 246. | |
| o-Tolyl urea<br>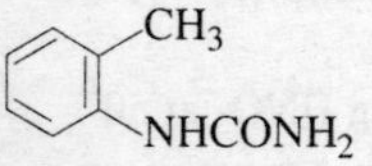 | 191 | Colourless, soluble in hot $H_2O$ | N-acetyl, 168<br>N-benzoyl, 210 | - On heating above its m.p. yields $NH_3$, $CO_2$ and di-o-tolyl urea, m.p. 250°C.<br>- On boiling with HCl, yields o-toluidine hydrochloride. |

| *Name and structure* | *M.P. (°C)* | *Properties* | *Derivatives* | *Specific test and derivatives* |
|---|---|---|---|---|
| Diacetyl m-phenylene-diamine 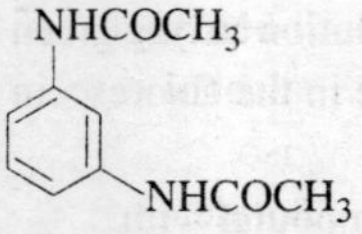 | 191 | Colourless, soluble in hot $H_2O$ | Dinitro, 228 | - On boiling with HCl yields acetic acid and m-phenylene diamine hydrochloride.<br>- With conc. $HNO_3$ at 0° in presence of urea, yields dinitro derivative, m.p. 228°C. |
| Biuret $NH_2CONHCONH_2$ | 192 | Colourless, insoluble in $H_2O$, soluble in hot $H_2O$, alcohol | Diphenyl biuret, 210<br>Xanthyl-amide, 260 | - Addition of one drop of aq. $CuSO_4$ solution to an alkaline solution of the compd gives red-violet colour.<br>- On heating with aniline yields diphenylbiuret, m.p. 210°C. |
| p-Nitrobenzamide 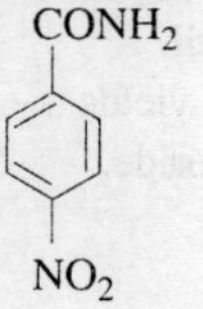 | 201 | Colourless, sparingly soluble in $H_2O$, soluble in alcohol, ether | | - On heating with NaOH soln. yields p-nitrobenzoic acid, m.p. 240°C.<br>- On reduction (Sn + HCl, boiling) gives p-aminobenzoic acid, m.p. 186°C.<br>- On heating with $P_2O_5$ gives p-nitrobenzonitrile, 148°C. |
| Isatin 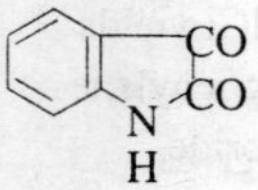 | 203 | Yellowish-red prisms, soluble in dil. NaOH, methanol, ethanol | N-Acetyl, 141<br>3-Oxime, 225(d)<br>3-Semicarba-zone, 266(d) | |
| Phthalanil 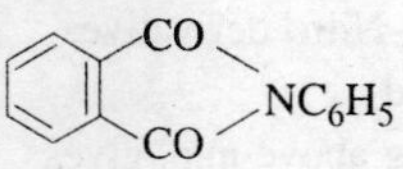 | 210 | | | - On boiling with HCl yields phthalic acid, m.p. 195°C and aniline hydrochloride.<br>- On boiling with $Ba(OH)_2$ solution and acidifying yields phthalanilic acid, m.p. 169°C. |
| p-Nitroacetanilide 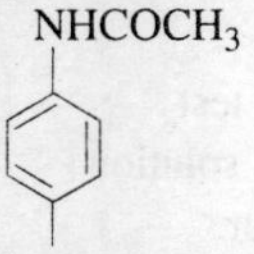 | 210 | Colourless or pale yellow, soluble in hot $H_2O$, alcohol, ether | | - On reduction (Sn + HCl) heat, yields p-phenylenediamine, m.p. 140°C. |
| Phthalamide 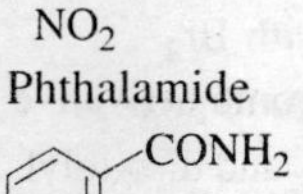 | 220 (d) | Colourless, sparingly soluble in $H_2O$, alcohol | Anil, 205<br>Xanthyl derivative 177 | - On heating above m.p. yields phthalimide, m.p. 238°C.<br>- On heating with aniline yields anil, m.p. 205°C.<br>- Gives fluorescein test (See phthalic acid) |

| *Name and structure* | *M.P. (°C)* | *Properties* | *Derivatives* | *Specific test and derivatives* |
|---|---|---|---|---|
| | | | | - On heating with NaOH solution gives phthalic acid, m.p. 195°C |
| o-Sulphobenzimide (Saccharin) 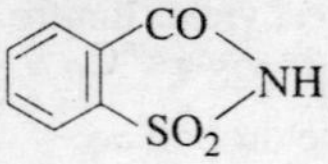 | 224 | Colourless, slightly soluble in hot $H_2O$, chloroform, dil. alkali | Benzyl, 118<br>Xanthyl dervaitve, 198 | - Gives red solution having green fluorescence in the fluorescein test.<br>- Heat the compound with KOH pellet until melting, cool, dissolve in dil. HCl and add $H_2O$ then a drop of $FeCl_3$—violet colour. |
| Succinanilide $C_6H_5NHCOCH_2CH_2CONHC_6H_5$ | 228 | | | - On boiling with HCl yields succinic acid, m.p. 185°C, and aniline hydrochloride.<br>- On heating above m.p. yields aniline and succinanil, m.p. 156°C. |
| Phthalimide 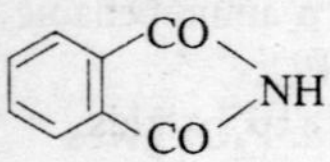 | 238 | Colourless, insoluble in $H_2O$, soluble in absolute alcohol | Xanthyl-amide, 177<br>Anil 210<br>N-acetate, 133<br>N-Benzoate, 115 | - Gives Fluorescein test (see phthalic acid). |
| sym. Diphenylurea (Carbanilide) $(C_6H_5NH)_2CO$ | 238 | Colourless, sparingly soluble in $H_2O$, alcohol, soluble in ether | Acetyl derivative, 106<br>Nitro, 248 | - On heating with HCl gives aniline.<br>- On boiling (15 min) with $Ac_2O$-acetic acid gives acetanilide, m.p. 114°C.<br>- With conc. $HNO_3$ (room temp.)—m-Nitro derivative, m.p. 248(d) |
| Succinamide $CH_2CONH_2$ \| $CH_2CONH_2$ | 260 | Colourless, insoluble in $H_2O$, alcohol, ether, soluble in hot $H_2O$ | Anil, 156 | - On heating above m.p. gives succinimide, m.p. 126°C.<br>- On hydrolysis (heating with NaOH soln) gives succinic acid, m.p. 185°C.<br>- Gives fluorescein test |
| Barbituric acid (malonyl urea) 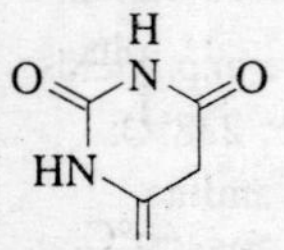 | 245 (d) | Colourless, soluble in hot $H_2O$, ether, sparingly soluble in alcohol, soluble in alkali | Dibromo, 235(d) | - With cold $NaNO_2$ solution gives purple colour.<br>- On treatment with $Br_2$ water gives dibromo derivative m.p. 235(d). |

| *Name and structure* | *M.P. (°C)* | *Properties* | *Derivatives* | *Specific test and derivatives* |
|---|---|---|---|---|
| Oxanilide<br>$C_6H_5NHCOCONHC_6H_5$ | 246 | Colourless, insoluble in hot $H_2O$, sparingly soluble in alcohol, ether | | - On boiling with HCl yields Oxalic acid and aniline hydrochloride.<br>- On boiling with $Ac_2O$ + Na acetate, yields vinylideneoxanilide, m.p. 208°C.<br>- With fuming $HNO_3$ in acetic acid, yields di-p-nitro deriv., m.p. 260°C, which on hydrolysis yields p-nitroaniline, m.p. 147°C. |
| sym. Di-o-tolyl urea<br>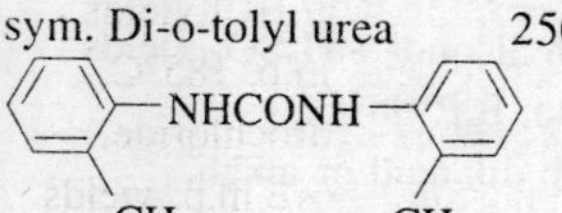 | 250 | Colourless, insoluble in hot $H_2O$ | o-acetotoluelidide, 112 | - On boiling with HCl yields o-toluidine hydrochloride.<br>- On boiling with $Ac_2O$ + Na acetate yields o-aceto-toluidide, m.p. 112°C. |
| sym Di-p-tolyl urea<br>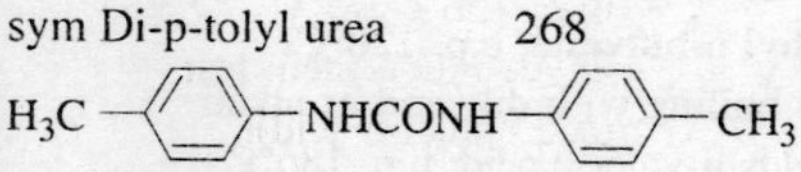 | 268 | Colourless, insoluble in $H_2O$ | p-acetotoluidide, 153 | - On boiling with $Ac_2O$ + Na acetate yields N-acetyl p-toluidine, m.p. 153°C. |
| Diacetyl-p-phenylene-diamine<br>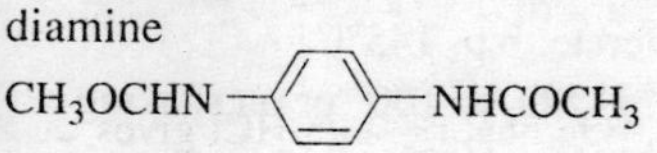 | 305 | Colourless, soluble in acetic acid | Dinitro derivative 258 | - On boiling with HCl yields acetic acid and p-phenylene diamine hydrochloride.<br>- With fuming $HNO_3$ at 0° yields dinitro deriv., m.p. 258°C. |
| Creatinine<br>HN=C·NH—CO<br>N($CH_3$)—$CH_2$ | 315 (d) | Soluble in hot $H_2O$ | Picrate, 220 | - On oxidation ($KMnO_4$) yields oxalic acid and methyl guanidine.<br>- On saturation of the solution with $NO_2$ and then adding $NH_4OH$ yield nitroso derivative m.p. 210°C.<br>- Reduces Fehlings solution.<br>- With nitroprusside + NaOH gives red colour changing to yellow. |
| Oxamide<br>$NH_2COCONH_2$ | 419 (sealed tube) | Colourless, insoluble in $H_2O$, sparingly soluble in hot alcohol, ether | Oxanilide, 246 | - Gives red colour with $CuSO_4$ + dil. NaOH solution.<br>- On boiling with NaOH soln. gives oxalic acid and $NH_3$.<br>- On heating with aniline gives oxanilide, m.p. 246°C. |

**Table 6.19(a)** Nitriles (Liquids)

| *Name and structure* | *B.P.(°C)* | *Properties* | *Specific tests and derivatives* |
|---|---|---|---|
| Acrylonitrile<br>$CH_2{=}CHCN$ | 78 | Colourless, characteristic odour | - β-Naphthol adduct, m.p. 142°C. |
| Acetonitrile<br>$CH_3CN$ | 81 | Colourless, pleasant smell, soluble in $H_2O$, alcohol, ether | - On heating with HCl gives acetic acid (smell of vinegar).<br>- With alcohol (1 mole) + HCl gas yields acetiminoethyl ether HCl, m.p. 98°(d) |
| Propionitrile<br>$CH_3CH_2CN$ | 97 | Colourless, soluble in $H_2O$ | - On boiling with dil. acid or alkali yields propionoc acid, b.p. 140°C.<br>- With alcohol (1 mole) + HCl gas yields propioiminoethyl ether HCl, m.p. 92(d)<br>- On boiling with alcohol + $H_2SO_4$ yields ethyl propionate, B.P. 98°C. |
| n-Butyronitrile<br>$CH_3CH_2CH_2CN$ | 118 | Colourless, soluble in $H_2O$ | - On boiling with dil. acid or alkali, yields butyric acid, B.P. 163°C.<br>- On boiling with alcohol + $H_2SO_4$ yields -ethyl n-butyrate, b.p. 120°C. |
| n-Valeronitrile<br>$CH_3CH_2CH_2CH_2CN$ | 141 | Colourless, sparingly soluble in $H_2O$ | - On boiling with dil. acid or alkali yields n-valeric acid, b.p. 186°C.<br>On boiling with alcohol + $H_2SO_4$ yields ethyl n-valerate, b.p. 145°C. |
| Benzonitrile<br>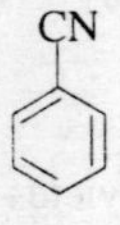 | 190 | Colourless, odour resembles that of nitrobenzene | - With conc. $H_2SO_4$ at 100° or on shaking with dil. NaOH + $H_2O_2$ in cold yields benzamide, m.p. 128°C.<br>- On boiling with 75% $H_2SO_4$ yields benzoic acid, m.p. 122°C.<br>- On boiling with $CH_3OH$ + conc. $H_2SO_4$ yields methyl benzoate, b.p. 198°C.<br>- With conc. $HNO_3$ + conc. $H_2SO_4$ yields m-nitro derivative, m.p. 118°C. |
| o-Tolunitrile<br>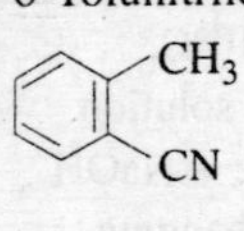 | 205 | Colourless, insoluble in $H_2O$ | - With conc. $H_2SO_4$ at 100° or on shaking with dil. NaOH + $H_2O_2$ in cold yields o-toluamide, m.p. 142°C.<br>- On boiling with 75% $H_2SO_4$ yields o-toluic acid, m.p. 104°C.<br>- On boiling with $CH_3OH$ + $H_2SO_4$ yields methyl o-toluate, b.p. 213°C.<br>- With conc. $HNO_3$ + conc. $H_2SO_4$ yields nitro derivative m.p. 105°C. |
| Ethylcyanoacetate<br>$CNCH_2COOC_2H_5$ | 207 | Colourless, sparingly soluble in $H_2O$, soluble in alcohol, ether | - On evaporation with $NH_4OH$ yields cyanoacetamide, m.p. 118°C.<br>- On boiling with alcohol + $H_2SO_4$ yields ethyl malonate, b.p. 198°C. |

| *Name and structure* | *B.P.(°C)* | *Properties* | *Specific tests and derivatives* |
|---|---|---|---|
| m-Tolunitrile<br>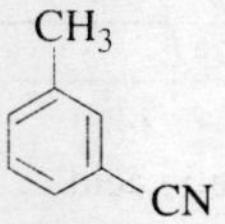 | 212 | Colourless, insoluble in $H_2O$ | - On hydrolysis with 75% $H_2SO_4$ yields m-toluic acid, m.p. 110°C.<br>- With conc. $H_2SO_4$ at 100° or on shaking with dil. NaOH + $H_2O_2$ in cold yields the amide, m.p. 95°C. |
| p-Tolunitrile<br>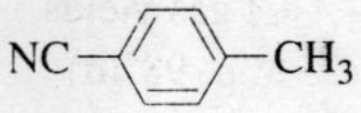 | 217 | Colourless, insoluble in $H_2O$ | - With conc. $H_2SO_4$ at 100° or on shaking with dil. NaOH + $H_2O_2$ in cold yield p-toluamide, m.p. 159°C.<br>- On boiling with 75% $H_2SO_4$ yields p-toluic acid, m.p. 178°C.<br>- On boiling with $CH_3OH$ + $H_2SO_4$ yields methyl p-toluate, m.p. 33°C. b.p. 217°C.<br>- With conc. $HNO_3$ + $H_2SO_4$ yields nitro-deriv., m.p. 107°C. |
| Phenylacetonitrile (benzyl cyanide)<br>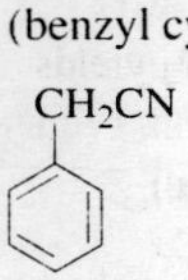 | 234 | Colourless, insoluble in $H_2O$, odour of bitter almonds | - With conc. $H_2SO_4$ at 100° yields phenyl-acetamide, m.p. 154°C.<br>- On boiling with 75% $H_2SO_4$ yields phenyl acetic acid, m.p. 76°C.<br>- With $CH_3OH$ + $H_2SO_4$ (boiling) yields methyl phenyl acetate, b.p. 218°C.<br>- With conc. $HNO_3$ + $H_2SO_4$ yields p-nitro deriv., m.p. 116°C. |
| Cinnamonitrile<br>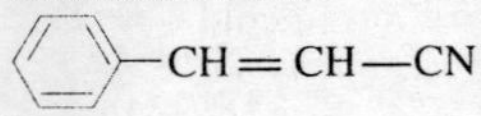 | 254 (m.p. 20°) | Colourless, insoluble in $H_2O$ | - On boiling with 75% $H_2SO_4$ yields cinnamic acid, m.p. 133°C.<br>- On shaking with dil. NaOH + $H_2O_2$ in cold yields cinnamoamide, m.p. 147°C. |
| α-Naphthonitrile<br>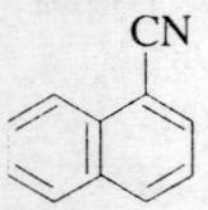 | 297 | Colourless, insoluble in $H_2O$ | - On heating rapidly with 75% $H_2SO_4$ yields α-naphthoamide, m.p. 202°C, this on prolonged boiling is converted into α-naphthoic acid, m.p. 162°C. |

**Table 6.19(b)** Nitriles (Solids)

| *Name and structure* | *M.P.(°C)* | *Properties* | *Specific tests and derivatives* |
|---|---|---|---|
| Succinonitrile<br>$CH_2CN$<br>\|<br>$CH_2CN$ | 55 | Colourless, insoluble in $H_2O$ | - On hydrolysis (75% $H_2SO_4$, refluxing) gives succinic acid, m.p. 185. |
| Cyanoacetic acid | 66 | See Carboxylic acids (Table 6.2b) | |
| β-Naphthonitrile<br>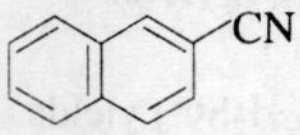 | 66<br>b.p. 305 | Colourless, insoluble in $H_2O$ | - On heating rapidly with 75% $H_2SO_4$ yields β-naphthoamide, m.p. 192°C, which on prolonged boiling is converted into β-naphthoic acid, m.p. 184°C. |
| Phthalonitrile<br>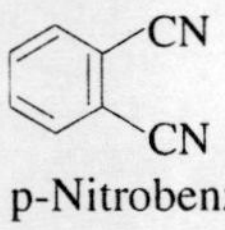 | 141 | Colourless, insoluble in $H_2O$ | - On hydrolysis with 75% $H_2SO_4$ (refluxing) yields phthalic acid, m.p. 195°(d). |
| p-Nitrobenzonitrile<br>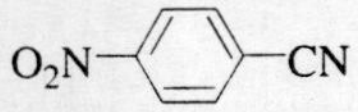 | 148 | Colourless, insoluble in $H_2O$ | - On boiling with 75% $H_2SO_4$ yields p-nitrobenzoic acid, m.p. 240°C.<br>- Amide, m.p. 201°C. |

**Table 6.20** Amino acids, esters

| *Name and structure* | *M.P. (°C)* | *Properties* | *Derivatives* | | *Specific tests and derivatives* |
|---|---|---|---|---|---|
| | | | *Acetyl* | *Benzoyl* | |
| Methylanthraniliate<br>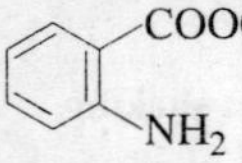 | 25 | Colourless, turns dark on exposure, orange blossom odour, insoluble in $H_2O$ | 101 | 100 | - Solution in ethanol gives blue fluorescence.<br>- Picrate, m.p. 106°C. |
| Urethane | 50 | See esters (Table 6.7b) | | | |
| Methyl carbamate<br>$H_2N\ COOCH_3$ | 52 | See esters (Table 6.7b) | | | |
| n-Propyl carbamate | 60 | See esters (Table 6.7b) | | | |
| Ethyl oxamate | 114 | See esters (Table 6.7b) | | | |
| N-Phenylglycine<br>$C_6H_5NHCH_2COOH$ | 126 | Colourless, soluble in $H_2O$, alcohol, sparingly soluble in ether | 194 | 63 | - On heating with a pellet of NaOH gives blue colour. |
| Anthranilic acid<br>(o-Aminobenzoic acid)<br>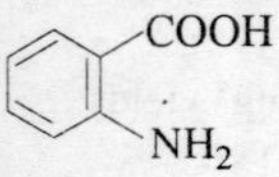 | 144 | Colourless or pale yellow, insoluble in $H_2O$, soluble in hot $H_2O$, ether, sparingly soluble in alcohol | 185 | 182 | - On heating with solid $CaCl_2$ and dissolving the mass in alcohol gives red solution. (violet fluorescence on standing).<br>- Anilide, m.p. 126°C.<br>p-Toluidide, m.p. 151°C. |
| m-Aminobenzoic acid<br>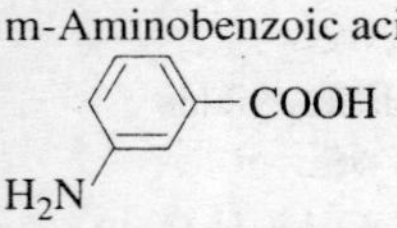 | 174 | Colourless, insoluble in $H_2O$, soluble in hot $H_2O$, sparingly soluble in alcohol, ether | 250 | 248 | - Amide, m.p. 75°C.<br>- Anilide, m.p. 129°C. |
| p-Aminobenzoic acid<br>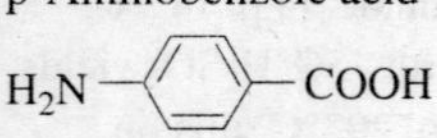 | 186 | Colourless or yellowish, insoluble in $H_2O$, soluble in hot $H_2O$, alcohol, ether | 252 | 278 | |
| Hippuric acid<br>(benzoyl glycine)<br>$CONH.CH_2COOH$<br>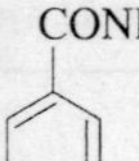 | 187 | Colourless, soluble in hot $H_2O$, hot alcohol, insoluble in ether | | | - Anilide, m.p. 208°C.<br>- Amide, m.p. 183°C.<br>- On boiling with HCl gives benzoic acid, m.p. 122°C.<br>- On treatment with conc. $H_2SO_4$ + conc. $HNO_3$ gives m-nitro derivative, m.p. 162°C. |
| $\beta$-Alanine<br>$NH_2CH_2CH_2COOH$ | 196 | Colourless, soluble in $H_2O$, alcohol, insoluble in ether | | 165 | - 3, 5-Dinitrobenzoyl, m.p. 202°C. |
| (D-orL Glutamic acid<br>($\alpha$-aminoglutaric acid)<br>$HOOC(CH_2)_2\ CH(NH_2)COOH$ | 198<br>(d) | Colourless, sparingly soluble in $H_2O$, alcohol | | 138 | - Gives deep blue colour on adding $CuSO_4$ solution to aq. solution of the compound.<br>- 3, 5-dinitrobenzoyl, deriv., m.p. 217°C<br>- Phthalyl derivative, 189°C. |
| p-Aminophenyl acetic acid<br> | 200(d) | Soluble in hot $H_2O$ | 168–70 | 198 | |

| *Name and structure* | *M.P. (°C)* | *Properties* | *Derivatives* | | *Specific tests and derivatives* |
|---|---|---|---|---|---|
| | | | *Acetyl* | *Benzoyl* | |
| dl-Proline 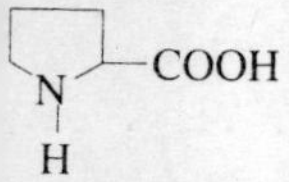 | 203 | | | | - Picrate m.p. 137°C.<br>- 3, 5-dinihobenzoate m.p. 217°C. |
| Sarcosine $CH_3NHCH_2COOH$ | 210 (d) | Soluble in $H_2O$, insoluble in ethanol | 135 | 104 | - Hydrochloride, m.p. 169°C.<br>- p-tosyl, m.p. 102°C.<br>- Anhydride, m.p. 149°C. |
| L (–) proline 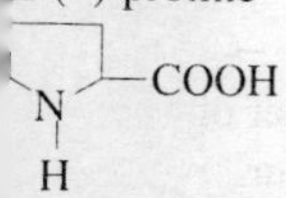 | 222 (d) | Colourless, hygroscopic, soluble in $H_2O$, slightly soluble in alcohol, insoluble in ether | – | 156 | - Picrate, m.p. 154°C.<br>-p-Tosyl, m.p. 132°C. |
| D or L-Lysine $NH_2(CH_2)_4CH(NH_2)COOH$ | 224 (d) | Colourless, soluble in $H_2O$, insoluble in alcohol | – | 150 (di) | - 3, 5-Dinitrobenzoyl, m.p. 169°C. |
| D or L-Asparagine $NH_2COCH_2CH(NH_2)COOH$ | 226 (Sealed tube) | Colourless, sparingly soluble in $H_2O$, insoluble in absolute alcohol | – | 189 | - 3, 5-Dinitrobenzoyl, 196°C.<br>- Picrate, m.p. 180°C.<br>- Gives blue colour with dil. NaOH + trace of $CuSO_4$ solution. |
| Glycine $NH_2CH_2COOH$ | 232 | | 206 | 187 | - Gives deep blue colour on addition of $CuSO_4$ solution to an aq. soln. of the compound.<br>- Gives red colour with aq. $FeCl_3$ solution.<br>-3, 5-Dinitrobenzoyl, m.p. 179°C. |
| dl-Threonine $CH_3CH(OH)CH(NH_2)COOH$ | 235 (d) | Colourless, soluble in $H_2O$, insoluble in alcohol, ether | – | 148 (mono) | |
| N-(p-Hydroxyphenyl) glycine  | 248 (d) | Colourless, sparingly soluble in cold $H_2O$, insoluble in ether | 175 (di) | 117 | With $AgNO_3$ solution gives black ppt. in cold; solution becomes purple on heating. |
| DL Serine $HOCH_2CH(NH_2)COOH$ | 246 (d) | Colourless, sparingly soluble in $H_2O$, insoluble in alcohol, ether | – | 171 | - 3, 5-Dinitrobenzoyl, m.p. 183°C. |
| L-Cystine 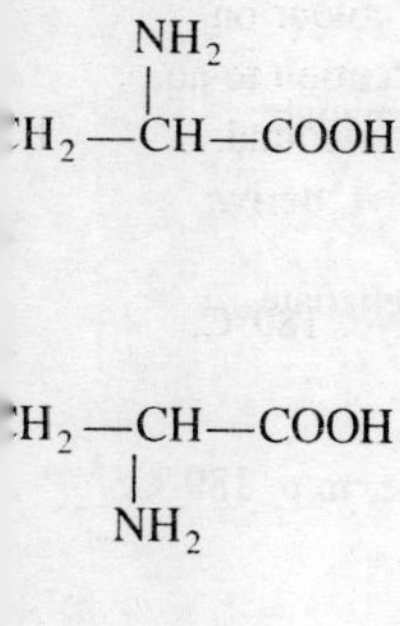 | 260 | Colourless, insoluble in $H_2O$ | | 180 | - Forms a blue insoluble Cu salt.<br>- With Sn + HCl yields cysteine (soluble in $H_2O$, gives blue colour with $FeCl_3$) |

| *Name and structure* | *M.P. (°C)* | *Properties* | *Derivatives* | | *Specific tests and derivatives* |
|---|---|---|---|---|---|
| | | | *Acetyl* | *Benzoyl* | |
| L-Hydroxylproline 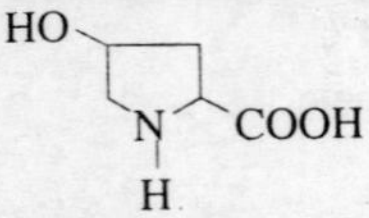 | 270 | | | | - P-Toluenesulphonyl-derivative, m.p. 153°C. |
| D or L Aspartic acid $HOOCCH_2CH(NH_2)COOH$ | 270 (sealed tube) | Colourless, sparingly soluble in $H_2O$, insoluble in alcohol | - | 185 | - Gives blue colour by adding $CuSO_4$ soln to aq. solution of the compound. |
| dl-Methionine $CH_3S(CH_2)_2CH(NH_2)COOH$ | 272 | | 114 | 151 | |
| dl-Phenylalanine $C_6H_5CH_2CH(NH_2)COOH$ | 273 (d) | Colourless, sparingly soluble in alcohol, soluble in $H_2O$, in insoluble in ether | – | 188 | - 3, 5-Dinitrobenzoyl, m.p. 93°C.<br>- On oxidation ($KMnO_4$)—benzoic acid, m.p. 122°C. |
| L-Histidine 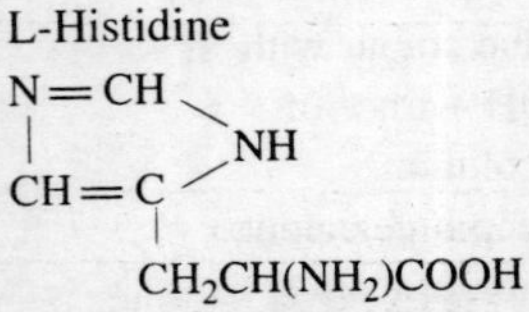 | 277 (d) | Colourless, sparingly soluble in alcohol, soluble in $H_2O$, insoluble in ether | – | 230 | - 3, 5-Dinitrobenzoyl, m.p. 189°C.<br>- Picrate, m.p. 86°C. |
| $\alpha$-Amino isobutyric acid $(CH_3)_2C(NH_2)COOH$ | 280 (Subl) | Soluble in cold $H_2O$, insoluble in alcohol | – | 198 | - On oxidation with NaOCl yields acetone. |
| L-Methionine $CH_3S(CH_2)_2CH(NH_2)COOH$ | 283 (d) | Colourless, soluble in $H_2O$, insoluble in alcohol, ether | – | 150 | - 3, 5-Dinitrobenzoyl, m.p. 95°C. |
| 5-Aminosalicylic acid 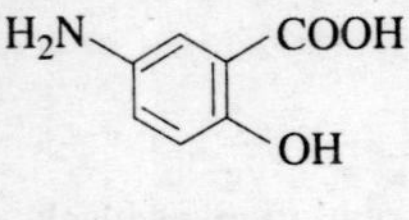 | 283 | Colourless, sparingly soluble in $H_2O$, insoluble in alcohol | 218 (mono) 184 (di) | 252 | - With $FeCl_3$ gives red Colour changing to brown ppt.<br>Azo-$\beta$-naphthol derivative m.p. 201°C. |
| L-Tryptophan 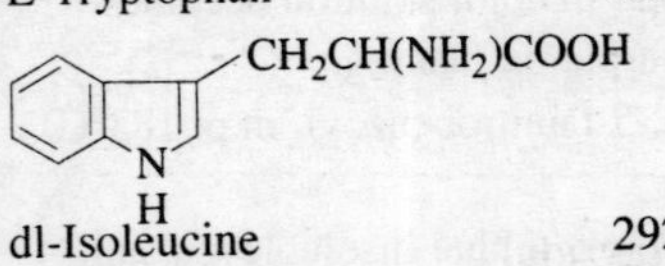 | 289 | Soluble in water-alcohol mixture | 189 | 104 | - Picrate, m.p. 195°C.<br>- 3, 5-Dinitrobenzoate, m.p. 233°C. |
| dl-Isoleucine $CH_3CH_2CH(CH_3)CH(NH_2)COOH$ | 292 (d) | Soluble in alcohol, ether | - | 118 | |
| L-Leucine $(CH_3)_2CHCH_2CH(NH_2)COOH$ | 294 (d) | Colourless, insoluble in alcohol, ether, sparingly soluble in $H_2O$, hot alcohol | – | 107 | - 3, 5-Dinitrobenzoate, m.p. 187°C. |
| dl-Alanine $CH_3CH(NH_2)COOH$ | 295 (d) | Colourless, or pale yellow, soluble in $H_2O$, insoluble in alcohol, ether | – | 166 | - 3, 5-Dinitrobenzoate, m.p. 177°C. |
| dl-Valine $(CH_3)_2CHCH(NH_2)COOH$ | 298 (d) | Colourless, sparingly soluble in $H_2O$ | 148 | 132 | - Hydrochloride, m.p. 189°C. |
| L-(+) Norleucine | 300 | Colourless, sparingly soluble | | | |

| Name and structure | M.P. (°C) | Properties | Derivatives: Acetyl | Derivatives: Benzoyl | Specific tests and derivatives |
|---|---|---|---|---|---|
| $CH_3(CH_2)_3CH(NH_2)COOH$ | (d) (sealed tube) | in $H_2O$ | | | |
| dl α-Amino-n-butyric acid $CH_3CH_2CH(NH_2)COOH$ | 307 (d) | Colourless, soluble in cold $H_2O$, insoluble in alcohol | – | 147 | |
| dl Tyrosine 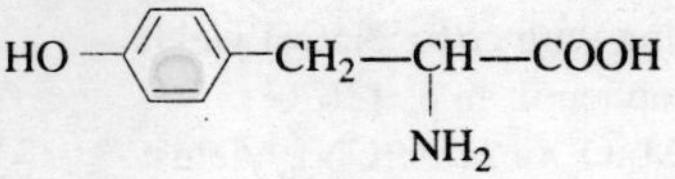 | 318 | Sparingly soluble in $H_2O$ | – | 197 | - On KOH fusion gives p-hydroxybenzoic acid, m.p. 213°C.<br>- Gives red colour with dil. solution of $CH_3CHO$ in conc. $H_2SO_4$. |
| dl-Norleucine $CH_3(CH_2)_3—CH(NH_2)—COOH$ | 327 (d) | | | | |
| dl-Leucin | 332 | | | | |
| L(–)Tyrosine | 344(d) | | | | |

**Table 6.21(a)** Nitroso compounds including nitrosamines (Liquids)

| Name and structure | B.P.(°C) | Properties | Specific tests and derivatives |
|---|---|---|---|
| Methylphenylnitrosamine 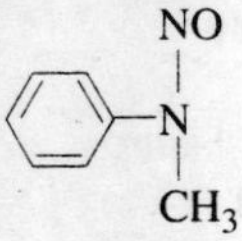 | 227<br>120/13 mm<br>(m.p. 12) | Pale yellow, insol. in $H_2O$, sol. in alcohol, ether | - With Sn + HCl yields methylaniline, B.P. 194.<br>- With Zn dust + acetic acid yields methylphenylhydrazine, b.p. 227<br>- With HCl gas in ether yields p-nitrosomethylaniline, m.p. 116 |
| Ethylphenylnitrosamine 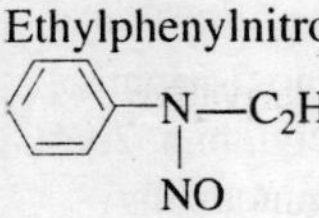 | 236 (d)<br>134/16 mm | Pale yellow, insol. in $H_2O$, sol. in alcohol, ether | - With Sn + HCl yields ethylaniline, b.p. 205.<br>- With Zn dust + acetic acid yields ethylphenylhydrazine, b.p. 237 (benzal deriv., m.p. 49).<br>- With HCl gas in ether yields p-nitrosoethylaniline, m.p. 78. |

**Table 6.21(b)** Nitroso compounds including nitrosamines (Solids)

| Name and structure | M.P.(°C) | Properties | Specific tests and derivatives |
|---|---|---|---|
| p-Nitroso-N-ethyl-N-methylaniline 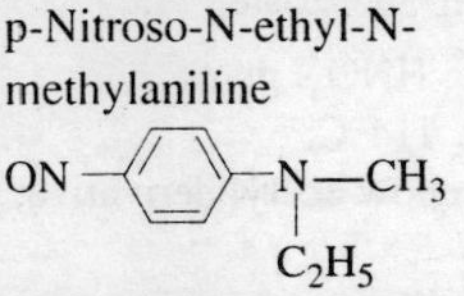 | 66 | | - Hydrochloride m.p. 114. |
| Diphenylnitrosoamine $(C_6H_5)_2N.NO$ | 67 | Pale yellow, insol. in $H_2O$ & dil. acids, sol. in alcohol, ether | - Gives blue colour on adding a trace of conc. $H_2SO_4$<br>- On boiling with Sn + HCl in alcohol yields diphenylhydrazine, m.p. 34.<br>- With HCl gas in ether yields p-nitrosodiphenylamine, m.p. 144. |

| *Name and structure* | *M.P.(°C)* | *Properties* | *Specific tests and derivatives* |
|---|---|---|---|
| Nitrosobenzene<br>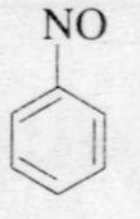 | 68 | Colourless, sharp odour, green in fused state, insoluble in $H_2O$, soluble in alcohol, ether | - On warming with p-bromoaniline in acetic acid gives p-bromoazobenzene, m.p. 88°C (crystallised from acetic acid).<br>- On reduction (Sn + HCl) gives aniline b.p. 184°C.<br>- With aniline in acetic acid yields azobenzene, m.p. 68°C. |
| p-Nitroso-N, N-diethylaniline<br>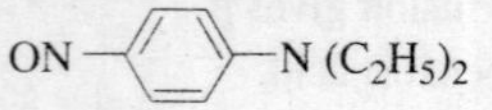 | 84 | Dark green, sparingly soluble in $H_2O$, soluble in alcohol, ether | - With Sn + HCl yields p-aminodiethylaniline b.p. 262°C.<br>- On boiling with conc. NaOH yields p-nitrosophenol, m.p. 126°C.<br>- With $KMnO_4$ or $K_3Fe(CN)_6$ yields p-nitrodiethylaniline, m.p. 77°C. |
| p-Nitroso-N, N-dimethyl-aniline<br>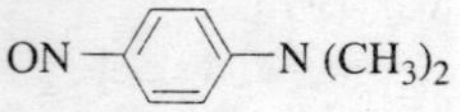 | 87 | Dark green, insoluble in $H_2O$, soluble in alcohol, ether | - With Sn + HCl yields p-amino dimethyl-aniline, m.p. 41°C.<br>- On boiling with conc. NaOH yields p-nitrosophenol, m.p. 126°C.<br>- With $KMnO_4$ or $K_3Fe(CN)_6$ yields p-nitrodimethylaniline, m.p. 163°C. |
| p-Nitroso-N, N-dibenzyl aniline<br>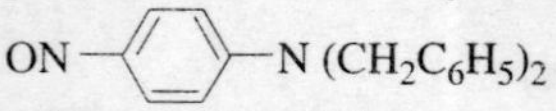 | 91 | | |
| 1-Nitroso-2-naphthol<br>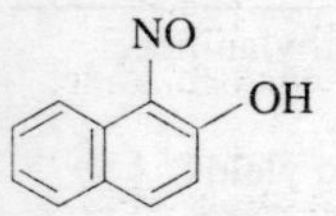 | 109 | Dull yellow, insoluble in $H_2O$, soluble in alcohol, ether | - Forms an addition product with 1 mole $Br_2$ in $CHCl_3$, m.p. 130°C.<br>- On boiling with HCl yields $\beta$-naphtha-quinone, m.p. 115-20°C (d) and hydroxylamine HCl.<br>- With dil. $HNO_3$ in cold yields 1-nitro-2-naphthol, m.p. 103°C.<br>- With fuming $HNO_3$ at 40-50° yields, 1, 6-dinitro-2-naphthol, m.p. 195°C(d)<br>- With $NH_4SH$ yields 1-amino-2-naphthol (picrate, m.p. 109°C, diacetyl, m.p. 206°C) |
| p-Nitroso-N-methylaniline<br>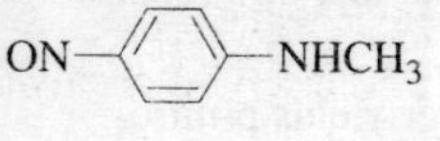 | 116 | Green, insoluble in $H_2O$, soluble in alcohol, ether, dil. acids | - With Sn + HCl yields p-aminomethy-laniline, m.p. 35°C, b.p. 258°C. (dibenzoyl derivative, m.p. 164°C).<br>- On boiling with conc. NaOH yields p-nitrosophenol, m.p. 126°C. |
| p-Nitrosophenol<br>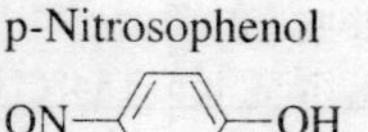 | 126 | Light pale yellow, soluble in hot $H_2O$, alcohol ether | - Soluble in dil. NaOH giving reddish brown solution.<br>- With $NH_4SH$ gives p-aminophenol, m.p. 184°C.<br>- On oxidation (conc. $HNO_3$) gives p-nitrophenol, m.p. 114°C.<br>- With $Ac_2O$ gives yellow acetyl derivative, m.p. 107°C.<br>- On warming with aniline yields p-hydroxyazobenzene, m.p. 152. |
| p-Nitrosodiphenylamine<br>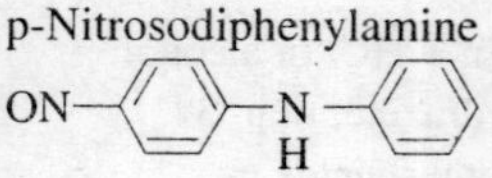 | 144 | Green, sparingly sol. in $H_2O$, soluble in alcohol ether, dil. NaOH | - With Zn dust + acetic acid yields p-aminodiphenylamine, m.p. 66<br>- On boiling with conc. NaOH yields p-nitrosophenol, m.p. 126. |

**Table 6.22** Azoxy-, Azo- and Hydrazo Compounds

| *Name and structure* | *M.P.(°C)* | *Properties* | *Specific tests and derivatives* |
|---|---|---|---|
| Azoxybenzene<br>$C_6H_5N{=}NC_6H_5$<br>↓<br>O | 36 | Light yellow, insoluble in $H_2O$, soluble in alcohol, petroleum ether | - With Zn dust + NaOH in dil. alcohol yields azobenzene, m.p. 68°C and then hydrazobenzene, m.p. 127°C.<br>- With $SnCl_2$ + HCl yields aniline, b.p. 184°C.<br>- On warming with conc. $H_2SO_4$ yields p-hydroxyazobenzene, m.p. 152°C. |
| o-Azotoluene<br>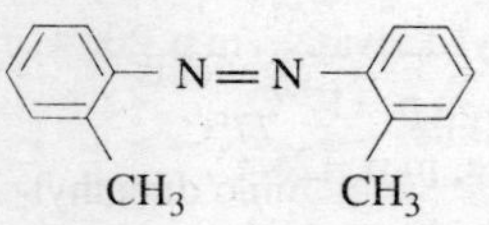 | 55 | Red, insoluble in $H_2O$ | - With Zn dust + NaOH in dil. alcohol, yields o-hydrazotoluene, m.p. 165, which with conc. HCl yields o-tolidine, m.p. 129°C.<br>- With $Na_2S_2O_4$ + NaOH in dil. alcohol yields o-toluidine, b.p. 199°C. |
| o-Azoxytoluene<br>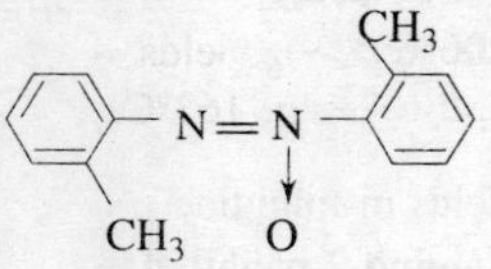 | 59 | Light yellow, insoluble in $H_2O$, soluble in alcohol | - With Zn dust + NaOH in dil. alcohol yields first o-azotoluene, m.p. 55°C and then o-hydrazotoluene, m.p. 165°C.<br>- With Sn + HCl yields o-toluidine, b.p. 199°C. |
| Azobenzene<br>$C_6H_5N{=}NC_6H_5$ | 68 | Orange red, insoluble in $H_2O$, sparingly soluble in alcohol | - With Zn dust+NaOH in dil. alcohol yields hydrazobenzene, m.p. 127°C which with conc. HCl yields benzidine, m.p. 127°C.<br>- With $Na_2S_2O_4$ + NaOH in dil. alcohol yields aniline, b.p. 184°C.<br>- With 2 mole $Br_2$ in acetic acid yields dibromo derivative, m.p. 187°C.<br>- With benzaldehyde at 200° yields benzanilide, m.p. 164°C. |
| p-Aminoazo-o-toluene<br>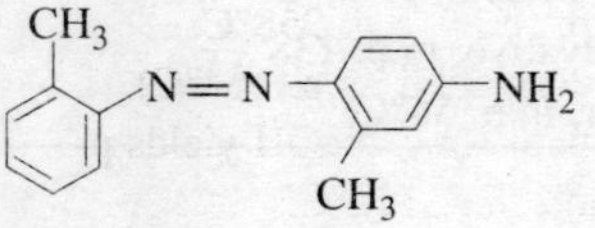 | 100 | Purple, insoluble in $H_2O$, soluble in alcohol and ether giving yellow solution. | - With Sn + HCl yields o-toluidine, b.p. 199°C and p-toluylenediamine, m.p. 64°C b.p. 273°C (diacetyl derivative, m.p. 220°C).<br>- Acetyl derivative, m.p. 185°C.<br>- Azo-$\beta$-naphthol derivative, m.p. 186°C. |
| p-Dimethylaminoazobenzene<br>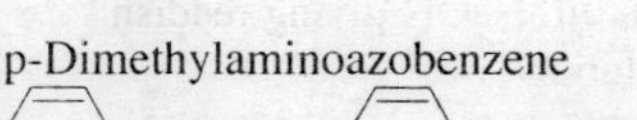 | 117 | Orange, soluble in alcohol, ether (yellow solution) and red solution in dil. acids. | - With Sn + HCl yields aniline, b.p. 184°C and p-aminodimethylaniline, m.p. 41°C, b.p. 262°C. |
| Hydrazobenzene<br>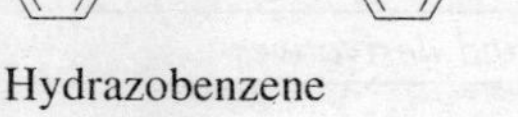 | 127 | See hydrazines (Table 6.23b) | |
| Benzeneazo-o-cresol<br>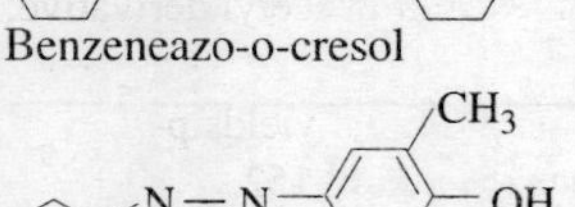 | 128 | Yellow, soluble in alcohol, ether and in NaOH (orange solution) | - With Sn + HCl yields aniline, b.p. 184°C and 5-amino-o-cresol, m.p. 176°C.<br>- Acetyl derivative, m.p. 87°C.<br>- Benzoyl derivative, m.p. 110°C. |

| *Name and structure* | *M.P.(°C)* | *Properties* | *Specific tests and derivatives* |
|---|---|---|---|
| o-Tolueneazo-β-naphthol 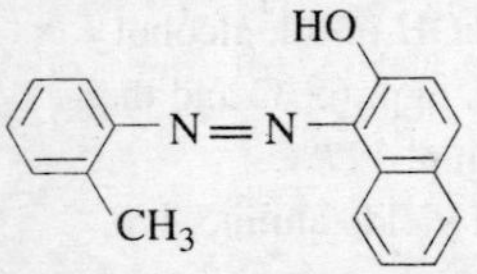 | 131 | | |
| 1-Benzeneazo-2-naphthol 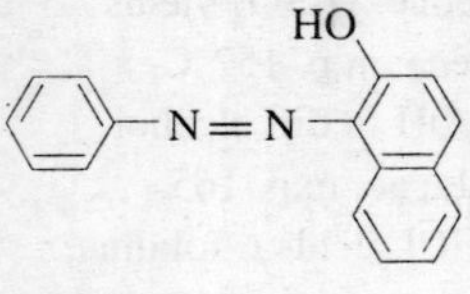 | 134 | Orange, insoluble in $H_2O$, soluble in alcohol, ether, insoluble in dil. NaOH | - Dissolves in conc. $H_2SO_4$ to give magenta colour.<br>- With Sn + HCl yields aniline, b.p. 184°C and 1-amino-2-naphthol (picrate, m.p. 109°C; diacetyl derivative, m.p. 206°C.)<br>- Acetyl derivative, m.p. 117°C.<br>- Benzoyl derivative, m.p. 125°C. |
| p-Tolueneazo-β-naphthol 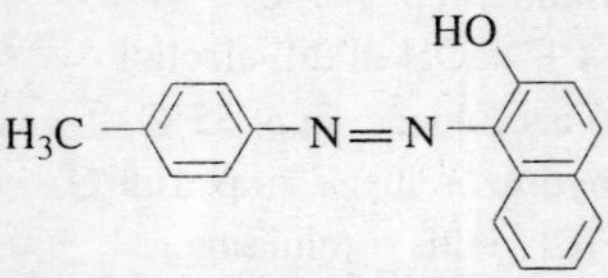 | 134 | Yellow, insoluble in $H_2O$, soluble in alcohol, ether | - With Sn + HCl yields p-toluidine, m.p. 44°C and 1-amino-2-naphthol (picrate, m.p. 109°C; diacetyl derivative, m.p. 206°C). |
| m-Tolueneazo-β-naphthol 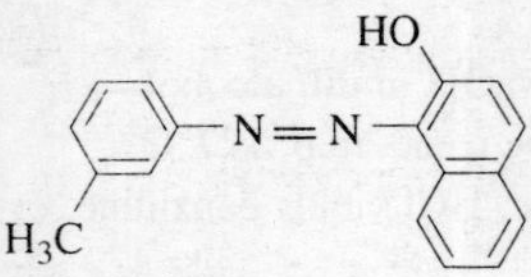 | 140 | Yellow, insoluble in $H_2O$, soluble in alcohol, ether | - With Sn + HCl yields m-toluidine, b.p. 202°C and 1-amino-2-naphthol (picrate, m.p. 109°C; diacetyl derivative, m.p. 206°C). |
| p-Azotoluene 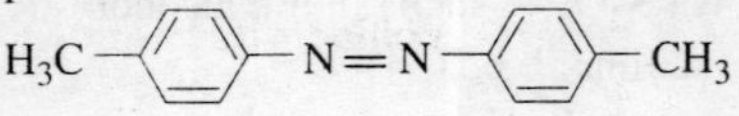 | 144 | Orange, insol. in $H_2O$, soluble in alcohol, ether | - With Zn dust + NaOH in dil. alcohol yields p-hydrazotoluene, m.p. 133°C.<br>- With Sn + HCl yields p-toluidine, m.p. 44°C. |
| p-Hydroxyazobenzene 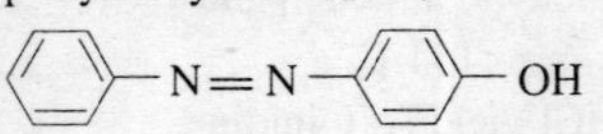 | 152 | Yellow, insoluble in $H_2O$, soluble in alcohol, ether (yellow solution) and in dil. NaOH (orange solution) | - With Sn + HCl yields aniline, b.p. 184°C and p-aminophenol, m.p. 184°C (d).<br>- Acetyl derivative, m.p. 84°C.<br>- Benzoyl derivative, m.p. 138°C.<br>- Methyl ether, m.p. 56°C. |

**Table 6.23(a)** Hydrazine derivatives (Liquids)

| *Name and structure* | *M.P.(°C)* | *Properties* | *Specific tests and derivatives* |
|---|---|---|---|
| Methylphenylhydrazine | 227 | See amines (Table 6.15a) | |
| Phenylhydrazine | 243 | See amines (Table 6.15a) | |

**Table 6.23(b)** Hydrazine derivatives (Solids)

| *Name and structure* | *M.P.(°C)* | *Properties* | *Specific tests and derivatives* |
|---|---|---|---|
| Diphenylhydrazine<br>$(C_6H_5)_2NNH_2$ | 34 | Insoluble in $H_2O$, soluble in alcohol | - Gives blue solution in conc. $H_2SO_4$.<br>- Reduces ammanical $AgNO_3$ solution.<br>- Hydrochloride, m.p. 167°C.<br>- With Sn + HCl yields diphenylamine, m.p. 54°C.<br>- Acetyl derivative, m.p. 184°C.<br>- Benzoyl derivative, m.p. 192°C.<br>- Benzal derivative, m.p. 122°C. |
| p-Tolylhydrazine<br>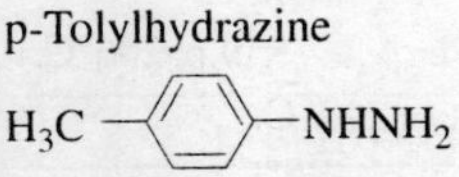 | 66 | | - Reduces Fehlings Solution in cold yielding toluene, (b.p. 110°C) and nitrogen.<br>- With excess iodine yields p-iodotoluene, m.p. 35°C, b.p. 211°C.<br>- Acetyl derivative, m.p. 121°C.<br>- Benzoyl derivative, m.p. 146°C.<br>- Benzal derivative, m.p. 125°C. |
| Semicarbazide<br>$NH_2CONHNH_2$ | 96 | Colourless, soluble in $H_2O$ and alcohol, insoluble in ether | - Hydrochloride, m.p. 173°C(d).<br>- Acid sulphate, m.p. 144°C.<br>- Nitrate, m.p. 123°C.<br>- Reduces Fehlings solution and ammonical $AgNO_3$ solution.<br>- With acetone yields isopropenyl derivative, m.,p. 187°C.<br>- Benzal derivative, m.p. 214°C. |
| 1-Phenyl-3-methyl-5-pyrazolone<br>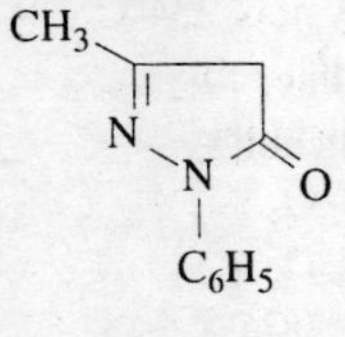 | 127 | Insoluble in cold $H_2O$, ether soluble in hot $H_2O$, alcohol | - Gives red colour on boiling with acetone.<br>- On boiling with aq. $FeCl_3$ gives purple ppt of pyrazole blue.<br>- Benzoyl derivative, m.p. 75.<br>- Benzal derivative, m.p. 107.<br>- With 2 moles of $Br_2$ in hot acetic acid, gives dibromo derivative, m.p. 80°C. |
| Acetyl phenylhydrazine<br>$NHNHCOCH_3$<br>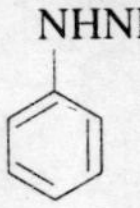 | 127 | Colourless, soluble in hot $H_2O$ | - On boiling with conc. HCl yields phenyl hydrazine hydrochloride and acetic acid.<br>- Reduces Fehling's solution.<br>- With 2 moles $Br_2$ in conc. HCl yields dibromo derivative, m.p. 146°C. |
| Hydrazobenzene<br>$NHNHC_6H_5$<br>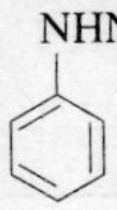 | 127 | Insoluble in $H_2O$, soluble in alcohol, ether | - Its alkaline soln. is rapidly oxidized by air to azo benzene, m.p. 68°C.<br>- With hot conc. HCl yields benzidine hydrochloride.<br>- Diacetyl derivative, m.p. 105°C (di).<br>- Dibenzoyl derivative, m.p. 162°C (d). |

| *Name and structure* | *M.P.(°C)* | *Properties* | *Specific tests and derivatives* |
|---|---|---|---|
| Benzoyl phenylhydrazine 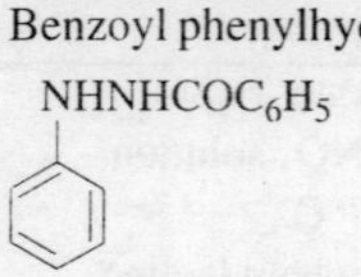 | 168 | Sparingly soluble in hot $H_2O$ and ether | - Readily soluble in warm dil. KOH.<br>- On oxidation by HgO or $N_2O_3$ yields benzoyl azobenzene, m.p. 80°C.<br>- On boiling with conc. HCl yields phenylhydrazine HCl and benzoic acid, m.p. 122°C. |
| 2, 4-Dinitrophenyl-hydrazine | 198 (d) | See Primary amines (Table 6.15b) | |

**Table 6.24** Purines

| *Name and structure* | *M.P.(°C)* | *Properties* | *Specific tests and derivatives* |
|---|---|---|---|
| Caffeine (1, 3, 7-Trimethylxanthine) 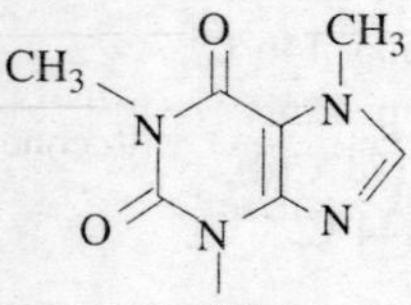 | 234 | White powder, soluble in $H_2O$ and $CHCl_3$, sparingly soluble in ether, $CCl_4$ | - Yields Prussian blue on warming with $K_4Fe(CN)_6$ and $HNO_3$.<br>- On boiling with alcoholic KOH yields $CO_2$, $NH_3$, $CH_3NH_2$, HCOOH and sarcosine ($CH_3NHCH_2COOH$), m.p. 210°C(d).<br>- Boil 0.1 g of the compound with 2 ml $H_2O$ and 1 ml HCl (conc.), add 10 ml saturated bromine water. Boil until colourless and dilute with water. On adding to 2 ml of this soln. 1 drop of 5% $FeSO_4$ and 2-3 drops of $NH_4OH$, blue colour results.<br>- With $I_2$ in KI gives ppt, m.p. 215°C. |
| Theophylline (1, 3-Dimethylxanthine) 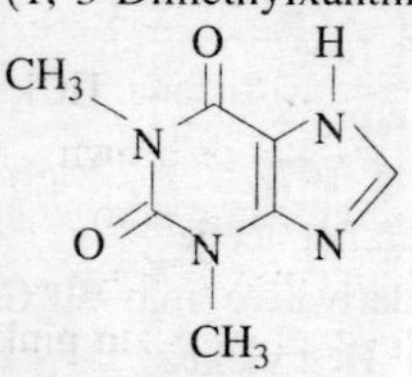 | 264 | Soluble in hot $H_2O$ | - On boiling in $CH_3OH$ with KOH and $CH_3I$ yields caffeine, m.p. 234°C.<br>- Gives blue colour with HCl + $Br_2$ + $FeSO_4$ + $NH_4OH$ etc as in the case of caffeine (see above). |
| Theobromine (3, 7-Dimethylxanthine) 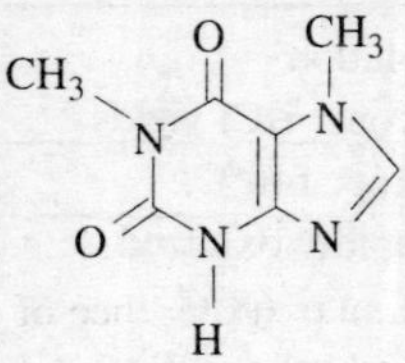 | 290 (sublimes) 329 (in sealed tube) | White solid, slightly soluble in hot water | - On warming in $CH_3OH$ with KOH and $CH_3I$ yields caffeine m.p. 234°C.<br>- Gives blue colour with HCl + $Br_2$ + $FeSO_4$ + $NH_4OH$ etc. as in the case of caffeine (see above).<br>- Yields crystalline ppt. with $AgNO_3$ in dil. $HNO_3$. |
| Uric acid 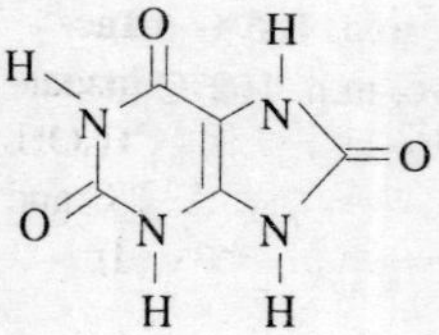 | decomposes on heating without melting | Insoluble in $H_2O$, alcohol, ether, soluble in sodium acetate and in glycerol | - Reduces acidic $KMnO_4$ in cold.<br>- Reduces Fehling's solution on long boiling.<br>- Reduces Ammonical $AgNO_3$.<br>- With $CuSO_4$ + $NaHSO_3$ gives white ppt. of cuprous urate. |

| *Name and structure* | *M.P.(°C)* | *Properties* | *Specific tests and derivatives* |
|---|---|---|---|
| Adenine (6-amino purine) 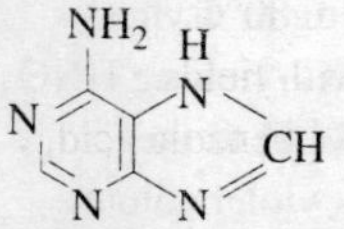 | 360-65 | | - Benzoyl m.p. 234°C.<br>- Picrate m.p. 280°C. |
| (2-Amino-6-hydroxypurine) Guanine 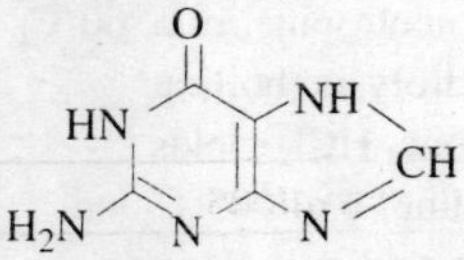 | > 360 (d) | | |

**Table 6.25(a)** Alkaloids (Liquids)

| *Name and structure* | *B.P.(°C)* | *Properties* | *Derivatives* | *Specific tests* |
|---|---|---|---|---|
| (+)-Coniine 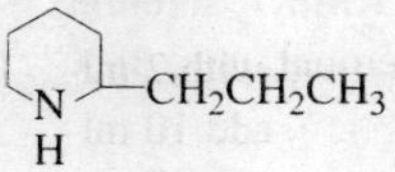 | 166 | Colourless, sparingly soluble in $H_2O$, alcohol, ether<br>$[\alpha]^{19}_{D} = +16°$ | Hydrochloride, m.p. 220°C<br>Picrate, m.p. 75°C | - Gives green colour with conc. $H_2SO_4$ + powdered $K_2Cr_2O_7$.<br>- After standing for 5 min. with alcoholic $CS_2$, the mixture gives a brown ppt. with $CuSO_4$ solution.<br>- Colours phenolphthalein pink in 50% alcohol.<br>- Gives white ppt. with $HgCl_2$.<br>- With HCl gas gives red colour changing to blue. |
| (–) Nicotine 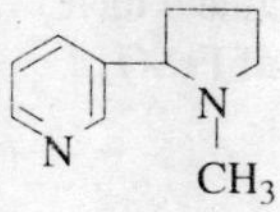 | 247 | Colourless, soluble in $H_2O$, alcohol, ether<br>$[\alpha]^{20}_{D} = -169°$ | Picrate, m.p. 218°C<br>Tartrate, m.p. 90°C | - On warming with conc. HCl gives light violet or brown colour, which changes to orange with conc. $HNO_3$.<br>- Colours phenolphthalein pink only in presence of $H_2O$ or very dil. alcohol. |

**Table 6.25(b)** Alkaloids (Solids)

| *Name and structure* | *M.P.(°C)* | *Properties* | *Derivatives* | *Specific tests* |
|---|---|---|---|---|
| Quinine hydrate | 57 | See Quinine, m.p. 177. | | |
| (–) Cocaine 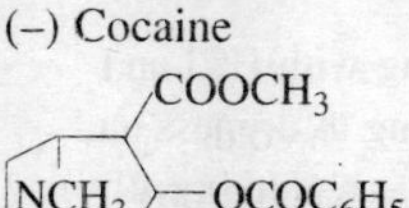 | 98 | Colourless, sparingly soluble in $H_2O$, soluble in alcohol, ether, chloroform<br>$[\alpha]^{20}_{D} = -16°$<br>(C = 4 in $CHCl_3$) | Hydrochloride, m.p. 195°C<br>Chromate, m.p. 127°C<br>Methiodide, m.p. 164°C | - With $K_2Cr_2O_7$ in presence of HCl gives yellow ppt.<br>- On warming with ethyl alcohol + conc. $H_2SO_4$ gives characteristic odour of ethyl benzoate.<br>- On hydrolysis yields $CH_3OH$, benzoic acid, m.p. 122°C and ecognine, m.p. 198°C(d). |
| Brucine hydrated | 105 | See Brucine, m.p. 178°C. | | |

| *Name and structure* | *M.P.(°C)* | *Properties* | *Derivatives* | *Specific test and derivs* |
|---|---|---|---|---|
| Atropine (dl-Hyoscyamine) 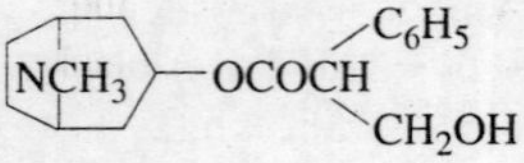 | 115 | Colourless, sparingly soluble in $H_2O$, soluble in alcohol, ether, chloroform | Hydrochloride, m.p. 165°C<br>Picrate, m.p. 175°C<br>Sulphate, m.p. 194°C | - Addition of aq. $HgCl_2$ to alcoholic solution gives yellow ppt<br>- On evaporating to dryness ($H_2O$ bath) with fuming $HNO_3$ and adding KOH to the residue gives violet colour changing to red.<br>- On warming with $Ac_2O$ at 85°, yields apotropine, m.p. 60°C. |
| Piperine 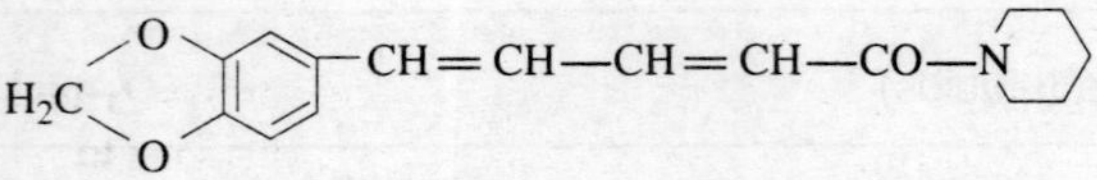 | 130 | Colourless, sparingly soluble in $H_2O$, soluble in alcohol, ether, chloroform | | - On hydrolysis (boiling with conc. HCl) yields piperidine, b.p. 105°C and piperic acid m.p. 216°C<br>- Solution in conc. $H_2SO_4$ gives yellow colour changing to brown.<br>- With acid $KMnO_4$ solution yields piperonal, m.p. 38°C. |
| Cevadine hydrated | 146 | See Cevadine, m.p. 213-14°C | | |
| Papaverine 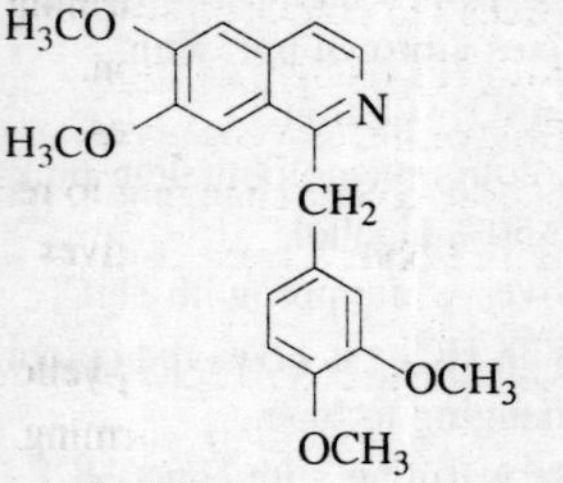 | 147 | Colourless, insoluble in $H_2O$, sol. in hot alcohol, hot chloroform | Hydrochloride, m.p. 230°C<br>Picrate, m.p. 179°C<br>Bromo derivative, m.p. 144°C | - Solution in conc. $H_2SO_4$ colourless, changes to violet on warming.<br>- On adding $K_2Cr_2O_7$ to solution in dil. acid yields yellow or orange ppt.<br>- With $Br_2$ water yields bromo derivative, m.p. 144°C. |
| (–) Codeine 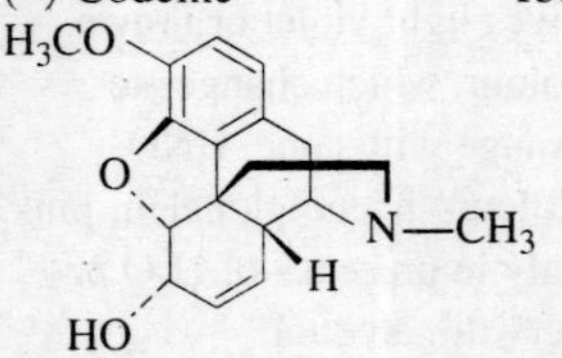 | 155 | Colourless, soluble in hot alcohol, ether, chloroform form<br>$[\alpha]^{15}_D = -136°$(alcohol) | Hydrochloride, m.p. 280°C<br>Sulphate, m.p. 278°C | - With conc. $H_2SO_4$ and then adding a crystal of $FeSO_4$ gives blue colour. |
| Quinidine 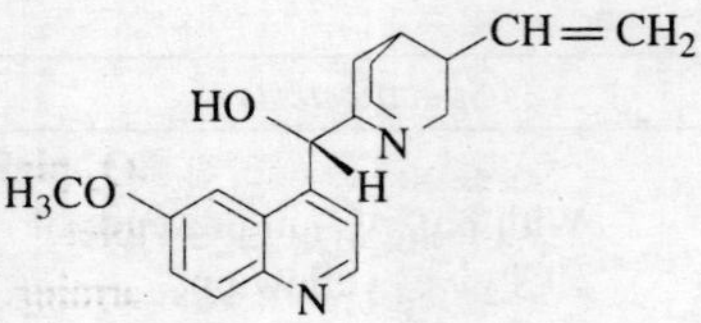 | 174 | Colourless, insoluble in $H_2O$, soluble in alcohol, ether, chloroform<br>$[\alpha]^{15}_D = +230°$<br>(C = 1.8 in $CHCl_3$) | - Hydrochloride, m.p. 258°C | - Gives dull ppt. with KI.<br>- With small quantity of $Br_2$ water and then adding a few drops of $NH_4OH$ gives green colour.<br>- On heating with HCl and evaporating to dryness and heating the residue gently, gives violet colour with evolution of violet vapours.<br>- Solution in dil. $H_2SO_4$ fluoresces blue. |
| (–) Quinine | 177 | Colourless, sparingly | - Hydro | - Solution of sulphate gives |

| *Name and structure* | *M.P.(°C)* | *Properties* | *Derivatives* | *Specific test and derivs* |
|---|---|---|---|---|
| 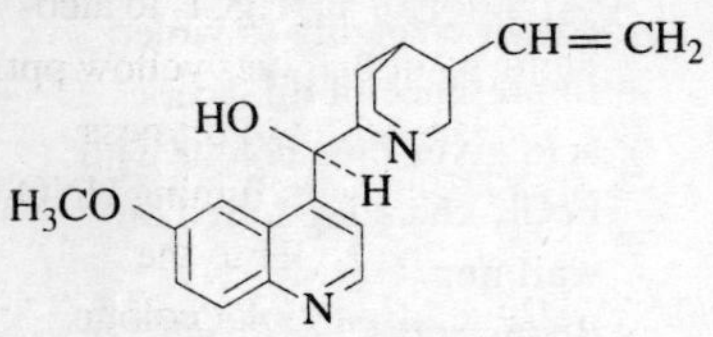 | | soluble in $H_2O$, soluble in organic liquids<br>$[\alpha]^{15}_{D} = -169°$ (97% alcohol) | chloride, m.p. 158°C | white ppt with ammonium oxalate.<br>- Solution in acetic acid and alcohol mixture gives black ppt on warming with alcoholic $I_2$ solution.<br>- With small quantity of $Br_2$ water and then adding a few drops of $NH_4OH$ gives green colour changing to red on addition of $K_3Fe(CN)_6$.<br>- Hydrochloride on heating alone assumes violet colour and gives off violet vapours. |
| Brucine<br>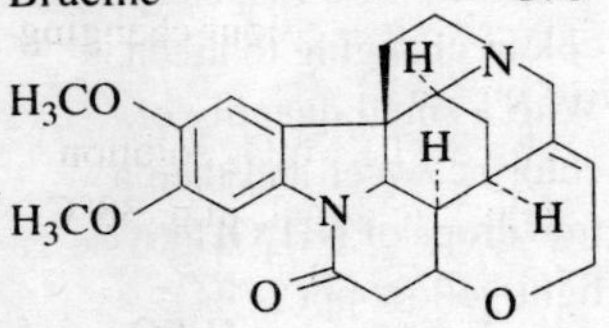 | 178 | Colourless, very sparingly soluble in $H_2O$, sparingly soluble in ether, soluble in alcohol, chloroform<br>$[\alpha]_D$ = approx. – 127°. (in $CHCl_3$) | Nitrate, m.p. 230°C(d) (anhydrous) | - Solution in conc. $H_2SO_4$–pink colour slowly changing to yellow.<br>- Red colour with conc. $HNO_3$ changing to yellow brown on evaporating at 100°–residue gives violet colour with $SnCl_2$.<br>- Gives red colour on warming with $Hg_2(NO_3)_2$ solution.<br>- Erdmann's reagent gives yellow colour changing to red.<br>- With Fohde's reagent gives red colour. |
| Cevadine<br>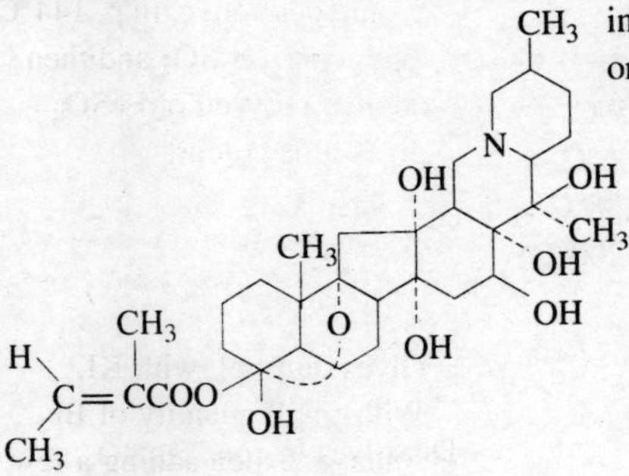 | 213-214 | Colourless, insoluble in $H_2O$, soluble in organic liquids | Benzoyl derivative m.p. 170–180°C | - Solution in conc. $H_2SO_4$-yellow changing to red on warming.<br>- On distilling alone yields β-picoline, m.p. 143°C. |
| Morphine<br>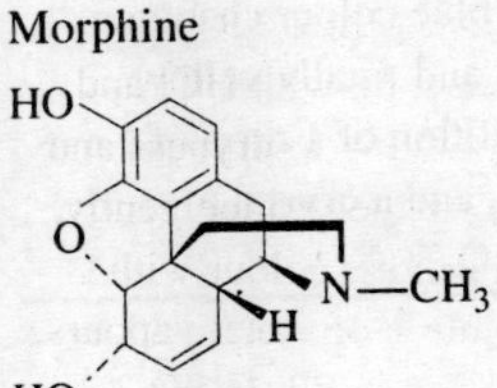 | 254 | Colourless, insoluble in $H_2O$, soluble in alcohol.<br>$[\alpha]^{25}_{D} = -132°$ (MeOH) | - Acetate, m.p. 200°C<br>- Hydro chloride, m.p. 200°C<br>- Sulphate, m.p. 250°C | - Solution in conc. $H_2SO_4$ pink, changing to orange-violet and then brown on warming.<br>- Red colour with conc. $HNO_3$ changing to yellow on warming.<br>- On heating to 100° with conc. $H_2SO_4$ and adding a small crystal of $FeSO_4$ gives red |

| Name and structure | M.P.(°C) | Properties | Derivatives | Specific test and derivs |
|---|---|---|---|---|
| | | | | colour changing to violet.<br>- In presence of dil. acetic acid gives blue colour with $FeCl_3$-changing to red on warming.<br>- Reduces $FeCl_3$ to $FeCl_2$ resulting solution gives blue ppt. with $K_3Fe(CN)_6$.<br>- With a solution of 1 drop of formalin in 1 ml conc. $H_2SO_4$ gives purple colour changing to blue.<br>- With Erdmann's reagent-pink colour.<br>- With Frohde's reagent-purple colour changing to green. |
| (+) Cinchonine<br>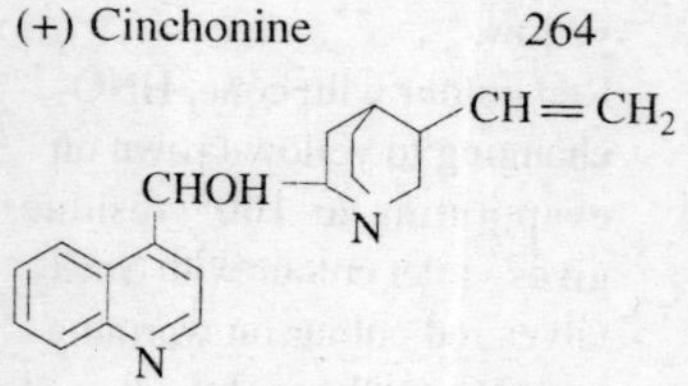 | 264 | Colourless, insoluble in $H_2O$, ether, sparingly soluble in alcohol, chloroform $[\alpha]_D + 229°$ (0.2% in alcohol) | - Hydro chloride, m.p. 217°C<br>- sulphate, m.p. 198°C | - With a small quantity of chlorine water and then a few drops of $NH_4OH$ gives light yellow ppt.<br>- On adding $K_4Fe(CN)_6$ to solution of salts gives yellow ppt. (soluble in hot $H_2O$, crystallising on cooling).<br>- Hydrochloride on heating assumes violet colour and evolves violet vapours. |
| Reserpine<br>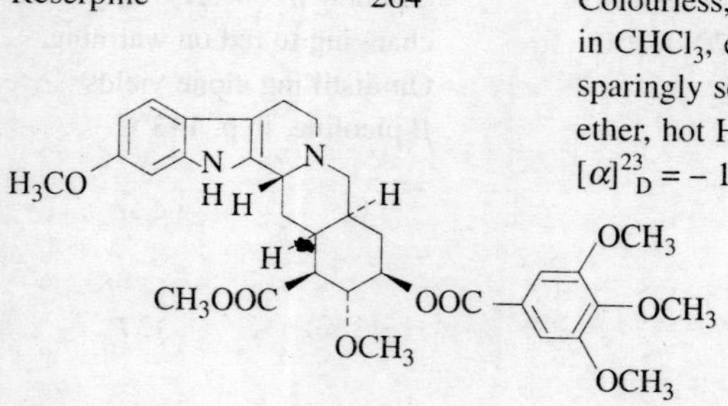 | 264 | Colourless, soluble in $CHCl_3$, ethyl acetate, sparingly soluble in ether, hot $H_2O$, acetone $[\alpha]^{23}_D = -118°$ ($CHCl_3$) | Hydrochloride, m.p. 224°C<br>Nitrate, m.p. 235°C<br>Picrate, m.p. 185°C<br>Oxalate, m.p. 206°C<br>Sulphate, m.p. 238°C (anhdrous, m.p. 200°C) | |
| (–) Strychnine<br> | 268 – 290 | Colourless, insoluble in $H_2O$, sparingly soluble in ether, soluble in hot alcohol, hot chloroform $[\alpha]^{18}_D = -139.3°$ (chloroform) | | - Dissolves in conc. $H_2SO_4$ to a colourless solution which gives blue colour changing to red and finally yellow on addition of 1 drop of $HNO_3$ and a crystal of $K_2Cr_2O_7$. |

**Table 6.26** Sulphonic Acids

| *Name and structure* | *M.P.(°C)* | *Properties* | *Amide* | *Anilide* | *S-Benzyl isothiouronium Salt* |
|---|---|---|---|---|---|
| Methanesulphonic acid $CH_3SO_3H$ | 20 | | 90 | 99 | – |
| Benzenesulphonic acid 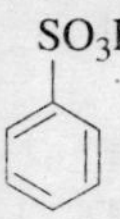 | 43 (anhydrous 65) | Colourless, deliquescent, soluble in $H_2O$, alcohol | 153 | 110 | 148 |
| m-Nitrobenzenesulphonic acid 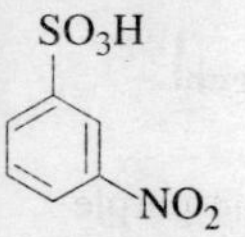 | 48 | Soluble in cold $H_2O$ | 167 | 126 | 146 |
| o-Toluenesulphonic acid 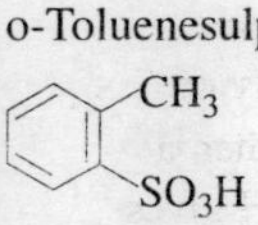 | 57 | Colourless, hygroscopic, soluble in $H_2O$, alcohol | 156 | 136 | 170 |
| m-Xylene-4-sulphonic acid 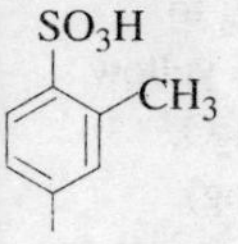 | 60 | | 137 | 110 | 146 |
| p-Chlorobenzenesulphonic acid 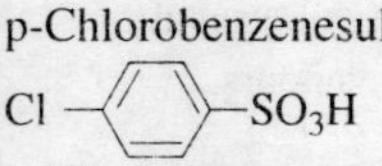 | 68 | | 144 | 104 | 175 |
| o-Nitrobenzenesulphonic acid 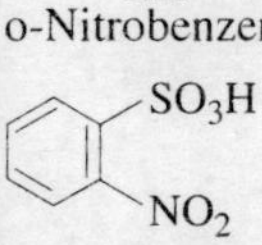 | 70 | | 193 | 115 | — |
| p-Nitrobenzenesulphonic acid 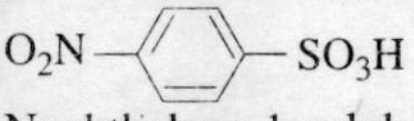 | 70 | | 179 | 136 | — |
| Naphthalene-1-sulphonic acid 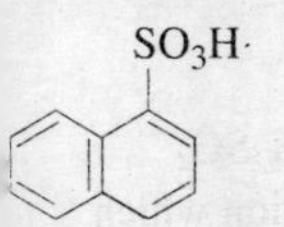 | 88 (anhydrous 90) | | 150 | 112(160) | 137 |
| p-Toluenesulphonic acid 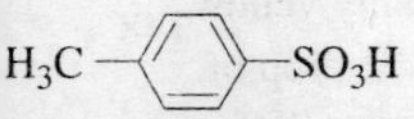 | 92 (anhydrous) 105 (hydrated) | | 137 (anhydrous 105 (hydrated) | 103 | 182 |
| p-Bromobenzenesulphonic acid 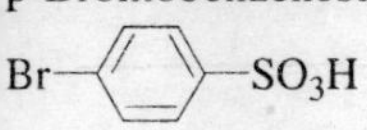 | 103 | | 166 | 119 | 170 |

| *Name and structure* | *M.P.(°C)* | *Properties* | *Amide* | *Anilide* | *S-Benzyl isothiouronium Salt* |
|---|---|---|---|---|---|
| 5-Sulphosalicylic acid 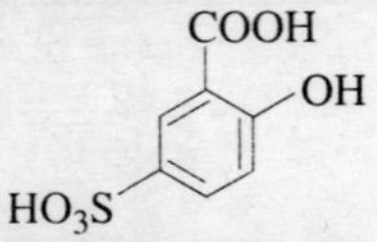 | 120 | | – | – | 204 |
| 2-Naphthol-6-sulphonic acid (Schaffer's acid) 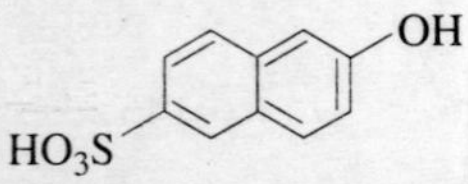 | 122 (d) | | 238 | – | – |
| Naphthalene-2-sulphonic acid 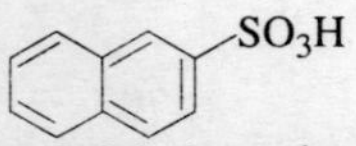 | 124 (anhydrous 91) | | 212 (217) | 132 | 191 |
| o-Sulphobenzoic acid | 134 | See Carboxylic acids (Table 6.2b) | | | |
| m-Sulphobenzoic acid | 141 | See Carboxylic acids (Table 6.2b) | – | – | 103 |
| 1-Naphthol-4-sulphonic acid 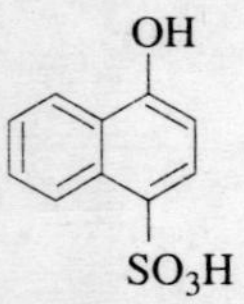 | 170 | | | | |
| (+)-Camphor 10-sulphonic acid 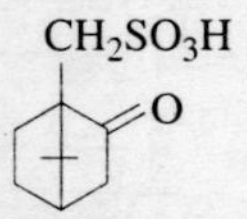 | 193 | Colourless, soluble in $H_2O$ $[\alpha]_D + 21°$ (10% aqueous solution) | 132 | 119 | 210 |
| Naphthalene-1, 5-disulphonic acid 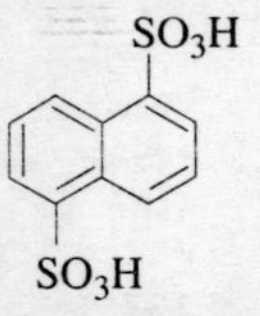 | 245 | | 310 | 249 | 254 |
| p-Sulphobenzoic acid 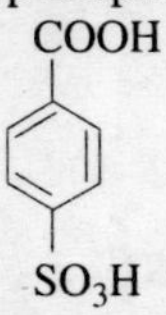 | 259 (anhyd.) 94 (hydrated) | Colourless, soluble in $H_2O$ | 236 (di) | 253 (di) | 213 |
| Metanilic acid 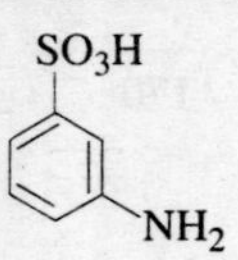 | > 300 | Sparingly soluble in cold $H_2O$ | 142 | – | 148 |

| *Name and structure* | *M.P.(°C)* | *Properties* | *Amide* | *Anilide* | *S-Benzyl isothiouronium Salt* |
|---|---|---|---|---|---|
| Sulphanilic acid<br>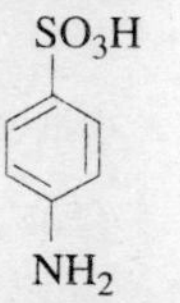 | > 300 | Sparingly soluble in cold $H_2O$ | 165 | 200 | 182 |
| Benzene m-disulphonic acid<br>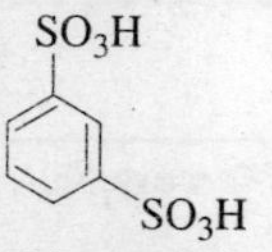 | – | Colourless, hygroscopic | 229 | 114 | 214 |
| Ethanesulphonic acid<br>$CH_3CH_2SO_3H$ | – | | 60 | 58 | 115 |
| Naphthalen-2, 6-disulphonic acid<br>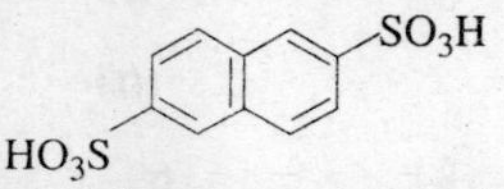 | – | Colourless, hygroscopic | 305 | – | 256 |
| Naphthalene acid 2, 7-disulphonic acid<br>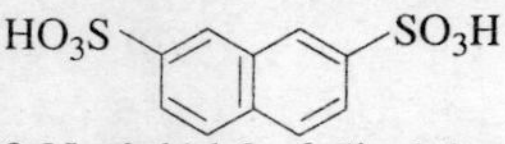 | – | Hygroscopic | 242 | – | 205 |
| 2-Naphthol-3, 6-disulphonic acid (R-acid)<br>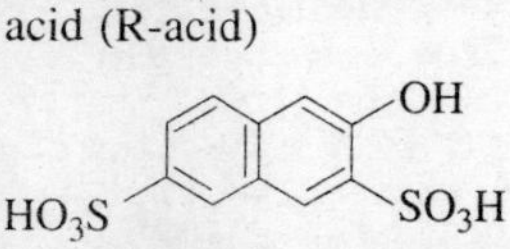 | – | Colourless, deliquescent, needles, soluble in $H_2O$ | – | 202 | 213 |
| 2-Naphthol 6, 8-disulphonic acid<br>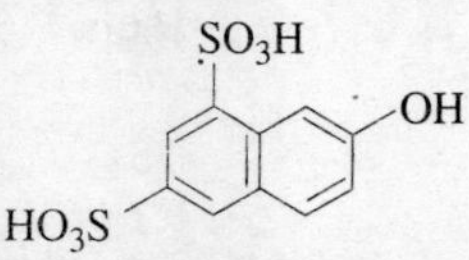 | – | Hygroscopic | – | 195 | 228 |
| Phenol p-sulphonic acid<br>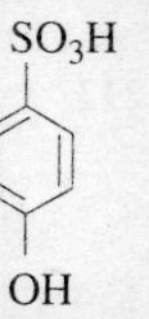 | – | Hygroscopic liquid | 177 | 141 | 169 |
| Naphthionic acid<br>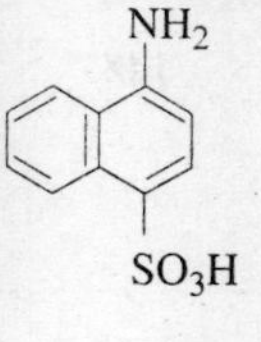 | 300 | | 261 | – | 195 (d) |

| *Name and structure* | *M.P.(°C)* | *Properties* | *Amide* | *Anilide* | *S-Benzyl isothiouronium Salt* |
|---|---|---|---|---|---|
| Orthanilic acid 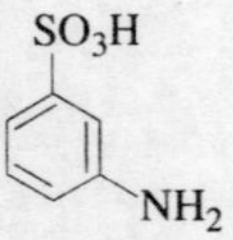 | > 300 (d) | Sparingly soluble in cold $H_2O$ | 142 | – | 148 |

**Table 6.27(a)** Thiols, Mercaptan (thioalcohols and thiophenols)—Liquids

| *Name and structure* | *B.P. (°C)* | *Properties* | *2, 4-dinitro phenyl thioether* | *3, 5-dinitro thiobenzoate* | *Mercury Salt* |
|---|---|---|---|---|---|
| Methylmercaptan $CH_3SH$ | 6 | Unpleasant odour | – | – | 175 |
| Ethylmercaptan $C_2H_5SH$ | 36 | Colourless, unpleasant garlic odour, soluble in dil NaOH | 115 | 62 | 76 |
| Isopropylmercaptan $(CH_3)_2CHSH$ | 56 | Colourless, unpleasant odour | 95 | 84 | 63 |
| n-Propyl mercaptan $CH_3CH_2CH_2SH$ | 67 | Colourless, insoluble in $H_2O$, soluble in alcohol, NaOH | 81 | – | 70 |
| Isobutyl mercaptan $(CH_3)_2CHCH_2SH$ | 88 | Soluble in dil. NaOH solution | 76 | 64 | 95 |
| Allyl mercaptan $CH_2{=}CHCH_2SH$ | 90 | | 72 | – | – |
| n-Butylmercaptan $CH_3CH_2CH_2CH_2SH$ | 97 | Soluble in dil. NaOH solution | 66 | – | 86 |
| Isoamyl mercaptan $(CH_3)_2CHCH_2CH_2SH$ | 117 | Colourless, soluble in dil. alkali | 59 | – | 100 |
| n-Amylmercaptan $CH_3CH_2CH_2CH_2CH_2SH$ | 127 | | 80 | – | 75 |
| Cyclohexyl mercaptan 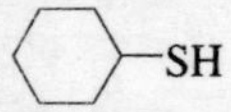 | 159 | – | 148 | – | – |
| Thiophenol (Phenyl mercaptan) 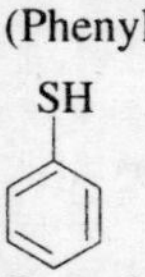 | 169 | Colourless, sparingly soluble in $H_2O$, soluble in alcohol, ether | 121 | – | – |
| Benzyl mercaptan 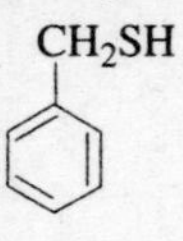 | 194 | Colourless, odour of onions, soluble in dil. NaOH, insoluble in $H_2O$, soluble in alcohol | 130 | – | – |

| Name and structure | B.P. (°C) | Properties | 2, 4-dinitro phenyl thioether | 3, 5-dinitro thiobenzoate | Mercury $SO_4$ |
|---|---|---|---|---|---|
| o-Tolylmercaptan (Thio o-cresol) 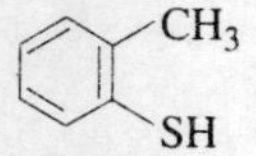 | 194 (m.P. 15°) | | 101 | – | – |
| m-Tolyl mercaptan (Thio m-cresol) 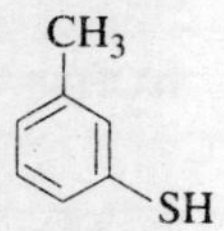 | 200 | | 91 | – | – |

**Table 6.27(b)** Thiols, Mercaptans (thioalcohols and thiophenols)—Solids

| Name and structure | M.P. (°C) | Properties | 2, 4-dinitro phenyl thioether | 3, 5-dinitro thiobenzoate | Mercury Salt |
|---|---|---|---|---|---|
| p-Tolyl mercaptan (Thio p-cresol)  | 43 (b.p. 195°) | | 103 | – | – |
| β-Thionaphthol 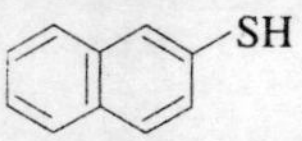 | 81 | | -2, 4-Dinitrophenyl sulphide, m.p. 145°C. | | |
| Thiosalicylic acid 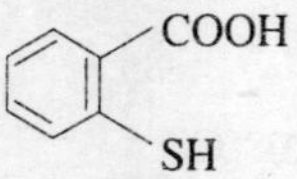 | 165 | See carboxylic acids (Table 2.6b) | | | |

**Table 6.28** Sulphonamides

| Name and structure | M.P.(°C) | Properties | Specific tests and derivatives |
|---|---|---|---|
| p-Toluenesulphonanilide 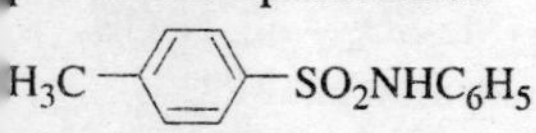 | 103 | Insoluble in $H_2O$, soluble in alcohol, ether | - On boiling with 20% HCl gives p-toluenesulphonic acid. and aniline, b.p. 184. |
| Benzenesulphonanilide $C_6H_5SO_2NHC_6H_5$ | 110 | Insoluble in $H_2O$, soluble in alcohol, ether | - On boiling with 20% HCl gives benzenesulphonic acid, m.p. 43 and aniline b.p. 184°C.<br>- On heating with benzyl chloride in alcoholic sodium hydroxide gives benzyl deriv., m.p. 119°C. |
| (+) Camphor 10-sulphonamide 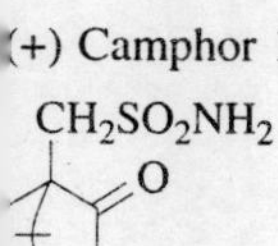 | 132 | | |

| *Name and structure* | *M.P.(°C)* | *Properties* | *Specific tests and derivatives* |
|---|---|---|---|
| p-Toluenesulphonamide 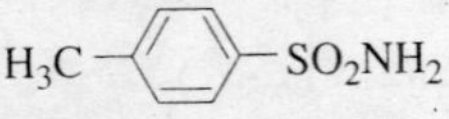 | 137 | Sparingly soluble in cold $H_2O$, soluble in alcohol, ether, dil. NaOH | - On methylation ($Me_2SO_4$+ dil. NaOH) yields dimethyl ether, m.p. 80.<br>- Xanthyl derivative, m.p. 197. |
| Naphthalene 1-sulphonamide 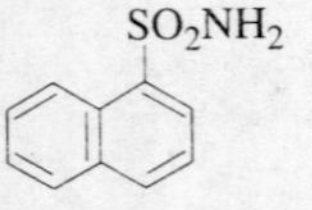 | 150 | Soluble in alcohol, ether, dil. NaOH | - On methylation ($Me_2SO_4$ + dil. NaOH) yields dimethyl ether, m.p. 137°C.<br>- N-Acetyl, m.p. 185°C.<br>- N-Benzoyl, m.p. 194°C. |
| Benzenesulphonamide 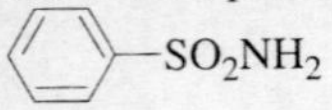 | 153 | Soluble in alcohol, ether, dil. NaOH | - On methylation ($Me_2SO_4$ + dil. NaOH) yields dimethyl ether, m.p. 47°C<br>- N-Acetyl, m.p. 125°C.<br>- N-Benzoyl, m.p. 147°C.<br>- Xanthyl derivative, m.p. 206°C. |
| o-Toluenesulphonamide 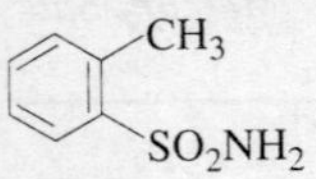 | 153 | Soluble in alcohol, ether, dil. NaOH | - With $K_3Fe(CN)_6$ solution yields o-sulphonamidobenzoic acid, m.p. 154°C, which on heating gives the imide (Saccharin), m.p. 224°C.<br>- On heating with strong $H_2SO_4$ yields toluene, b.p. 110.<br>- Xanthyl derivative, m.p. 183°C. |
| m-Nitrobenzene sulphonamide 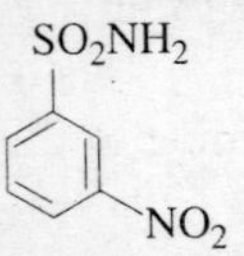 | 162 | | - On methylation ($Me_2SO_4$ + dil. NaOH) yields dimethyl ether, m.p. 122°C.<br>- N-acetyl, m.p. 189°C. |
| Sulphanilamide 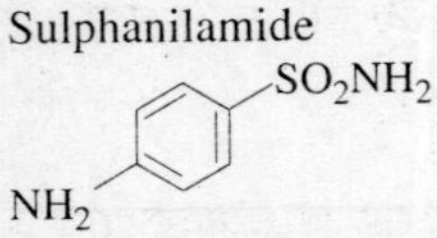 | 165 | | - Diacetyl, m.p. 217°C.<br>- Dibenzoyl, m.p. 268°C.<br>- Xanthyl derivative, m.p. 208°C. |
| Naphthalene-2-sulphonamide 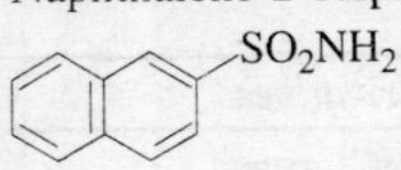 | 212 (217) | Soluble in alcohol, ether, dil. NaOH | - On methylation ($Me_2SO_4$ + dil. NaOH) yields dimethyl ether, m.p. 94°C.<br>- N-Acetyl, m.p. 146°C. |
| Sulphobenzoic imide (Saccharin) 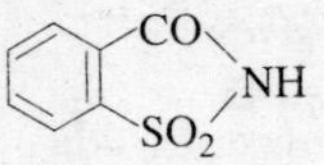 | 224 | Colourless, almost insoluble in $H_2O$, dil. NaOH or $Na_2CO_3$. Sodium derivative has intense sweet taste | - On methylation ($Me_2SO_4$ + dil. NaOH) gives N-methyl ether, m.p. 131°C.<br>- Xanthyl derivative, m.p. 198°C.<br>- On warming with resorcinol + trace of conc. $H_2SO_4$ gives sulphonfluorescein (red solution in dil. NaOH with green fluorescence). |
| m-Benzenedisulphonamide 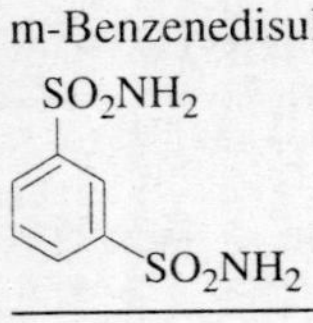 | 229 | Insol. in $H_2O$, soluble in dil. NaOH | - Xanthyl derivative, m.p. 170°C. |

# 7

# Preparation of Reagents and Indicators

## 7.1 PREPARATION OF REAGENTS

**Barfoed's Reagent.** Dissolves 6.5 g of crystallised neutral copper acetate in 100 ml of 1% acetic acid solution. Filter if necessary. The reagent is moderately stable.
The reagent is used for detecting the presence of monosaccharides.
**Baeyer's Reagent.** $KMnO_4$ solution in water (2%).
**Benedict's Reagent.** Dissolve 86.5 g of sodium citrate and 50 g of anhydrous sodium carbonate in 350 ml of water. To this add a solution of 8.65 g of hydrated crystalline copper sulphate ($CuSO_4 \cdot 5H_2O$) in 50 ml of water. Make up the total volume to 500 ml.
**Bials' Reagent.** Dissolve 1.2 g of orcinol in 400 ml of concentrated hydrochloric acid and add 15–20 drops of 10% aqueous ferric chloride solution.
**Bromine water.** Shake well 5 ml of bromine with 100 ml of water and allow the excess bromine to remain in solution (concentration approximately 3%).
**Ceric Ammonium Nitrate Reagent.** Dissolve 40 g of ceric ammonium nitrate in 100 ml of warm (2N) nitric acid.
This reagent is used for detecting the presence of alcohols.
**Chlorine water.** Pass chlorine gas in 100 ml of cold water until a saturated solution is obtained (concentration approximately 0.7%).
**Chromic Anhydride Reagent.** Add water (60 ml) slowly with stirring to a suspension of 20 g of chromic anhydride in 20 ml of concentrated sulphuric acid.
**Copper-Benzidine Reagent.** Dissolve 0.5 g of benzidine in 100 ml of 5% aqueous acetic acid and then add 100 ml of 0.2N cupric acetate solution. Store the solution in a dark bottle. A darkening of the reagent may occur after about 15 days but this does not affect its usefulness (*Kreshov, Vilborg and Filippova, Anal, Abs., 1954, 1(6), 1294*).
**Denige's Solution.** Dissolve 5 g of yellow mercuric oxide into a warm solution of 20 ml of concentrated sulphuric acid in 100 ml of water. After dissolution, the solution is cooled and filtered.
**2, 4-Dinitrophenylhydrazine Reagent (Brady's Reagent).** Treat 2g of 2, 4-dinitrophenylhydrazine with 4 ml of concentrated sulphuric acid and cautiously add 35 ml of methanol with cooling. Warm to effect solution and then add 10 ml of water. Allow the solution to stand for 4-5 hours and filter.
**Dragendorff's (Thresh's) Reagent.** Dissolve 8 g of bismuth nitrate in 20 ml of dilute nitric acid (1 part $HNO_3$ + 2 parts water) and add to it with stirring a solution of 27 g of potassium iodide in 40 ml

of water. The mixture is stirred and allowed to cool. The solution is made up to 100 ml with distilled water.

**Erdmann's Reagent.** Mix 100 ml of concentrated sulphuric acid with 0.5 ml of concentrated nitric acid.

**Fehling's solution.** *Fehling A*-dissolve 34.6 g of crystalline copper sulphate in water containing a few drops of dilute sulphuric acid. The total solution is made to 500 ml.

*Fehling B*-Dissolve 173 g of sodium potassium tartrate (Rochelle salt) and 70 g of sodium hydroxide in water and make up to 500 ml.

Mix equal volumes of solutions A and B just before use.

**Ferric chloride solution.** Dissolve hydrated ferric chloride (135 g) in a mixture of water (500 ml) and concentrated hydrochloric acid (20 ml). Make the total volume to 1 litre by adding distilled water.

For making neutral ferric chloride solution, treat the above ferric chloride solution (5 ml) with dilute ammonium hydroxide solution, adding dropwise, until a faint but permanent precipitate is obtained. Filter the solution.

**Ferri-Ferrocyanide Reagent.** Dissolve 0.1 g of ferric chloride and 0.12 g of potassium ferricyanide in 125 ml of water.

**Frohde's Reagent.** Dissolve 1 g of sodium or ammonium molybdate in 100 ml of concentrated sulphuric acid.

**Hydroxylamine hydrochloride solution (0.5 N).** Dissolve 8.75 g of hydroxylamine hydrochloride in 40 ml of water. Dilute the solution to 250 ml with methanol.

**Iodine-potassium iodide solution (for iodoform reaction).** Dissolve 10 g of potassium iodide and 5g iodine in 50 ml of water.

**Litmus solution.** Dissolve 1 g of blue litmus in 100 ml of water.

Red litmus solution is prepared by treating the above blue litmus solution with a few drops of dilute hydrochloric acid.

**Lucas reagent.** Dissolve 16 g of anhydrous zinc chloride in 10 ml of concentrated hydrochloric acid and cooling.

**Mayer's reagent.** Dissolve 7 g of mercuric chloride and 25 g of potassium iodide in 500 ml of water.

**Millions Reagent.** Dissolve 10 g of mercury in 10 ml of concentrated nitric acid and dilute the solution with 20 ml of water.

**Molisch's Reagent.** Dissolve 10 g of $\alpha$-naphthol in 100 ml of alcohol.

**Nessler's Reagent.** Dissolve 5 g of potassium iodide in a minimum amount of cold water and add a saturated solution of mercuric chloride until a faint permanent precipitate is obtained. Add 40 ml of 50% aqueous potassium hydroxide solution. Allow the precipitate to settle and dilute to 100 ml with water. Allow to stand over-night and decant the reagent solution from the precipitate.

**Nickel chloride-carbon disulphide Reagent.** Dissolve 0.5 g of crystallised nickel chloride in 100 ml of water. Saturate the above solution by adding carbon disulphide drop by drop. Shake the solution thoroughly until the globules of carbon disulphide remain in solution.

**Nickel chloride-5-nitrosalicylaldehyde Reagent.** Dissolve 0.5 g of 5-nitrosalicylaldehyde in a mixture of 15 ml of pure triethanolamine (free from ethanolamine) and 25 ml of water by stirring. A solution of 0.5 g of crystallised nickel chloride in 5 ml of water is added and total volume is made up to 100 ml with water.

**Ninhydrin Reagent.** Dissolve 0.2 g ninhydrin in 100 ml of rectified spirit.

**Nylander's Reagent.** Dissolve 2 g of basic bismuth nitrate and 4 g of sodium potassium tartrate in 100 ml of 10% sodium hydroxide solution.

**Pinoff's Reagent.** Dissolve 4 g of ammonium molybdate in 100 ml of water.
**Schiff's Reagent.** Dissolve 0.5 g of rosaniline hydrochloride (fuchsine) in 250 ml of warm water. Cool, saturate with sulphur dioxide. The solution becomes colourless or pale yellow. Dilute to 500 ml with water.
**Selewonoff's Reagent.** Dissolve 0.05 g of resorcinol in 100 ml of dilute hydrochloric acid (4N).
**Sodium nitroprusside solution.** Dissolve 0.1 g of sodium nitroprusside in 20 ml of water. It is advisable to prepare a fresh solution, since it deteriorates on keeping and becomes green.
**Modified Sodium Nitroprusside Reagent.** Dissolve 1.95 g of sodium nitropusside in a solution containing 40 ml of dimethylsulphoxide and 10 ml of water. The reagent should be stored in a refrigerator.
**Sodium iodide in acetone.** Dissolve sodium iodide (15 g) in acetone (100 ml).
**Tollen's Reagent.** Dissolve 5 g of silver nitrate in 100 ml of water. Just before use add a drop or two of dilute sodium hydroxide solution (5%) till a permanent turbidity remains. Dissolve it in minimum amount of ammonium hydroxide solution. Do not store.

## 7.2 Preparation of Indicators

**Bromophenol blue.** Dissolve 0.2 g of bromophenol blue in 100 ml of rectified spirit.
**Diphenylamine.** Dissolve 5 g of diphenylamine in 50 ml of concentrated sulphuric acid.
**Methyl orange.** Dissolve 0.2 g of methyl orange in 200 ml of water.
**Methylene blue.** Dissolve 0.1 g of methylene blue in 50 ml of water.
**Methyl red.** Dissolve 0.1 g of methyl red in 100 ml of rectified spirit.
**Phenolphthalein.** Dissolve 1g of phenolphthalein in 100 ml of rectified spirit.
**Starch solution.** Suspend 2 g of soluble starch in 10 ml of water and add into 90 ml of boiling water, all the while stirring the solution. Boil to get a clear solution.

# Summary of the Scheme Used for Identification of Unknown Organic Compounds



## Carbohydrates

### *Tests*



### *Derivatives*

## Carboxylic Acids

### Tests



### Derivatives



## Carbonyl Compounds—Aldehydes and Ketones

### Tests



### Derivatives

## Phenols

### Tests



### Derivatives



## Quinones

### Tests



### Derivatives



## Alcohols

### Tests

### *Derivatives*



## Esters

### *Tests*



### *Derivatives*



## Lactones

### *Tests*



### *Derivatives*



## Anhydrides

### *Tests*

## Acid Chlorides

### *Tests*



### *Derivatives*



## Nitro Compounds

### *Tests*



### *Derivatives*



## Amines

### *Tests*

## Amino Acids

### Tests



### Derivatives



## Sulphonic Acids

### Tests



### Derivatives



## Thiols and Mercaptans

### Tests



### Derivatives

## Sulphonamides

### Tests



### Derivatives



## SUGGESTED BOOKS FOR FURTHER READING


1. J.R. Dyer. Applications of absorption spectroscopy of organic compounds. New Jersey. Prentice Hall.
2. D.H. Williams and I. Fleming. Spectroscopic methods in organic chemistry. London. McGraw-Hill.
3. R.M. Silverstein, G.C. Bassler and T.C. Morrill. Spectroscopic identification of organic compounds. New York. John Wiley & Sons.
4. P.S. Kalsi, Spectroscopy of organic compounds, New Delhi. Willey Eastern Limited.
5. B.S. Furniss, A.J. Hannoford, V. Rogers, P.W.G. Smith and A.R. Tatchnell. Vogel's textbook of practical organic chemistry. U.K. ELBS.
6. H.T. Clarke (Revised by B. Haynes, E.C. Brick and G.G. Shone). A handbook of organic analysis (qualitative and quantitative). Arnold Heinemann (India).
7. R.L. Shriner, R.C. Fusion and D.V. Curtin. The systematic indentification of organic compounds. A laboratory manual. London. John Wiley & Sons.

# Index

*Note:* Number in the brackets are melting or boiling points. The boiling points have been underlined.